Skin Sensitization Testing Using New Approach Methodologies

Skin Sensitization Testing Using New Approach Methodologies

Guest Editor

Judy Strickland

Basel • Beijing • Wuhan • Barcelona • Belgrade • Novi Sad • Cluj • Manchester

Guest Editor
Judy Strickland
Inotiv, Inc.
Morrisville
NC
USA

Editorial Office
MDPI AG
Grosspeteranlage 5
4052 Basel, Switzerland

This is a reprint of the Special Issue, published open access by the journal *Toxics* (ISSN 2305-6304), freely accessible at: https://www.mdpi.com/journal/toxics/special_issues/3CR8V99L28.

For citation purposes, cite each article independently as indicated on the article page online and as indicated below:

Lastname, A.A.; Lastname, B.B. Article Title. *Journal Name* **Year**, *Volume Number*, Page Range.

ISBN 978-3-7258-4383-1 (Hbk)
ISBN 978-3-7258-4384-8 (PDF)
https://doi.org/10.3390/books978-3-7258-4384-8

Contents

About the Editor

Judy Strickland

Dr. Judy Strickland previously served as a Principal Predictive Toxicologist, accumulating 22 years of experience in evaluating alternative test methods that reduce, refine, or replace the use of animals in regulatory test methods under a contract to support the U.S. National Toxicology Program Interagency Center for the Evaluation of Alternative Toxicological Methods (NICEATM). She was most recently an employee of Inotiv, Inc., from which she retired in late 2023. She served on the Organization for Economic Cooperation and Development (OECD) Expert Group on Skin Sensitization and the Expert Group on Defined Approaches for Skin Sensitization. Prior to these positions, Dr. Strickland provided technical support to the U.S. Interagency Coordinating Committee on the Validation of Alternative Methods' evaluations of multiple modifications of the local lymph node assay, an animal reduction method for regulatory skin sensitization assessment, and served on the OECD Consultation on the Local Lymph Node Assay. She is excited about recent developments in alternative approaches for skin sensitization assessment because non-animal methods are now accepted to meet some regulatory requirements. Methods for integrating multiple sources of non-animal data have effectively increased the confidence in relying on such information. Dr. Strickland received the U.S. National Institutes of Health Award of Merit for validating the use of computational tools to replace a regulatory requirement for animal-based testing, including demonstrating the utility of Tox21 efforts in 2016. In 2022, Dr. Strickland received the Rosalind Franklin Society Award in Science for "Application of Defined Approaches to Assess Skin Sensitization Potency of Isothiazolinone Compounds," which was published in Applied In Vitro Toxicology. She is a member of the Society of Toxicology and a Diplomate of the American Board of Toxicology.

Preface

The following Special Issue addresses new approach methodologies for the assessment of chemicals and products to determine their potential to induce skin sensitization. New approach methodologies are methods that do not involve the use of animals, typically including in vitro or in silico methods or a combination of the two. This Special Issue was motivated by my experience of more than 20 years in evaluating such alternative toxicity test methods to reduce, refine, and replace the use of animals for regulatory toxicity testing. Skin sensitization assessments of chemicals and products have made greater progress in reducing animal use than many other regulatory toxicity evaluations. I am grateful for the broad range of submissions received for this Special Issue. It is conceivable that some of the methods and approaches presented can serve as inspiration for the development and evaluation of new approach methodologies for additional toxicity evaluations.

Judy Strickland
Guest Editor

toxics

MDPI

Editorial

Skin Sensitization Testing Using New Approach Methodologies

Judy Strickland †

Inotiv, Inc., Morrisville, NC 27560, USA; judystrickland@frontier.com
† Retired.

Chemical regulatory authorities require skin sensitization information for a number of chemical sectors, such as cosmetics, pesticides, newly marketed chemicals, household products, and topical pharmaceuticals [1]. If regulatory agencies need such information, then chemical and product developers and distributors also need it. Usually, the information needed is skin sensitization hazard, a binary classification to denote whether the substance has the potential to induce skin sensitization or not, and, if the substance has skin sensitization potential, skin sensitization potency is needed. Potency information is provided as a multilevel categorical classification or a point estimate, sometimes referred to as a "point of departure" (POD), which is a toxicity value used for risk assessment. Such information is typically determined in animal tests. However, skin sensitization is one of the regulatory toxicity endpoints seeing much progress in the uptake of new approach methodologies, which are non-animal skin sensitization evaluations, for regulatory and developmental purposes. The mechanisms of skin sensitization initiated by binding to skin proteins is well-characterized by an adverse outcome pathway published by the Organization for Economic Cooperation and Development (OECD) [2]. Test method developers have used the key events in the adverse outcome pathway as the targets for developing non-animal tests that assess whether a chemical or product may activate them.

This Special Issue, "Skin Sensitization Testing Using New Approach Methodologies", shows that there is much research and application activity in this area, as evidenced by the 10 articles contributed.

These articles provide a broad coverage of the area of new approach methodologies for skin sensitization assessments. A number of articles involve test method development. Fritch et al., Contribution 10, documents the problem formulation stage of test method development for T-cell activation, which is key event 4 of the adverse outcome pathway for skin sensitization produced by chemicals that bind to skin proteins [2]. Although there aresome non-animal tests that measure T-cell activation to assess a substance's potential for skin sensitization [3,4], there are no regulatory tests established in harmonized international test guidelines. Contribution 10 shows that known approaches to assess T-cell activation are challenged by chemical-mediated toxicity, inefficient or unknown generation of T-cell epitopes, and low throughput. Contribution 10 reviews the available solutions and strategies to confirm in vitro T-cell signals and concludes that application and standardization are necessary to define chemical applicability domains and to strengthen the role of T-cell tests in regulatory risk assessment.

In Contribution 1, Burla et al. report work to further the development of a non-animal test for respiratory sensitizers, which are classified as substances of very high concern under the European Commission Regulation for the Registration, Evaluation, Authorization and restriction of Chemicals (REACH). There are currently no animal or non-animal test methods approved for regulatory use in evaluating chemicals for respiratory sensitization hazards. Respiratory sensitizers test positive in the existing animal skin sensitization tests,

 https://doi.org/10.3390/toxics13050326

but these tests do not provide enough information to distinguish between skin and respiratory sensitizers. ALIsens®, a three-dimensional in vitro alveolar model, had previously been evaluated for the classification of respiratory sensitizers [5]. Contribution 1 continues the evaluation of ALIsens® by establishing its efficacy in discriminating respiratory sensitizers from skin sensitizers and non-sensitizers.

Contributions 8, Ahmed et al., and 9, Martins-Ribeiro et al., describe a laboratory test for skin sensitization that has been adapted to detect the immunotoxicity of therapeutic biologics (Contribution 8), or antibodies damaged by heat, which produces antibody aggregation (Contribution 9). These tests are very human-relevant. Both investigations expose human skin explants to evaluate histopathology, and also expose peripheral blood mononuclear cells from the same donor to evaluate T-cell proliferation or cytokine levels. Contributions 8 and 9 show that the adapted skin sensitization tests [3,4] were applicable to identifying adverse immune reactions in addition to skin sensitization.

A number of contributions to this Special Issue build and evaluate computational models to classify test substances for skin sensitization hazard or to place sensitizers in potency categories. In Contribution 2, Tieghi et al. develop and evaluate an in silico model that predicts human skin sensitization hazard and potency categories per the Globally Harmonized System of Classification and Labelling of Chemicals [6]. Human skin sensitization data were used to build this model and the model is freely available. Such a model can be easily implemented in the early stages of drug or chemical development and testing to minimize resource waste and facilitate the early stages of molecule development while reducing animal testing. Thus, these models are useful even if they do not progress towards regulatory acceptance.

Mohoric et al. (Contribution 3) and Tourneix et al. (Contribution 6) report on Bayesian models that combine data streams from in vitro and in silico results to estimate substance potency for inducing skin sensitization. Both models were based on the work of Jaworska et al. [7,8]. Animal data from the murine local lymph node assay (LLNA) were also used to build these models. Both contributions build and evaluate Bayesian models that predict LLNA potency in four categories as well as PODs for sensitizers that can be used for risk assessment. However, the models reported in Contributions 3 and 6 use different potency categories and different methods for predicting the LLNA potency value. The models from both contributions provide confidence indications for each prediction. Contributions 3 and 6 show that risk assessment can be performed without animal data.

Gradin et al. does as well, in Contribution 4, and provides a much simpler and less data intensive approach to developing a POD for skin sensitization risk assessment. The approach uses the information from only one in vitro test, which is based on gene expression, the GARDskin Dose–Response assay. The best performing model, which estimates PODs, is a regression of no-expected-sensitization-induction levels based on human and LLNA data (in vivo PODs) on the assay readout using only an intercept parameter.

In Contribution 5, Wareing et al. provides an in vitro test method evaluation. It evaluates the applicability of three key event assays included in the existing OECD test guideline methods [9–11], one for each of the first three key events, for testing inorganic nanomaterials for skin sensitization potential. Although the in vitro methods were modified in an attempt to adjust their suitability to nanomaterials, they were originally developed and evaluated for small organic molecules and were not a good fit for testing inorganic nanomaterials.

One contribution, 7, by Juncan et al., provides a case study showing how new approach methodologies for skin sensitization can be applied to toxicity evaluation during the product development of cosmetic ingredients. Skin sensitization was one of the toxicity endpoints assessed. Evaluations of stability, physicochemical properties, and microbiology were also performed. Contribution 7 reports on the development and evaluation processes all the way from the assessment of several candidate ingredients to be included in the new formulation to assessments of the final formulation. Contribution 7 shows how a novel proprietary in silico model that predicts skin sensitization can be used to evaluate ingredients during product development. The in silico evaluation was verified with clinical testing.

The contributions to this Special Issue, "Skin Sensitization Testing Using New Approach Methodologies", highlight a number of data gaps. Skin sensitization test results or model predictions are typically not considered alone to make a decision on how to classify a substance. Usually, all information about the substance in question is considered, including, but not limited to, physicochemical properties, results for other toxic endpoints, and the known skin sensitization potential of similar substances. A decision tree on how to apply the results of new approach methodologies for skin sensitization hazard and potency is needed in order to integrate the skin sensitization information with other pertinent chemical information. This especially applies to the in silico methods described in Contribution 2 and the individual in vitro test methods described in Contributions 4 and 5. However, the Bayesian network models described in Contributions 3 and 6 already use a number of different data streams. Would these model predictions be sufficient as stand-alone hazard or potency classifications? More research and evaluation is needed to identify the best approach for the application of these methods.

Contribution 7 reminds us that the existing OECD test guidelines are not applicable to all types of substances. The existing guidelines are intended to target mechanisms relevant to organic substances that induce skin sensitization by binding to skin proteins. Inorganic nanomaterials (as is the case in Contribution 7), metals such as nickel, and biological agents may induce skin sensitization by other mechanisms [2]. Thus, research and development of applicable test methods to detect substances that induce sensitization by means other than binding to skin proteins is needed. Such test development and application of new approach methodologies for skin sensitization will continue to be a focus of the regulatory and regulated communities in order to reduce animal use for these assessments and provide faster and more human-relevant results.

Data Availability Statement: Not applicable.

Conflicts of Interest: The author declares no conflicts of interest.

List of Contributions:

1. Burla, S.; Chary, A.; Serchi, T.; Cambier, S.; Sullivan, K.; Baker, E.; Sadekar, N.; Gutleb, A.C. Responses of an In Vitro Coculture Alveolar Model for the Prediction of Respiratory Sensitizers (ALIsens®) Following Exposure to Skin Sensitizers and Non-Sensitizers. *Toxics* **2025**, *13*, 29. https://doi.org/10.3390/toxics13010029.
2. Tieghi, R.S.; Moreira-Filho, J.T.; Martin, H.-J.; Wellnitz, J.; Otoch, M.C.; Rath, M.; Tropsha, A.; Muratov, E.N.; Kleinstreuer, N. A Novel Machine Learning Model and a Web Portal for Predicting the Human Skin Sensitization Effects of Chemical Agents. *Toxics* **2024**, *12*, 803. https://doi.org/10.3390/toxics12110803.
3. Mohoric, T.; Wilm, A.; Onken, S.; Milovich, A.; Logavoch, A.; Ankli, P.; Tagorti, G.; Kirchmair, J.; Schepky, A.; Kühnl, J.; et al. Increasing Accessibility of Bayesian Network-Based Defined Approaches for Skin Sensitisation Potency Assessment. *Toxics* **2024**, *12*, 666. https://doi.org/10.3390/toxics12090666.
4. Gradin, R.; Tourneix, F.; Mattson, U.; Andersson, J.; Amaral, F.; Forreryd, A.; Alépée, N.; Johansson, H. In Vitro Prediction of Skin-Sensitizing Potency Using the GARDskin Dose–Response Assay: A Simple Regression Approach. *Toxics* **2024**, *12*, 626. https://doi.org/10.3390/toxics12090626.

5. Wareing, B.; Aktalay Hippchen, A.; Kolle, S.N.; Birk, B.; Funk-Weyer, D.; Landsiedel, R. Limitations and Modifications of Skin Sensitization NAMs for Testing Inorganic Nanomaterials. *Toxics* **2024**, *12*, 616. https://doi.org/10.3390/toxics12080616.
6. Tourneix, F.; Carron, L.; Jouffe, L.; Hoffmann, S.; Alépée, N. Deriving a Continuous Point of Departure for Skin Sensitization Risk Assessment Using a Bayesian Network Model. *Toxics* **2024**, *12*, 536. https://doi.org/10.3390/toxics12080536.
7. Juncan, A.M.; Rus, L.-L.; Morgovan, C.; Loghin, F. Evaluation of the Safety of Cosmetic Ingredients and Their Skin Compatibility through In Silico and In Vivo Assessments of a Newly Developed Eye Serum. *Toxics* **2024**, *12*, 451. https://doi.org/10.3390/toxics12070451.
8. Ahmed, S.S.; Ahmed, M.M.; Ishaq, A.; Freer, M.; Stebbings, R.; Dickinson, A.M. An In Vitro Human Skin Test for Predicting Skin Sensitization and Adverse Immune Reactions to Biologics. *Toxics* **2024**, *12*, 401. https://doi.org/10.3390/toxics12060401.
9. Martins-Ribeiro, A.; Kizhedath, A.; Ahmed, S.S.; Glassey, J.; Ishaq, A.; Freer, M.; Dickinson, A.M. A Human Skin Explant Test as a Novel In Vitro Assay for the Detection of Skin Sensitization to Aggregated Monoclonal Antibodies. *Toxics* **2024**, *12*, 332. https://doi.org/10.3390/toxics12050332.
10. Fritsch, N.; Aparicio-Soto, M.; Curato, C.; Riedel, F.; Thierse, H.-J.; Luch, A.; Siewert, K. Chemical-Specific T Cell Tests Aim to Bridge a Gap in Skin Sensitization Evaluation. *Toxics* **2024**, *12*, 802. https://doi.org/10.3390/toxics12110802.

References

1. Daniel, A.B.; Strickland, J.; Allen, D.; Casati, S.; Zuang, V.; Barroso, J.; Whelan, M.; Régimbald-Krnel, M.J.; Kojima, H.; Nishikawa, A.; et al. International Regulatory Requirements for Skin Sensitization Testing. *Regul. Toxicol. Pharmacol.* **2018**, *95*, 52–65. [CrossRef] [PubMed]
2. OECD. *The Adverse Outcome Pathway for Skin Sensitisation Initiated by Covalent Binding to Proteins*; OECD Series on Testing and Assessment; OECD: Paris, France, 2014; ISBN 978-92-64-22144-4.
3. Ahmed, S.S.; Whritenour, J.; Ahmed, M.; Bibby, L.; Darby, L.; Wang, X.; Watson, J.; Dickinson, A. Evaluation of a human in vitro skin test for predicting drug hypersensitivity reactions. *Toxicol. Appl. Pharmacol.* **2019**, *369*, 39–48. [CrossRef] [PubMed]
4. Ahmed, S.S.; Wang, X.N.; Fielding, M.; Kerry, A.; Dickinson, I.; Munuswamy, R.; Kimber, I.; Dickinson, A.M. An in vitro human skin test for assessing sensitization potential. *J. Appl. Toxicol.* **2016**, *36*, 669–684. [CrossRef] [PubMed]
5. Chary, A.; Serchi, T.; Moschini, E.; Hennen, J.; Cambier, S.; Ezendam, J.; Blomeke, B.; Gutleb, A.C. An in vitro coculture system for the detection of sensitization following aerosol exposure. *Altern. Anim. Exp.* **2019**, *36*, 403–418. [CrossRef] [PubMed]
6. United Nations Economic Commission for Europe. *Globally Harmonized System of Classification and Labelling of Chemicals (GHS Rev.9)*; United Nations: New York, NY, USA; Geneva, Switzerland, 2021. Available online: https://unece.org/transport/documents/2021/09/standards/ghs-rev9 (accessed on 13 April 2025).
7. Jaworska, J.; Dancik, Y.; Kern, P.; Gerberick, F.; Natsch, A. Bayesian Integrated Testing Strategy to Assess Skin Sensitization Potency: From Theory to Practice. *J. Appl. Toxicol.* **2013**, *33*, 1353–1364. [CrossRef] [PubMed]
8. Jaworska, J.S.; Natsch, A.; Ryan, C.; Strickland, J.; Ashikaga, T.; Miyazawa, M. Bayesian Integrated Testing Strategy (ITS) for Skin Sensitization Potency Assessment: A Decision Support System for Quantitative Weight of Evidence and Adaptive Testing Strategy. *Arch. Toxicol.* **2015**, *89*, 2355–2383. [CrossRef] [PubMed]
9. OECD. *Test No. 442C: In Chemico Skin Sensitisation*; OECD: Paris, France, 2024.
10. OECD. *Test No. 442D: In Vitro Skin Sensitisation*; OECD: Paris, France, 2024.
11. OECD. *Test No. 442E: In Vitro Skin Sensitisation*; OECD: Paris, France, 2024.

toxics

MDPI

Article

Responses of an *In Vitro* Coculture Alveolar Model for the Prediction of Respiratory Sensitizers (ALIsens®) Following Exposure to Skin Sensitizers and Non-Sensitizers

Sabina Burla [1,2,3,*], Aline Chary [1], Tommaso Serchi [1], Sébastien Cambier [1], Kristie Sullivan [4,5], Elizabeth Baker [4], Nikaeta Sadekar [6] and Arno C. Gutleb [1,3,*]

1 Luxembourg Institute of Science and Technology (LIST), 4422 Belvaux, Luxembourg
2 Faculty of Food Science and Technology, University of Agricultural Sciences and Veterinary Medicine of Cluj-Napoca, 400372 Cluj-Napoca, Romania
3 Invitrolize Sarl, 4422 Belvaux, Luxembourg
4 Physicians Committee for Responsible Medicine (PCRM), Washington, DC 20016, USA
5 Institute for In Vitro Sciences, Inc. (IIVS), Gaithersburg, MD 20878, USA
6 Research Institute for Fragrance Materials (RIFM), Woodcliff Lake, NJ 07430, USA
* Correspondence: sabina.burla@invitrolize.com (S.B.); arno.gutleb@invitrolize.com (A.C.G.)

Abstract: In recent years, a global increase in allergy incidence following chemical exposure has been observed. While the process of skin sensitization is well characterized through the adverse outcome pathway (AOP) framework, the immunological mechanisms underlying respiratory sensitization remain less well understood. Respiratory sensitizers are classified as substances of very high concern (SVHC) under the European Union (EU) regulation for the registration, evaluation, authorization and restriction of chemicals (REACH), emphasizing the importance of evaluating respiratory tract sensitization as a critical hazard. However, the existing new approach methodologies (NAMs) for the identification of skin sensitizers lack the capacity to differentiate between skin and respiratory sensitizers. Thus, it is imperative to develop physiologically relevant test systems specifically tailored to assess respiratory sensitizers. This study aimed to evaluate the efficacy of ALIsens®, a three-dimensional (3D) *in vitro* alveolar model designed for the identification of respiratory sensitizers and to determine its ability to correctly identify sensitizers. In this study, we used a range of skin sensitizers and non-sensitizers to define the optimal exposure dose, identify biomarkers, and establish tentative thresholds for correct sensitizer classification. The results demonstrate that ALIsens® is a promising *in vitro* complex model that could successfully discriminate respiratory sensitizers from skin sensitizers and non-sensitizers. Furthermore, the thymic stromal lymphopoietin receptor (TSLPr) cell surface marker was confirmed as a reliable biomarker for predicting respiratory sensitization hazards.

Keywords: chemical allergy; new approach methodologies; respiratory sensitization; air-liquid-interface exposure; dendritic cell activation; cell surface markers

1. Introduction

Allergic reactions arising from chemical exposure represent a significant concern for human health and can manifest in various ways. Exposure to respiratory sensitizers, in particular, can lead to long-term health conditions, including allergic rhinitis [1] and asthma [2]. Respiratory sensitization resulting from chemical exposure occurs less frequently than skin sensitization, largely attributable to a limited number of identified chemical compounds [3], compared to over 3000 molecules identified as skin sensitizers [4], and exposure context

https://doi.org/10.3390/toxics13010029

from occupational settings to everyday environments. Despite its lower prevalence, respiratory sensitization is critically important due to its significant impact and severity, being associated with considerable morbidity on the affected individuals and related increased socioeconomic costs [5,6].

Regulatory agencies enforce specific guidelines and regulations for the classification and labeling of substances identified as respiratory and skin sensitizers. Early identification of respiratory sensitizers and distinguishing respiratory from skin sensitizers provides the necessary data to evaluate the hazards associated with exposure and for appropriate risk management measures to be implemented for the protection of human health [7]. This information is essential for making informed decisions regarding the safe handling, use and management of these substances. As many respiratory sensitizers are encountered in occupational settings, testing is essential for workers' protection. Currently, regulatory agencies do not request the testing of chemicals for respiratory sensitization potential during product registration or authorization due to the lack of validated testing methods.

The mechanisms underlying respiratory sensitization are not fully understood [8,9], but the proposed adverse outcome pathway (AOP) for respiratory sensitization [10] shows points of convergence with the AOP for skin sensitization [11], suggesting common mechanisms involved in the development of the two types of sensitization [12]. These relationships between respiratory and skin sensitization are important for the hazard and risk assessment process of chemical sensitizers. It has been observed that all identified respiratory sensitizers yield positive results in the *in vivo* assays designed for skin sensitization and exhibit dermal sensitization, while the opposite is not universally true [13]. Only a minority of skin sensitizers are also respiratory sensitizers. This suggests that if a compound tests negative in validated tests for the identification of skin sensitizers, the likelihood of it being a respiratory sensitizer may be low, but finally, no decision on its respiratory sensitization hazard can be taken [3,14]. However, if the assay for skin sensitization evaluation tests positive, further investigation is required to ascertain the respiratory sensitization hazard [15]. Distinguishing respiratory from skin sensitizers is essential for their correct classification to implement effective risk management, apply targeted safety measures, and reduce health risks.

Identification of low molecular weight (LMW) chemicals that may induce respiratory sensitization is of paramount importance as respiratory sensitizers have been categorized as substances of very high concern (SVHC) due to their severe impacts on human health [16]. When a chemical compound is classified as a SVHC under the Registration, Evaluation, Authorisation and Restriction of Chemicals (REACH) regulation [17], several regulatory consequences follow. The first step is the inclusion of the substance in the candidate list of SVHC [18], which triggers obligations for both manufacturers and importers, such as providing information on the presence of the SVHC in products to downstream users and consumers. The identification as an SVHC often leads to initiatives for substituting it with safer alternatives or phasing out the use of these substances where feasible. In case the European Chemicals Agency (ECHA) moves a substance from the Candidate List to the Authorization List (Annex XIV of REACH Regulation), its use becomes strictly controlled [17]. These measures have a significant impact on the manufacturers, and identification of respiratory sensitizers in the early development phases is very important for a large array of industries for several reasons: to reduce risks associated with SVHCs, to ensure better protection for human health and the environment by promoting safe handling practices to minimize exposure and mitigate risks, ensuring safer work environments and consumer products.

At present, the classification of chemicals as respiratory sensitizers requires evaluating their sensitization potential through the analysis of scientific evidence derived from occupational data, clinical and experimental studies, and read-across approaches [9,19–21]. A significant challenge in the process of risk assessment of chemicals inducing respiratory sensitization is the absence of validated or widely recognized methods for identifying and characterizing chemical respiratory sensitizers [8], as well as for differentiating between respiratory and skin sensitizers [12,22].

Animal models, such as the local lymph node assay (LLNA), have been employed to evaluate respiratory sensitizers. Although primarily designed for skin sensitization testing, this model has offered some insights into the respiratory sensitization potential, but despite the adaptation to a respiratory LLNA [23], the data obtained could not be used to gain a deep understanding of the mechanism leading to respiratory sensitization. Additionally, *in vivo* tests have been criticized for their limitations in accurately predicting human responses, as species differences play a significant role in data interpretation [24]. In response to ethical concerns and the stringent need for accurate tests, many countries from various regions around the globe have phased out or implemented restrictions on animal testing on cosmetics. The regulatory agencies, therefore, promote the development of alternative *in vitro* methods to animal testing to evaluate the impacts of chemicals on human health. The European Union led this global movement by adopting the Cosmetics Regulation in 2009 [25]. Alternative methods that are utilized to assess the safety of cosmetic ingredients and final products fall under the category of new approach methodologies (NAMs) [26,27], a comprehensive descriptor for any non-animal assay, methodology, approach, or combination thereof that can be employed to provide input on chemical hazard and risk assessment [28]. NAMs consist of *in chemico, in silico* and *in vitro* methods, which can be used as stand-alone methodologies or can be comprised of integrated approaches to testing and assessment (IATAs), defined approaches (DAs) for data interpretation, and performance-based evaluation of test methods [29].

While significant progress has been made in the identification of skin sensitizers, resulting in the development and publication of a completely described AOP [11] in which every key event (KE) is supported by one or more NAMs, the progress in the development of NAMs for the identification of respiratory sensitizers has been comparatively slower. However, recent proposals have introduced notable additions to the AOP for respiratory sensitization [30].

The lung is a complex organ characterized by cellular heterogenicity, and the type of cells, culture conditions and exposure methodology are key factors that need to be properly addressed when designing and developing physiologically relevant *in vitro* human lung models for the assessment of xenobiotics' effects upon inhalation [31]. Respiratory sensitization is a complex immunological process in which dendritic cells, the most effective antigen-presenting cells, play a central role. Their activation, identified as a common key event (KE3) in both the established skin sensitization and proposed respiratory sensitization AOPs [12], is crucial. Incorporating dendritic cells, either in monocultures or cocultures with bronchial cells and in the alveolo-capillary barrier *in vitro* models, is essential for advancing the understanding of the immunological mechanisms underlying respiratory sensitization. Dendritic cells can be derived from human peripheral blood or THP-1, a human monocytic leukemia-derived immortalized cell line, using granulocyte-macrophage colony-stimulating factor GM-CSF and interleukin 4 (IL-4) [32,33]. Established cell lines are commonly used as surrogates for dendritic cells in sensitization studies [34], and several *in vitro* assays have been developed with the aim of correctly classifying respiratory sensitizers [9]. Undifferentiated THP-1 cells, in particular, have been utilized in monoculture for both the development [35,36] and validation of the human cell line activation test (h-CLAT) assay for the evaluation of skin sensitizers [37]. Additionally, THP-1 cells in monoculture have been used in studies evaluating respiratory sensitizers [38–41]. The cluster of differentiation (CD)34$^+$ human acute myeloid leukemia cell line (MUTZ-3) used undifferentiated in monoculture has also served as a dendritic cell model for the assessment of respiratory sensitizers [42]. In coculture systems, naïve THP-1 cells have been incorporated to study respiratory sensitizers [43–45].

Expression of CD86 and CD54 markers is measured on the surface of dendritic cells (DCs) to assess the skin sensitizing potential of chemicals in the h-CLAT assay [37]. Since skin and respiratory sensitization share certain mechanisms [12], measuring the expression of CD86 and CD54 cell surface markers was found suitable for the detection of respiratory sensitizers due to their roles in immune response modulation. As skin and respiratory sensitizers differ primarily in the type of T cell response they induce, with skin sensitizers triggering a T helper 1 (Th1) response and respiratory sensitizers eliciting a Th2 response, defining specific markers for the correct identification of respiratory sensitizers and differentiation from skin sensitizers is essential. Thymic stromal lymphopoietin (TSLP), a cytokine produced primarily by epithelial cells at barrier surfaces and also by dendritic cells [46,47], has been studied in the context of asthma [48]. It has been identified as a key regulator of the Th2 immune response. Dendritic cells are also direct targets of TSLP, as its receptor is highly expressed in specific populations of myeloid dendritic cells. When naïve allogeneic T cells are cocultured with TSLP-conditioned dendritic cells, they acquire an inflammatory Th2-like phenotype characterized by the production of cytokines associated with respiratory sensitization [46].

ALIsens®, a complex three-dimensional (3D) *in vitro* coculture alveolar model based on alveolar type II cells secreting lung surfactant (A549) [49,50], endothelial cells (EA.hy926) [51], dendritic-like cells (undifferentiated THP-1) [52] and macrophage-like cells (differentiated THP-1) was developed by Chary and collaborators, 2019 [43]. The model has been shown to correctly identify respiratory sensitizers from the class of acid anhydrides based on a set of biomarkers: thymic stromal lymphopoietin receptor (TSLPr), CD86 and CD54 cell surface marker, soluble mediators release in cell culture medium (cytokines and chemokines) and gene expression [43]. An increased expression of CD86, CD54, TSLPr and of the ligand for OX40 (OX40L) on the surface of dendritic cells from exposed *in vitro* models has been shown to correctly identify respiratory sensitizers [41,43,50,53]. An enhanced expression of CD54 and TSLPr cell surface markers was detected 24 h post-exposure to respiratory sensitizers. TSLPr expression increased upon respiratory sensitizers exposure, but remained unchanged after exposure to irritants, highlighting its specificity in detecting respiratory sensitizers [43].

Since skin sensitizers were not tested in ALIsens®, the specificity of this model for identifying respiratory sensitizers remains uncertain, including its ability to differentiate between respiratory and skin sensitizers [9]. Additionally, the most relevant biomarkers from the panel proposed in the development stage of the model needed to be identified. The increased expression of CD86, CD54 and TSLPr markers in the context of an alveolo-capillary barrier enhances the model's ability to detect respiratory sensitizers, offering a reliable method to differentiate between sensitizers and other chemical exposures [43].

The aim of this study was to evaluate the response of ALIsens® by analyzing the expression of CD86, CD54 and TSLPr cell surface markers following exposure to known skin sensitizers and non-sensitizers. Additionally, this investigation sought to identify specific marker(s) and establish a preliminary threshold that could enable accurate classification of respiratory sensitizers.

2. Materials and Methods

All materials (cell lines, cell culture reagents and ware, chemicals, flow cytometry reagents), equipment and software are presented in Table S1 in the Supplementary Material, indicating the catalog number, supplier and location information.

2.1. Reagents and Chemicals

2.1.1. Cell Culture Reagents and Chemicals

Dulbecco's Modified Eagle's Medium (DMEM), high glucose, with GlutaMAX^{TM} supplement, Roswell Park Memorial Institute 1640 (RPMI-1640), high glucose, with GlutaMAX^{TM} supplement, Iscove's Modified Dulbecco's Medium (IMDM), Phosphate buffered saline (PBS) and TrypLE^{TM} Express Enzyme (1×) were purchased from Gibco^{TM} (Erembodegem, Belgium). Accutase was purchased from Invitrogen^{TM} ThermoFischer Scientific (Carlsbad, NM, USA). Fetal bovine serum (FBS) was purchased from Sigma-Aldrich (Overijse, Belgium). 4-(2-Hydroxyethyl) piperazine-1-ethanesulfonic acid (HEPES) (Chemical Abstracts Service Number (CAS No) 7365-45-9), phorbol-12-myristate-13-acetate (PMA) (CAS No 16561-29-8), resazurin sodium salt (CAS No 62758-13-8) were purchased from Sigma-Aldrich (Overijse, Belgium). 2-mercaptoethanol (β-ME) (CAS No 60-24-2) was purchased from Bio-Rad (Temse, Belgium).

2.1.2. Chemicals Tested in the Study

Brij® 35 surfactant was purchased from Merck Life Sciences, Hoeilaart, Belgium. Citric acid and lactic acid were purchased from VWR Chemicals, Leuven, Belgium. Dimethyl sulfoxide (DMSO) (CAS No 67-68-5), α-terpineol, 1,2-benzisothiazol-3(2H)-one, 2-mercaptobenzothiazole, benzylideneacetone, *trans*-cinnamaldehyde, chloroxylenol, diphenylcyclopropenone, ethylene glycol dimethacrylate, eugenol, imidazolidinyl urea, D-limonene, *p*-phenylenediamine were purchased from Sigma-Aldrich (Overijse, Belgium). The CAS No for the tested chemicals is provided in Table 1.

2.1.3. Selection of Chemicals

The selection criteria for the tested chemicals were based on the availability of LLNA reference data, their use as proficiency chemicals in the Organisation for Economic Co-operation and Development Test Guidelines (OECD TGs) for skin sensitization, and the availability of results from *in chemico* and *in vitro* methods for skin sensitization, specifically the direct peptide reactivity assay (DPRA) [57] and the h-CLAT [37], respectively. Among the 11 selected skin sensitizer chemicals, 4 are classified as weak sensitizers, 4 are classified as moderate sensitizers, 1 as strong sensitizer and 1 as extreme sensitizer, according to LLNA (Table 1).

Table 1. Chemicals tested in the study.

Chemical	Acronym	CAS No [a]	Molecular Weight (g/mol)	Vehicle for Stock Solution Preparation	Exposure Doses (µg/cm²)		Potency Category	h-CLAT [e]		Proficiency/Reference Chemical OECD TG [f]
					Minimum	Maximum		CD86	CD54	
α-Terpineol	α-Terp	98-55-5	154.25	DMSO	6.25	250	Non-sensitizer [b]	- [¥]	- [¥]	n/a
Citric acid	CA	77-92-9	192.12	Water+ 0.1% Brij® 35	50	500	No category	-	-	
Lactic acid	LA	50-21-5	90.08	DMSO	50	475	Non-sensitizer [c]	-	-	442C, 442D, 442E
Chloroxylenol	ChX	88-04-0	156.61		2.5	15	No category	NA	NA	n/a
Ethylene glycol dimethacrylate	EGD	97-90-5	198.22		50	300	Weak [c]	-	+	442D, 442E
Eugenol	Eug	97-53-0	164.20		5	75		+	+	
Imidazolidinyl urea	IU	39236-46-9	388.29		60	160		+	+	442E
D-Limonene	D-Lim	5989-27-5	136.23		112.5	750		-	+	
Benzylideneacetone	BA	122-57-6	146.19		10	100	Moderate [c]	+	+	442C
trans-Cinnamaldehyde	Cinn	14371-10-9	132.16		5	35		n/a *	n/a *	442C, 442D
2-Mercaptobenzothiazole	2-MBT	149-30-4	167.25		5	25		-	+	442D, 442E
1,2-Benzisothiazol-3(2H)-one	BZI	2634-33-5	151.19		10	25		-	+	n/a
p-Phenylenediamine	*p*PD	106-50-3	108.14		5	50	Strong [c]	+	-	442D, 442E
Diphenylcyclopropenone	DPC	886-38-4	206.24		1.5	9	Extreme [d]	-	+	n/a

[a] CAS No (Chemical Abstracts Service Number). [b] PLNA (Popliteal Lymph Node Assay) [54,55]. [c] LLNA (Local Lymph Node Assay) [56]. [d] Potency category classification [56]. [e] h-CLAT: +, sensitizer; -, non-sensitizer [56]. [f] OECD TG—Organisation for Economic Co-operation and Development Test Guideline: 442C [57], 442D [58], 442E [37]. [¥] CD86, CD54 data: [59]. * Cinnamic aldehyde (CAS 104-55-2): h-CLAT—CD54+ and CD86+ [56]. n/a—not applicable. NA—not available.

2.2. Cell Culture

The human cell lines A549, EA.hy926 and THP-1 were obtained from the American Type Culture Collection (ATCC®, Manassas, VA, USA). A549, EA.hy926 and THP-1 cells were cultivated as previously described [43]. Adherent cell lines, A549 and EA.hy926, were routinely subcultured twice a week using TrypLE™ Express Enzyme (1×) and were used up to passage number 25. A549 cells were cultivated in DMEM supplemented with 10% FBS. EA.hy926 cells were cultivated in DMEM, supplemented with 10% FBS and 25 mM HEPES. The suspension cell line, THP-1, was routinely subcultured twice a week and was used up to passage number 25. THP-1 cells were cultivated in RPMI-1640 supplemented with 10% FBS and 50 μM β-ME. All cells were cultivated in a humidified incubator at 37 °C, with 5% CO_2 and tested every 6 months for mycoplasma contamination using the MycoAlert® Mycoplasma Detection Kit (Lonza, Roermond, The Netherlands). The identity of the cell lines was confirmed via human short tandem repeat (STR) profiling cell authentication service provided by ATCC®. A volume of approximately 40 μL cell suspension, containing at least 1×10^6 cells, was spotted on the Flinders Technology Associates (FTA) sample collection card. The FTA cards were allowed to dry and then returned to ATCC® to genotype the cell lines. The identity of the cells was confirmed by ATCC®.

2.3. ALIsens® Test System Build-Up

The 3D alveolar *in vitro* model, ALIsens®, was set up following the methodology proposed by Klein et al., 2013 [60], with the modifications implemented by Chary et al., 2019 [43] for the *in vitro* test system designed for the identification of respiratory sensitizers. The workflow schema is presented in Figure 1.

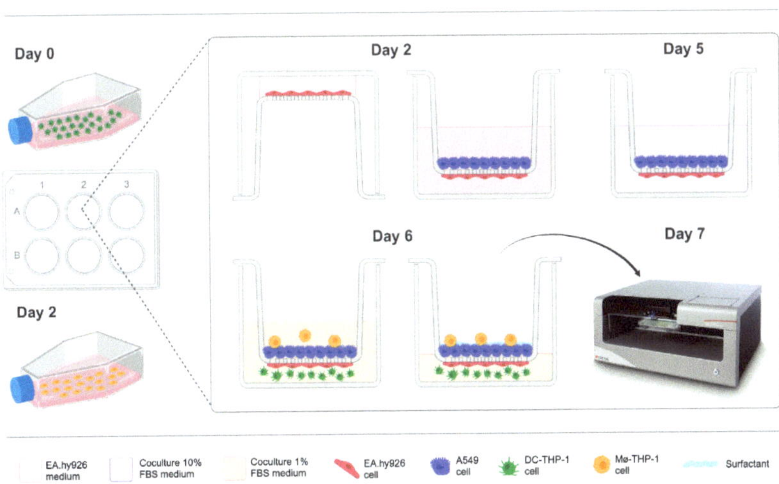

Figure 1. Schematic representation of the workflow for the build-up of ALIsens® for the detection of respiratory sensitizers (adapted and modified from [43]) [61].

On day 0, THP-1 cells were seeded in T175 flasks (ThermoFischer Scientific, Merelbeke, Belgium) at a density of 4×10^5 cells/mL in a complete THP-1 cell culture medium. Cells were induced to differentiate into macrophage-like THP-1 cells (MΦ-THP-1) by the addition of PMA at a concentration of 20 ng/mL. PMA was prepared as a stock solution (10 mg/mL) in ultrapure absolute ethanol and stored at −20 °C. This stimulation process was carried out for 48 h in a humidified incubator at 37 °C and 5% CO_2. Subsequently, the PMA-supplemented medium was replaced with fresh THP-1 complete cell culture medium on day 2.

On day 2, 6-well cell culture plates (Greiner Bio-One, Vilvoorde, Belgium) were filled with hanging cell culture inserts, provided with a high pore density membrane with a surface area of 4.5 cm^2 and 5 μm pore size (cellQART, Sabeu, Northeim, Germany). The plates with inserts were inverted to seed human endothelial cells, EA.hy926 (2.4×10^4 cells/cm^2), on the basolateral side of the insert semipermeable membrane. After the cells attached to the membrane, the plate with inserts was returned to its original orientation, and A549 epithelial cells were seeded on the apical side of the cell culture inserts (6×10^4 cells/cm^2). Complete EA.hy926 cell culture medium was added to the basolateral compartment, and both epithelial and endothelial cells were cultured to reach confluency over three days at 37 °C and 5% CO_2.

On day 5, the cell culture medium was removed from both compartments and coculture medium consisting of a mix of 25 mM HEPES buffered DMEM (75%), RPMI-1640 (15%) and IMDM (10%) supplemented with 10% FBS (coculture 10% FBS medium) was added.

On day 6, undifferentiated dendritic-like THP-1 (DC-THP-1) cells (1×10^6 cells/well) were seeded in the wells of new 6-well suspension cell culture plates (Greiner Bio-One, Vilvoorde, Belgium) using coculture medium consisting of a mix of HEPES buffered DMEM (75%), RPMI (15%) and IMDM (10%) supplemented with 1% FBS (coculture 1% FBS medium). The inserts containing the A549/EA.hy926 cells biculture were transferred to the plates containing the DC-THP-1 cells. MΦ-THP-1 cells were washed with PBS and were detached using Accutase for harvesting and seeding in the apical compartment (2.4×10^4 cells/cm^2) on the surface of the A549 cell monolayer. The medium for the tetraculture *in vitro* model was supplemented with 1% FBS to prevent excessive proliferation of DC-THP-1 cells. The attachment of MΦ-THP-1 cells to the epithelial monolayer was confirmed via light microscopy 4 h after seeding, and the medium was aspirated from the apical compartment. The test system was kept for 24 h at the air-liquid interface (ALI) before exposure.

2.4. Exposure

Exposure of ALIsens® was carried out using the TECAN D300e Digital Dispenser and T8+ dispensehead cassettes (Tecan, Mechelen, Belgium). The test items were dispensed onto the apical surface of the test system using the D300e Pattern software (version 1). The dispenser accurately and reproducibly delivers solutions of test items solubilized in pure DMSO or sterile water containing 0.1% Brij® 35 surfactant on the apical surface of the test system. The circular pattern was composed of 200 droplets of a volume of 30 nL each, arranged in a 0.5×0.5 mm grid format and spread over the apical surface area, with a total single dispense volume of 6 μL per insert. Prior to exposure to the selected test items, the impact of the two vehicles used for solution preparation on the viability of the test system was assessed. Pure DMSO and sterile water containing 0.1% Brij® 35 surfactant were dispensed on the apical surface, and cell viability was assessed after 24 h of incubation concurrently with non-exposed inserts (incubator control). ALIsens® was then exposed to increasing doses of the evaluated chemicals for 24 h to experimentally determine the individual dose-response curves for viability of the: (1) cells comprised in the apical compartment (A549+MΦ-THP-1), (2) EA.hy926 cells and (3) DC-THP-1 cells from the basolateral compartment and mathematically calculate the dose-response curves for cell viability of the (4) ALIsens® model as an average of the first three individual dose-response curves. The dose-response curve of the ALIsens® model was used to calculate the doses of investigated chemicals that reduce the cell viability of the test system by 25%, which were subsequently used to expose ALIsens® for the evaluation of endpoints for respiratory sensitization. The minimum and maximum doses dispensed for each test item are presented in Table 1.

2.5. Endpoint Measurements

2.5.1. Cell Viability

Cell viability of ALIsens® was measured using the resazurin assay. The test system was exposed for 24 h to increasing doses (μg/cm^2) of test items and the corresponding vehicle controls. Coculture 1% FBS medium was supplemented with 20 mM resazurin sodium salt stock solution prepared in PBS to reach a working concentration of 400 μM resazurin (resazurin working solution). The hanging cell culture inserts were transferred from the initial plates to new 6-well plates containing 1 mL of resazurin working solution to evaluate the viability of EA.hy926 cells from the tetraculture, and 1 mL of resazurin working solution was added in the apical compartment of the test system to evaluate the viability of A549 and MΦ-THP-1 cells from the tetraculture. To evaluate the viability of DC-THP-1 cells, 20 μL of resazurin stock solution was added to the cell suspension remaining in the wells of the plate and homogenized by horizontal manual shaking to reach a final concentration of 400 μM resazurin. The plates containing the test system exposed to test items, and the corresponding vehicle controls were incubated at 37 °C, 5% CO_2, in the dark for 1 h. Following the incubation period, aliquots of coculture 1% FBS medium ± metabolized resazurin, resorufin, were sampled from the apical (A549+ MΦ-THP-1) and basolateral (EA.hy926) side of the hanging cell culture insert and from the well of the plate containing the DC-THP-1 cell suspension, and pipetted to a 96-well, flat-bottom microplate (Greiner Bio-One, Vilvoorde, Belgium). The absolute fluorescence intensity of resorufin was measured using the TECAN Spark 20M multi-mode microplate reader (Tecan, Mechelen, Belgium) at 530 nm excitation and 590 nm emission. Relative cell viability was calculated using Equation (1):

$$\text{Cell viability} = \frac{\text{Absolute fluorescence intensity of the sample} - \text{Blank}}{\text{Absolute fluorescence intensity of the vehicle control} - \text{Blank}} \times 100 \quad (1)$$

The exposure dose reducing the viability of ALIsens® by 25%, resulting in a 75% residual cell viability (CV75) was calculated using Equation (2):

$$\text{LogCV75} = \frac{(75 - \mathrm{c}) \times \text{Log(b)} - (75 - \mathrm{a}) \times \text{Log(d)}}{\mathrm{a} - \mathrm{c}} \quad (2)$$

where a is the minimum value of cell viability over 75%, c is the maximum percentage of cell viability below 75%, and b and d are the concentrations showing the percentage of cell viability at a and c, respectively [37].

2.5.2. Flow Cytometry

DC-THP-1 cell suspensions were collected from the wells of the plates in conical centrifuge tubes (Greiner Bio-One, Vilvoorde, Belgium) and were centrifuged for 5 min at 300× *g*. The cell pellets were resuspended and washed in PBS at room temperature. The cells were centrifuged for 5 min at 300× *g*, the PBS wash was discarded, and the cells were resuspended in PBS supplemented with 2% FBS. Cells were stained at 4 °C for 45 min using commercially available fluorescent-labeled monoclonal antibodies: allophycocyanin (APC) mouse anti-human TSLPr (clone 1F11/TSLPR), Brilliant™ Blue 515 (BB515) mouse anti-human CD54 (clone HA58) and phycoerythrin (PE) mouse anti-human CD86 (clone 2331 (FUN-1)), and κ isotype-matched controls (mouse IgG1), APC (clone MOPC-21), BB515 (clone X40) and PE (clone MOPC-21) (BD Biosciences, Erembodegen, Belgium). 1 μM SYTOX™ Blue Nucleic Acid Stain (Invitrogen™ ThermoFischer Scientific, Carlsbad, NM,

USA) was used as a marker for non-viable cells. Expression of the cell surface markers was measured using the BD FACSCelesta™ Cell Analyzer (BD Biosciences, Erembodegen, Belgium), and data was collected using the BD FACSDiva™ software (version 9) provided with the instrument for data acquisition. A total of 10,000 events were acquired per sample, and further analysis was performed using FlowJo software (version 10) (BD Life Sciences, Ashland, OR, USA). Relative geometric mean fluorescence intensities (rMFI) were expressed as % compared to the vehicle control and calculated using Equation (3):

$$\text{rMFI} = \frac{\text{MFI sample} - \text{MFI isotype control of the sample}}{\text{MFI vehicle control} - \text{MFI isotype control of the vehicle control}} \times 100 \quad (3)$$

2.6. Graphics and Language

BioRender platform (Science Suite Inc., Toronto, ON, Canada) was used for the preparation of Figure 1. GraphPad Prism software (version 10) (GraphPad Software Inc., San Diego, CA, USA) was used for the preparation of Figures 2–6. ChatGPT-3.5 artificial intelligence (AI) language model (OpenAI, San Francisco, CA, USA) was used to improve the language and readability of the manuscript. Mendeley Reference Manager software (version 2) (Elsevier, Amsterdam, The Netherlands) was used to prepare the list of references.

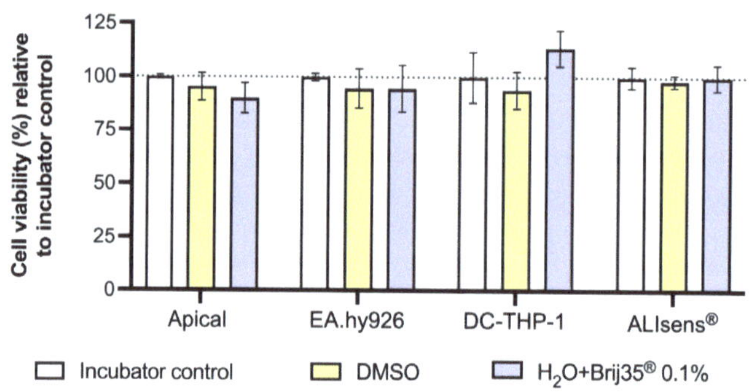

Figure 2. Viability of cells from the apical compartment (A549 and MΦ-THP-1), EA.hy926 and DC-THP-1 cells and ALIsens® following exposure to pure DMSO and sterile water containing 0.1% Brij® 35, relative to the incubator control. Data are expressed as mean ± SD of n = 1, N = 3. *p* values (>0.05) were determined by one-way ANOVA, Dunnett's post-hoc.

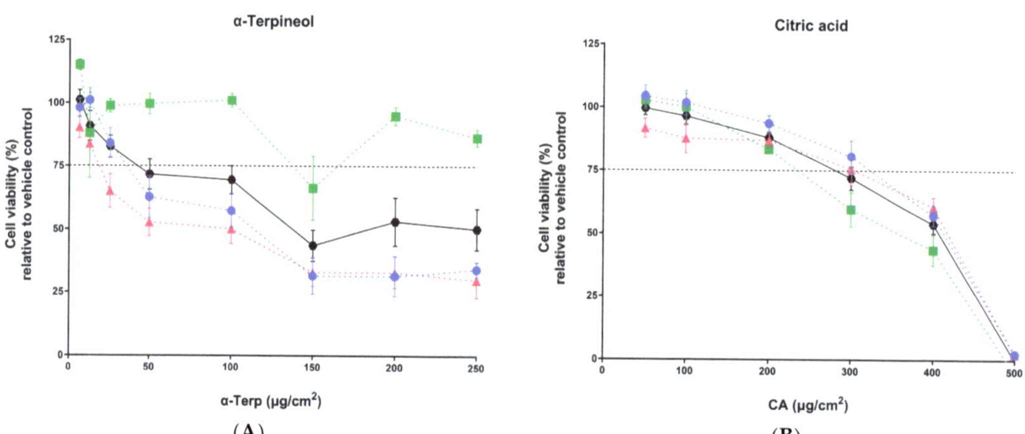

(A) (B)

Figure 3. *Cont.*

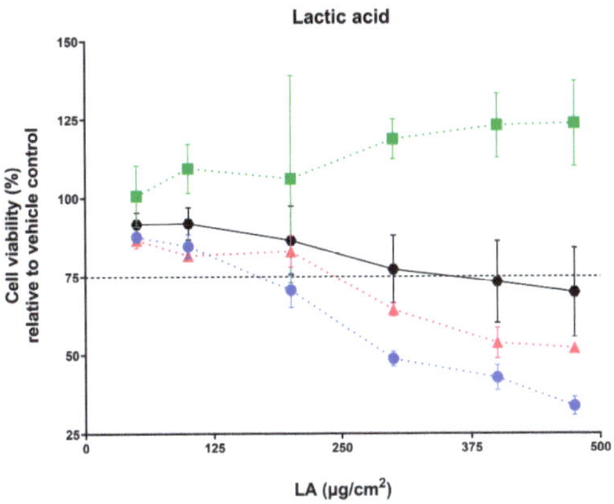

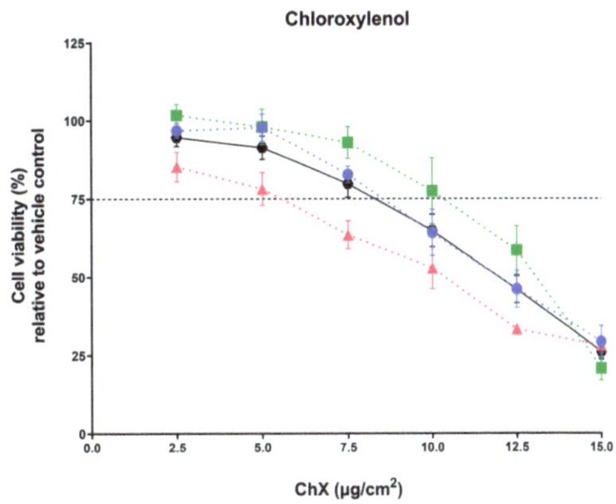

(C) (D)

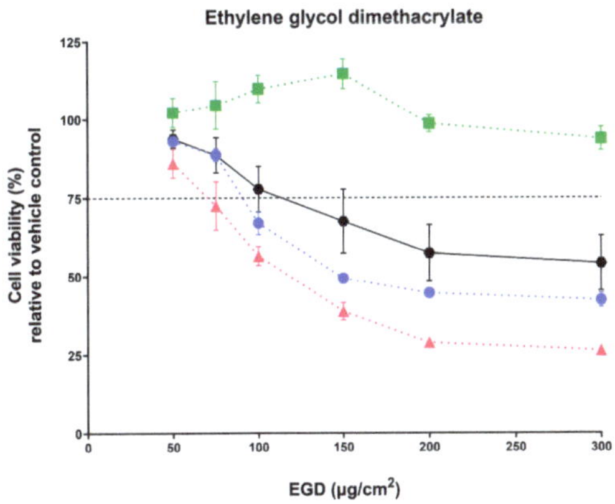

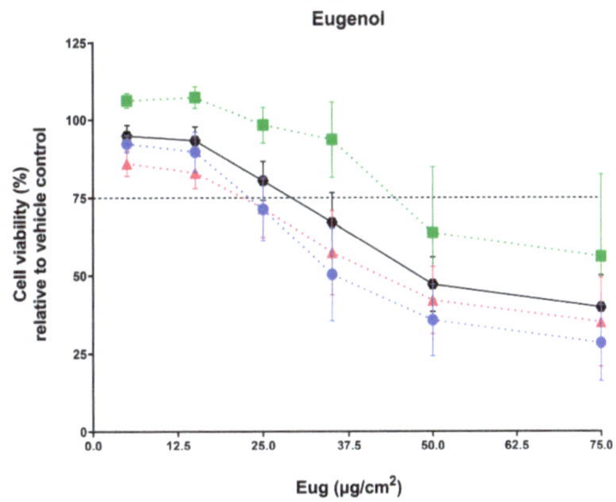

(E) (F)

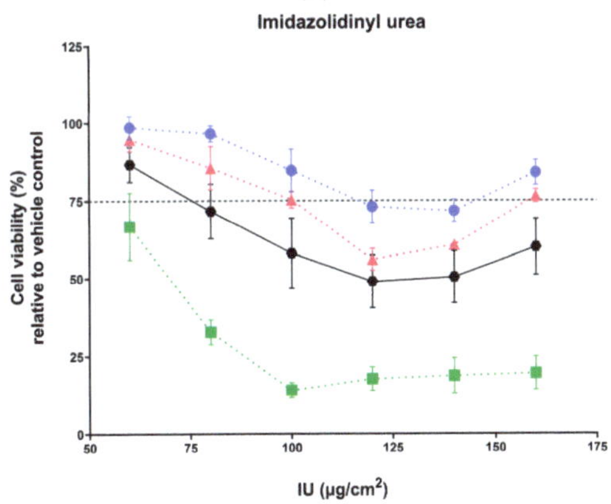

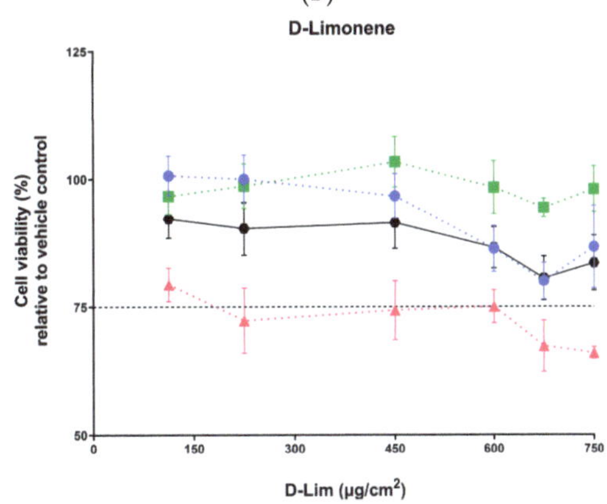

(G) (H)

Figure 3. *Cont.*

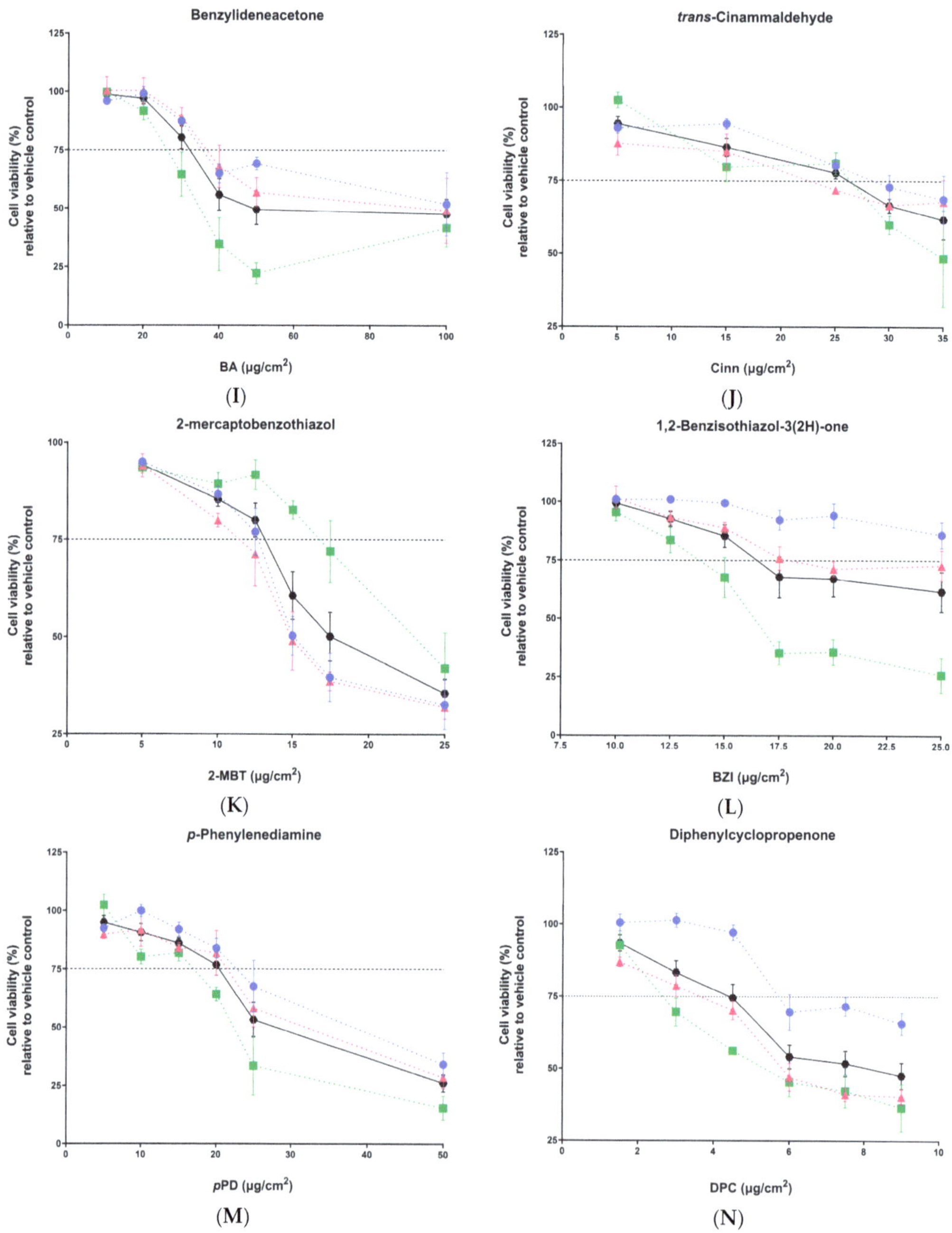

Figure 3. Viability of ALIsens® (black hexagons and line), A549 and MΦ-THP-1 cells (blue circles and dotted line), EA.hy926 cells (pink triangles and dotted line) and DC-THP-1 cells (green squares and dotted line) relative to the corresponding vehicle control following 24 h exposure to the investigated chemicals (**A**) α-terpineol, (**B**) citric acid, (**C**) lactic acid, (**D**) chloroxylenol, (**E**) ethylene glycol dimethacrylate, (**F**) eugenol, (**G**) imidazolidynil urea, (**H**) D-limonene, (**I**) benzylideneacetone, (**J**) *trans*-cinnamaldehyde, (**K**) 2-mercaptobenzothiazol, (**L**) 1,2-benzisothiazol-3(2H)-one, (**M**) *p*-phenylenediamine, (**N**) diphenylcyclopropenone. Data are expressed as mean ± SEM of n = 1, N ≥ 3.

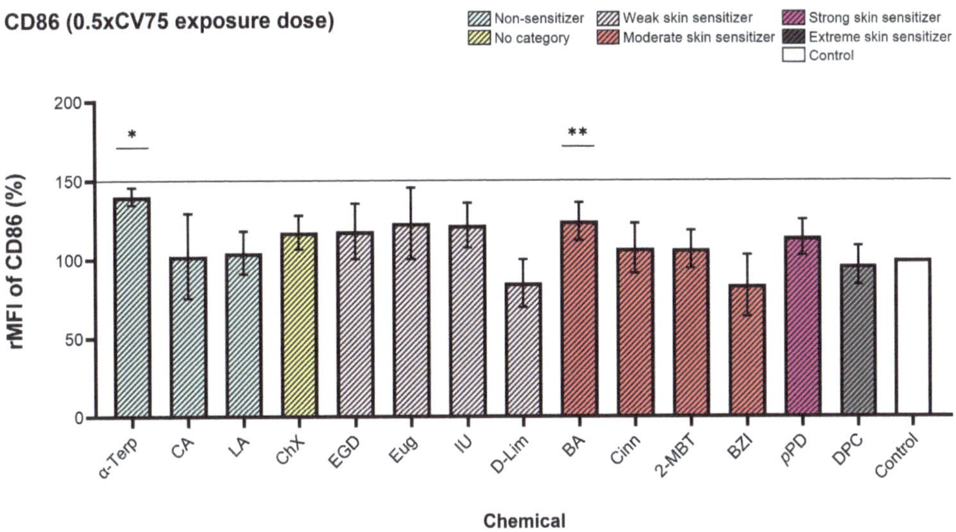

(**A**)

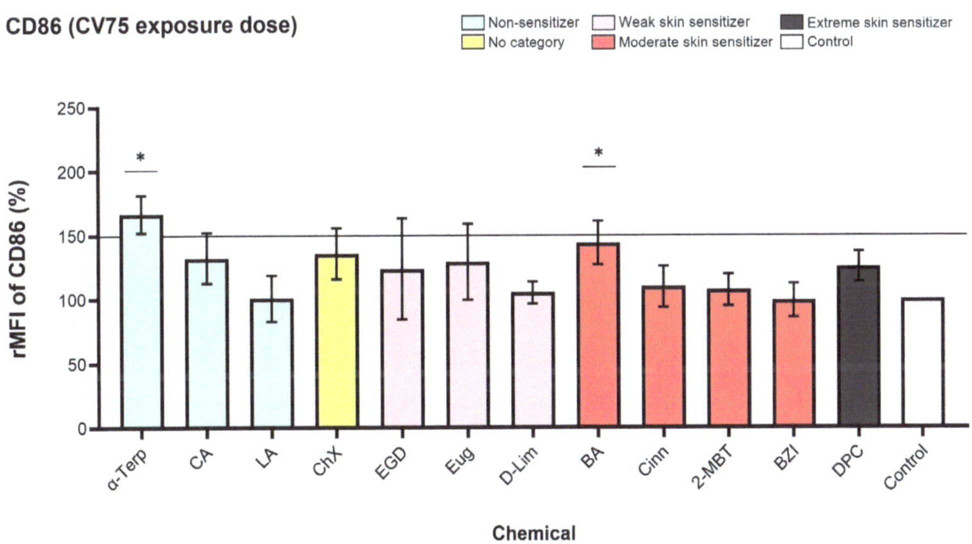

(**B**)

Figure 4. Expression of CD86 marker measured on the surface of DC-THP-1 cells following exposure of ALIsens® for 24 h to (**A**) 0.5× CV75 dose and (**B**) CV75 dose of the evaluated chemicals, assessed by flow cytometry. Data are expressed as mean ± SD of n = 1, of N = 3 (α-Terp, CA, ChX, IU, D-Lim, DPC), N = 4 (LA, EGD, Eug, Cinn, 2-MBT, *p*PD), N = 5 (BZI); N = 6 (BA) for the 0.5× CV75 exposure dose and of N = 3 (α-Terp, CA, ChX, D-Lim, BA, DPC), N = 4 (LA, EGD, Eug, Cinn, 2-MBT, BZI) for the CV75 exposure dose. p values were determined by one-sample t-test. $p \leq 0.05$ = *, $p \leq 0.01$ = **.

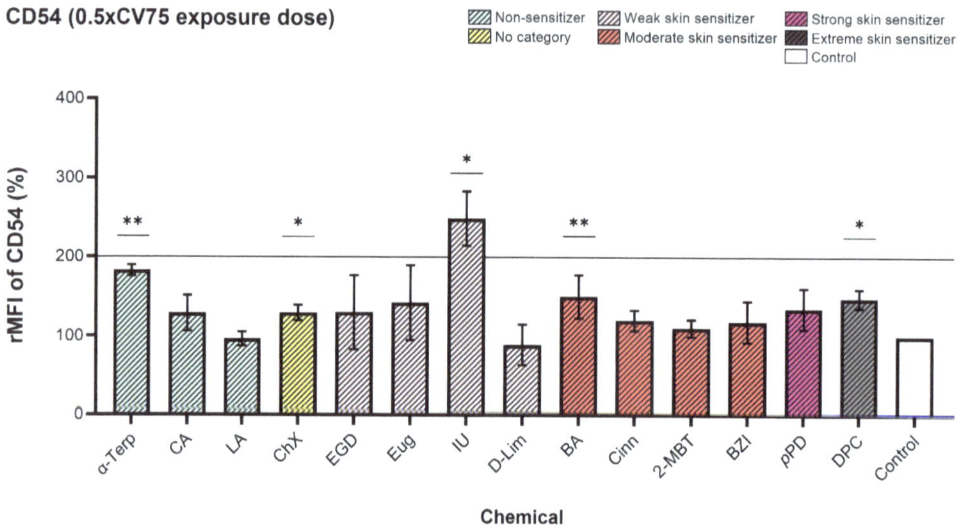

(A)

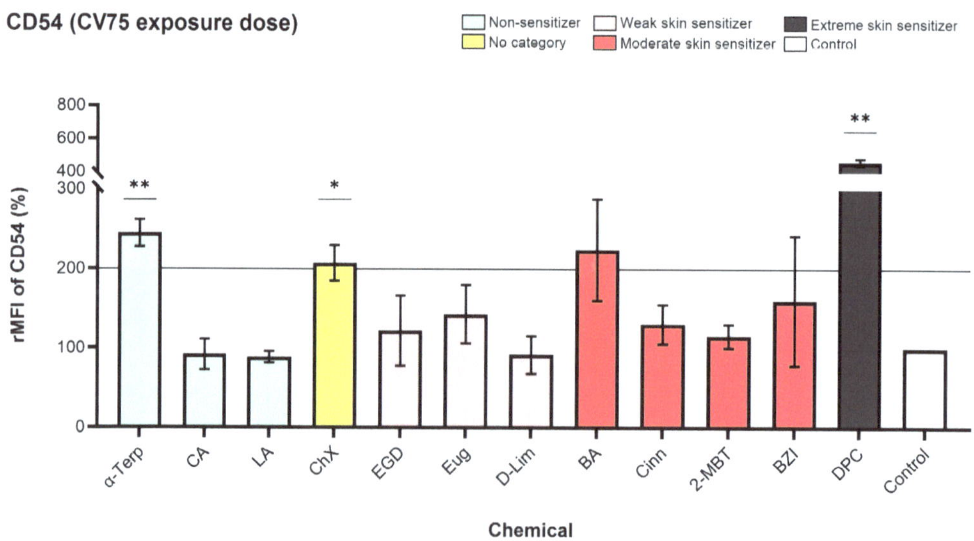

(B)

Figure 5. Expression of CD54 marker measured on the surface of DC-THP-1 cells following exposure of ALIsens® for 24 h to (**A**) 0.5× CV75 and (**B**) CV75 of the evaluated chemicals, assessed by flow cytometry. Data are expressed as mean ± SD of n = 1, of N = 3 (α-Terp, CA, ChX, IU, D-Lim, DPC), N = 4 (LA, EGD, Eug, Cinn, 2-MBT, *p*PD), N = 5 (B ZI); N = 6 (BA) for the 0.5× CV75 exposure dose and of N = 3 (α-Terp, CA, ChX, D-Lim, BA, DPC), N = 4 (LA, EGD, Eug, Cinn, 2-MBT, BZI) for the CV75 exposure dose. *p* values were determined by one-sample *t*-test. $p \leq 0.05$ = *, $p \leq 0.01$ = **.

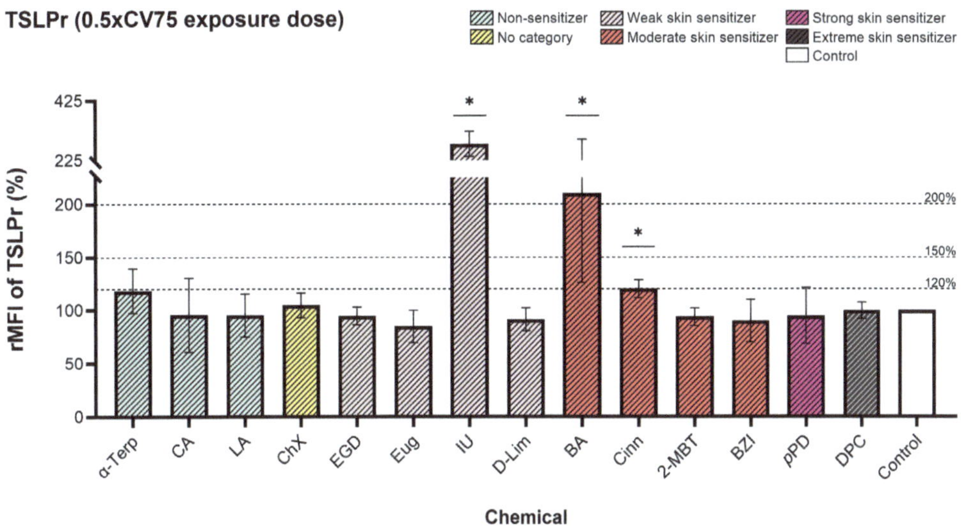

(A)

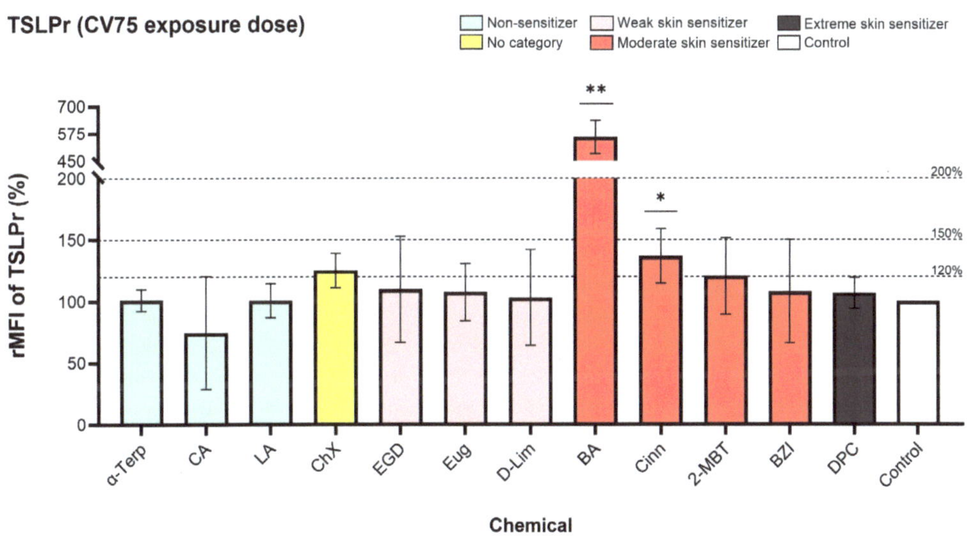

(B)

Figure 6. Expression of TSLPr marker measured on the surface of DC-THP-1 cells following exposure of ALIsens® for 24 h to (**A**) 0.5× CV75 and (**B**) CV75 of the evaluated chemicals, assessed by flow cytometry. Data are expressed as mean ± SD of n = 1, of N = 3 (α-Terp, CA, ChX, IU, D-Lim, DPC), N = 4 (LA, EGD, Eug, Cinn, 2-MBT, *p*PD), N = 5 (BZI); N = 6 (BA) for the 0.5× CV75 exposure dose and of N = 3 (α-Terp, CA, ChX, D-Lim, BA, BZI, DPC), N = 4 (LA, EGD, Eug, Cinn, 2-MBT) for the CV75 exposure dose. *p* values were determined by one-sample *t*-test. $p \leq 0.05$ = *, $p \leq 0.01$ = **.

2.7. Statistical Analysis

Data are presented as mean ± standard deviation (SD) or standard error of the mean (SEM). The number of technical replicates (n), biological replicates (N), and statistical tests are indicated in the figures' descriptions. Data from the cell viability dose-response curves was used to calculate the CV75 value. Differences between control and exposed groups were evaluated for the vehicle impact on the viability of ALIsens® and cell surface marker expression and were considered statistically significant if p-values were $p \leq 0.05$ = *, $p \leq 0.01$ = **. Data were analyzed using GraphPad Prism software (version 10).

3. Results

3.1. Cell Viability

3.1.1. Vehicle Impact on the Viability of ALIsens®

The test items solutions were prepared in pure DMSO and sterile water with 0.1% Brij® 35 surfactant and were delivered on the apical surface of ALIsens® using the TECAN D300e digital dispenser. DMSO is a vehicle frequently used for the dissolution of chemicals that are tested in cell culture and is usually used at a concentration that does not influence the test and the viability of the cell lines used in the method [62], being usually diluted in the exposure vehicle. Brij® 35, a nonionic surfactant, was added to the aqueous solution to be dispensed to reduce the surface tension. Surfactants are irritants [63], and they can lead to cytotoxic effects in *in vitro* studies employing cell lines [64]. In our study, the solutions of tested chemicals were directly applied to the apical compartment of ALIsens® in a pattern of 200 droplets, each containing 30 nL, covering the entire surface of the apical compartment. The impact of the vehicles used for the preparation of solutions of tested chemicals on the viability of the *in vitro* test system was evaluated against a non-exposed sample, the incubator control. Data reveals that direct application of pure DMSO and sterile water containing 0.1% Brij® 35 surfactant did not have a significant impact on the viability of any of the cell types comprised in the *in vitro* alveolar test system (Figure 2).

3.1.2. Dose-Response Curves of Evaluated Chemicals

The viability of ALIsens® was evaluated 24 h post-exposure using the resazurin assay. This allowed the determination of the dose-response curve of each tested chemical that provided the data to calculate the final exposure doses (μg/cm^2) for the analysis of sensitization potential. The standard ALIsens® protocol requires that the respiratory sensitization potential is tested 24 h post-exposure using an exposure dose of the test item, leading to a 25% reduction of cell viability (residual cell viability of 75%), CV75 [43].

The cell viability assay has shown that all tested chemicals (Figure 3), with the exception of D-limonene (Figure 3H), reduced the viability of cells with increasing exposure dose. This allowed the calculation of the CV75 dose for the chemicals which reduced the viability of the test system by more than 25%. The calculated doses that were used for exposures for the evaluation of respiratory sensitization markers are presented in Table 2. As D-limonene did not reduce the cell viability of ALIsens® under 75%, the highest exposure dose from the range of doses applied for the generation of the dose-response curve (750 μg/cm^2) and 0.5× 750 μg/cm^2 (375 μg/cm^2) were used for the exposures for the assessment of respiratory sensitization endpoints.

Following exposure to increasing doses of citric acid (Figure 3B), chloroxylenol (Figure 3D), eugenol (Figure 3F), *trans*-cinnamaldehyde (Figure 3J), 2-mercaptobenzothiazol (Figure 3K), 1,2-benzisothiazol-3(2H)-one (Figure 3L), *p*-phenylenediamine (Figure 3M) and diphenylcyclopropenone (Figure 3N), all cell types comprised in the ALIsens® model exhibit a uniform response, with the viability of each cell type steadily decreasing as the exposure dose increases. α-Terpineol (Figure 3A), chloroxylenol (Figure 3D), ethylene glycol dimethacrylate (Figure 3E) and D-limonene (Figure 3H) exposures had the highest impact on the EA.hy926 cells. The viability of DC-THP-1 cells does not consistently decrease with increasing doses of α-terpineol (Figure 3A), lactic acid (Figure 3C), ethylene glycol dimethacrylate (Figure 3E) and D-limonene (3H). Overall, these chemicals have minimal to no impact on the viability of DC-THP-1 cells. On the other hand, DC-THP-1 cells seem to be the most sensitive cells of the ALIsens® model upon exposure to imidazolidinyl urea (Figure 3G), benzylideneacetone (Figure 3I), 1,2-benzisothiazol-2(3H)-one (Figure 3L) and diphenylcyclopropenone (Figure 3N).

Table 2. Calculated 75% cell viability (CV75) exposure dose value and 0.5× CV75 exposure dose values for ALIsens® exposed for 24 h to the evaluated chemicals.

Chemical	CV75 [1] (μg/cm²)	0.5× CV75 [2] (μg/cm²)
α-Terpineol	41.4	20.7
Citric acid	278	139
Lactic acid	346.4	173.2
Chloroxylenol	8.25	4.13
Ethylene glycol dimethacrylate	112.9	56.5
Eugenol	28.5	14.2
Imidazolidynil urea	75.5	37.8
D-Limonene	750 *	375 **
Benzylideneacetone	31.9	15.9
Trans-Cinnamaldehyde	26.2	13.1
2-Mercaptobenzothiazol	13.1	6.6
1,2-Benzisothiazol-3(2H)-one	16.4	8.2
p-Phenylenediamine	20.4	10.2
Diphenylcyclopropenone	4.62	2.31

[1] The exposure doses resulting in 75% residual cell viability of the 24 h exposed ALIsens®, expressed in μg/cm².
[2] The exposure doses multiplied with a factor of 0.5 resulting in 75% residual cell viability of the 24 h exposed ALIsens®, expressed in μg/cm². *—maximum exposure dose; **—0.5× maximum exposure dose.

3.2. Cell Surface Markers Expression

The CD86, CD54, and TSLPr cell surface markers expression measured on DC-THP-1 cells from the exposed ALIsens® model are presented in Figure 4, Figure 5, and Figure 6, respectively.

α-Terpineol enhanced the expression of CD54 and CD86 markers on the surface of DC-THP-1 cells following exposure of the ALIsens® model. Expression of CD86 (140% for 0.5× CV75 exposure dose and 167% for CV75 exposure dose) (Figure 4) and CD54 (183% for 0.5× CV75 exposure dose and 245% for CV75 exposure dose) (Figure 5) markers was significantly induced at both tested doses of α-terpineol, exceeding the 150% and 200% thresholds, respectively, at the CV75 exposure dose (Figures 4B and 5B). No significant induction of the TSLPr cell surface marker was observed following exposure to the 0.5× CV75 and CV75 doses of α-terpineol, which resulted in an upregulation of 118% and 101% as compared to control, respectively (Figure 6A,B). Chloroxylenol significantly increased the expression of CD54 cell surface marker following exposure to both exposure doses (129% for 0.5× CV75 and 207% for CV75) (Figure 5A,B) and over the 200% threshold following exposure to CV75 exposure dose (Figure 5B). No significant increase in the expression of the other investigated markers, CD86 and TSLPr, was observed following exposure to the two doses of chloroxylenol (Figures 4 and 6).

Imidazolidinyl urea exposure significantly increased the expression of CD54 (249%) and TSLPr (279%) markers and over the 200% threshold for the CD54 marker at 0.5× CV75 dose. Benzylideneacetone significantly enhanced the expression of CD54 cell surface marker at the lower exposure dose, 0.5× CV75 (150%) (Figure 5A), with an increase in expression over the 200% threshold following exposure to CV75 dose (Figure 5B). Both evaluated exposure doses of benzylideneacetone (0.5× CV75 and CV75) significantly induced the expression of CD86 (Figure 4A,B) and TSLPr (Figure 6A,B) cell surface markers.

Trans-Cinnamaldehyde significantly upregulated TSLPr expression on the surface of DC-THP-1 cells following exposure to both doses of 0.5× CV and CV75, resulting in values of 120% and 137%, respectively, as compared to control (Figure 6A,B).

Diphenylcyclopropenone significantly induced the expression of CD54 cell surface marker up to 463%, over the 200% threshold, at the CV75 exposure dose (Figure 5B).

For two evaluated chemicals, imidazolidinyl urea and *p*-phenylenediamine, the expression of the cell surface markers following exposure to the CV75 doses was not calculated, as the cell viability of DC-THP-1 cells measured with SYTOX^{TM} Blue in the flow cytometric readings, was lower than 75%). A low cell viability of the cells can lead to potential diffuse labeling of cytoplasmic structures as a result of cell membrane destruction [35], and this could render false-positive or false-negative results. Imidazolidinyl urea reduced the cell viability of the DC-THP-1 cells from ALIsens® below 75% already at the first testing dose from the dose-response curve, 60 μg/cm^2. It also significantly increased the expression of the CD54 cell surface marker at the 0.5× CV75 exposure dose (37.8 μg/cm^2), surpassing the 200% threshold set by the h-CLAT assay. Additionally, this skin sensitizer significantly induced the expression of the TSLPr marker on the surface of DC-THP-1 cells up to the value of 279%.

The accuracy of the three proposed thresholds for TSLPr cell surface marker expression to correctly identify skin sensitizers and non-sensitizers as non-respiratory sensitizers was determined using data from all 14 evaluated chemicals at the 0.5× CV75 dose. For the CV75 dose, accuracy was calculated based on data from 12 chemicals, as the data from exposures to imidazolidinyl urea and *p*-phenylenediamine were not included due to low DC-THP-1 cell viability, which prevented the evaluation of cell surface marker expression. At the 0.5× CV75 dose, an accuracy of 79% was observed for the 120 threshold, while 86% accuracy was achieved for both the 150% and 200% thresholds. At the CV75 dose, the accuracy was 67% for the 120 thresholds and 92% for the 150% and 200% thresholds (Table 3).

Table 3. Summary data for TSLPr expression following exposure to the investigated chemicals to 0.5× CV75 and CV75 exposure doses.

Exposure Dose (μg/cm^2)	**0.5× CV75**			**CV75**		
Tentative Threshold (%)	**120**	**150**	**200**	**120**	**150**	**200**
α-Terp	-	-	-	-	-	-
CA	-	-	-	-	-	-
ChX	-	-	-	+	-	-
LA	-	-	-	-	-	-
EGD	-	-	-	-	-	-
Eug	-	-	-	-	-	-
IU	+	+	+	n/a	n/a	n/a
D-Lim	-	-	-	-	-	-
BA	+	+	+	+	+	+
Cinn	+	-	-	+	-	-
2-MBT	-	-	-	+	-	-
BZI	-	-	-	-	-	-
p-PD	-	-	-	n/a	n/a	n/a
DPC	-	-	-	-	-	-
Accuracy (%)	79	86	86	67	92	92

rMFIs (%) of TSLPr expression on the surface of DC-THP-1 cells were compared to three tentative criteria: 120%, 150% and 200%. + represents a rMFI value of TSLPr ≥ tentative criterion, and "-" means a rMFI value of TSLPr < tentative criterion. n/a represents not available data.

4. Discussion

AOP 40 outlines the mechanisms of skin sensitization, providing comprehensive insights into the skin sensitization pathway, making it the gold standard in the field [11]. This AOP serves as a starting point for understanding respiratory sensitization [30]. It informs AOP 39 [65] by offering a well-established framework for studying the molecular and cellular processes that lead to sensitization, including KEs such as hapten binding to skin proteins, cellular activation, and cytokine release. Since the immune mechanisms underlying skin and respiratory sensitization share similarities along the events of the two AOPs, with the main difference between skin and respiratory sensitization being the type of adaptive immune responses [12], the detailed guidance in AOP 40 helps researchers apply these principles to respiratory outcomes in AOP 39, supporting the development of non-animal testing strategies for respiratory sensitization. Gaining further knowledge of the mechanisms involved in the process of respiratory sensitization is critical for advancing the development of accurate assays [30].

The adverse outcomes (AOs) from exposure to different sensitizers vary, with respiratory sensitizers causing allergic reactions in the respiratory system, leading to conditions like asthma, rhinitis, or other respiratory tract issues [8], while skin sensitizers trigger allergic reactions in the skin, resulting in allergic contact dermatitis (ACD) or eczema. The severity of AOs is closely linked to their effect on an individual's health and quality of life. Since the impact of these two factors varies, along with the type of treatment required, it is important to distinguish between the two classes of sensitizers. In addition, sensitization in one system can potentially exacerbate sensitization in the other, leading to more severe allergic reactions overall.

The exposure route is another aspect that needs to be considered in the context of sensitization, as skin sensitizers typically induce sensitization upon skin contact, while respiratory sensitizers primarily induce sensitization through inhalation. There is considerable evidence suggesting that respiratory sensitization can be acquired as a result of skin exposure [10,66–68]. Accurate identification of respiratory sensitizers is crucial for implementing effective preventive measures and minimizing exposure risks in occupational and environmental settings. Over-classifying chemicals as respiratory sensitizers can lead to unnecessary costs from additional workplace controls or regulatory bans, while under-classification compromises worker safety. Therefore, distinguishing respiratory sensitizers from skin sensitizers and non-sensitizers/respiratory irritants is essential for balanced protection and cost management [9].

The validated *in vitro* test systems for the evaluation of skin sensitizers cannot differentiate between skin and respiratory sensitizers [22], and moreover, some respiratory sensitizers which have been classified as sensitizers in the LLNA, yield false-negative results in some of the methods used for the evaluation of skin sensitizers, such as phthalic anhydride, a respiratory sensitizer [13,21], which tested negative in the h-CLAT assay [56]. The primary focus of ALIsens® is to correctly identify the respiratory sensitization hazard of LMW chemicals. The test system proved to correctly identify acid anhydrides as respiratory sensitizers based on the selection of biomarkers, either expressed on the surface of DCs or secreted into the cell culture medium from the basolateral compartment of the model [43]. Nevertheless, further chemicals from the reference lists of respiratory sensitizers [21,69,70] need to be evaluated to define the capacity and the accuracy of the model to identify such allergens.

In the development of ALIsens®, the decision of positive or negative with respect to respiratory sensitization was based on statistically significant differences in the expression of evaluated cell surface markers (CD54, CD86, TSLPr) measured on at least three biological replicates. In the h-CLAT prediction model, each test item is evaluated in at least two independent runs to derive a single prediction. If the first two runs are not concordant for at least one of the two markers, a third run is needed, and the final prediction is based on the majority result of the three individual runs. The upregulation of CD54 or/and CD86 markers of at least or greater than 200% and 150%, respectively, confirms THP-1 cell activation, representing KE3 [37], a common KE for skin and respiratory sensitization [10,12].

This study aimed to assess the response of ALIsens® by analyzing changes in the expression of cell surface markers when exposed to skin sensitizers and non-sensitizers to evaluate the specificity of the model to correctly identify respiratory sensitizers and differentiate between skin and respiratory sensitizers. In particular, we sought to identify the most indicative marker(s) and propose a preliminary threshold that could accurately identify skin sensitizers, non-sensitizers and non-respiratory sensitizers. To achieve this, we tested a panel of chemical substances for their potential to increase the expression of CD86, CD54 and TSLPr cell surface markers on DC-THP-1 cells. The study included skin sensitizers of varying potency (weak, moderate, strong, and extreme) and non-sensitizers, most of which are proficiency chemicals from the OECD TG for the evaluation of skin sensitizers (lactic acid, ethylene glycol dimethacrylate, eugenol, imidazolidinyl urea, D-limonene, benzylideneacetone, *trans*-cinnamaldehyde, 2-mercaptobenzothiazole, *para*-phenylenediamine) [37,57,58]. TSLPr has been previously identified as a promising and reliable marker for correctly detecting respiratory sensitizers and distinguishing them from non-sensitizers [43]. However, since no skin sensitizers were included in previous studies, it remained unclear whether ALIsens® could correctly identify respiratory sensitizers and potentially differentiate them from skin sensitizers and non-sensitizers [9,43].

Among the tested skin sensitizers, α-terpineol led to a significant increase in the expression of CD86 and CD54 cell surface markers at both exposure doses applied to ALIsens® for 24 h. For terpineol, a mixture of the four isomers, α-, β-, γ- and terpin-4-ol, with the α-terpineol as principal constituent, there is no concern for skin sensitization [71]. In the h-CLAT assay, α-terpineol was found positive, as it enhanced the expression of CD54 cell surface marker, but in the "2 out of 3" integrated approach, it was classified as a non-sensitizer, as it scored negative [59]. Our findings for α-terpineol are in accordance with the results from the h-CLAT assay. α-terpineol did not enhance the expression of the TSLPr cell surface marker, neither in a significant manner nor over the tentative thresholds, indicating that ALIsens® can correctly identify α-terpineol as a non-respiratory sensitizer. Imidazolidinyl urea, tested at the 0.5× CV75 exposure dose, significantly increased the expression of CD54 and TSLPr cell surface markers, surpassing the established threshold of 200% in the h-CLAT assay and the proposed tentative criterion of 200%, respectively. This skin sensitizer yielded positive results in the h-CLAT assay for both evaluated cell surface markers, CD86 and CD54 [56]. Benzylideneacetone elevated the expression of TSLPr on the surface of DC-THP-1 cells at both tested exposure doses, surpassing the tentative threshold of 200% for TSLPr at both the 0.5× CV75 and CV75 doses. Chloroxylenol increased TSLPr expression beyond the proposed 120% threshold at the CV75 exposure dose, and *trans*-cinnamaldehyde increased the expression of the TSLPr cell surface marker at both exposure doses, over 120%, though it remained under 150% rMFI value. 2-mercaptobenzothiazole increased the expression of TSLPr over the 120% proposed threshold, but only at the low tested dose.

In the development of the h-CLAT assay, three arbitrary thresholds, relative fluorescence intensity (RFI) values of 120%, 150%, and 200%, were selected to evaluate the assay's ability to correctly distinguish skin sensitizers from non-sensitizers. Two independent experiments were conducted at two laboratories, each testing four doses for each test item. The mean RFI values for CD86 and CD54 cell surface markers and cell viability were determined from these two independent experiments to assess accuracy across the selected thresholds [36]. A similar approach was applied in this study, and the obtained data facilitated the calculation of ALIsens® accuracy in identifying skin sensitizers and non-sensitizers based on the proposed thresholds. The ALIsens® model was exposed to two evaluated doses of each chemical, with at least three biological replicates per condition, to assess the rMFI values for TSLPr, CD54, and CD86 cell surface markers.

For the tentative criterion rMFI of 120%, the lowest accuracy was observed for both exposure conditions (0.5× CV75 and CV75) with values of 79% and 67%, respectively. For the 150% and 200% tentative criteria, an accuracy of 86% was observed for the 0.5× CV75 exposure dose and 92% for the CV75 exposure dose. It has been previously observed that activation of dendritic cells requires a certain level of toxicity [72,73], and at the same time, respiratory sensitization is a threshold-based phenomenon [66]. The accuracy for the thresholds of 150% and 200% was consistent across both exposure doses, with the 150% threshold emerging as the most suitable for effectively identifying skin sensitizers and non-sensitizers as non-respiratory sensitizers in the ALIsens® model.

Cell viability is the gatekeeper assay in the evaluation of immunotoxic effects, and as a subtoxic dose was successfully used for the evaluation of skin sensitizers [74], a robust approach should be used. The expression for the evaluated cell surface markers was not evaluated for two tested chemicals at the CV75 exposure dose due to the low cell viability of the DC-THP-1 cells in the flow cytometric analysis. Therefore, an alternative approach should be developed for the testing of chemicals that significantly reduce the viability of THP-1 cells used as a model for dendritic cells in the test system but have minimal to no impact on the other cell types (i.e., A549, EA.hy926) used for the build-up of ALIsens®. Calculating the CV75 solely from the DC-THP-1 cell viability data acquired following exposure of ALIsens® to such chemicals would represent a potential solution to ensure, on the one hand, that the dendritic cells were exposed to the test item indirectly, through the alveolo-capillary barrier, while on the other hand, the population of viable cells is more or less 75%, in order not to hamper the measurements of cell surface markers.

The cell viability information is important in evaluating the potential of substances to induce sensitization in the respiratory tract since ALIsens® addresses a biological mechanism, the modulation of the expression of cell surface phenotypic markers associated with the maturation of DC-THP-1 cells. Despite its simplicity, the cell viability assay showed the complexity of the tetraculture model, as the different types of cells of ALIsens® respond in a particular way following exposure to low molecular weight chemicals belonging to different chemical classes. Delivery of test items in the apical compartment did not necessarily reduce the cell viability of the combination of epithelial and macrophage-like THP-1 cells to a greater extent than reducing the viability of cells from the basolateral compartment, endothelial or dendritic-like THP-1 cells. These data suggest that the different cellular components of a complex alveolar *in vitro* test system may have certain implications in the mechanism of respiratory sensitization. The dissection of the model to evaluate the independent contribution of each cell type of the alveolo-capillary barrier and their individual and orchestrated impact on the dendritic cells could help connect some dots in the mechanism of respiratory sensitization.

In this study, we evaluated the third KE of the respiratory sensitization AOP, highlighting that ALIsens® is a promising complex 3D NAM for the evaluation of respiratory sensitization hazard. The cellular components and the architecture of the model designed for the identification of respiratory sensitization could also contribute to the differentiation of respiratory sensitizers from skin sensitizers and non-sensitizers based on the second KE (KE2), the activation of danger signals through the release of inflammatory cytokines and chemokines [10,43]. Secreted markers GM-CSF, macrophage inflammatory protein-3-alpha (MIP-3α)/CCL20 and IL-10 have been identified as predictive markers for the correct classification of acid anhydrides in a linear discriminant analysis [43]. Several other *in vitro* NAMs developed for the evaluation of chemicals to induce respiratory sensitization also addressed the KE2 of the AOP and measured the expression or secretion of cytokines and chemokines, including IL-6, IL-8, IL-18, monocyte chemoattractant protein-1 (MCP-1)/chemokine ligand (CCL)2, growth regulated oncogene-a (GRO-α)/C-X-C motif (CX)CL1, and regulated on activation, normal T cell expressed and secreted (RANTES)/CCL5, IL-1a, tumor necrosis factor (TNF-α) [41,75–77]. As respiratory sensitization can be acquired following skin exposure, NAMs dedicated to the detection of skin sensitizers could be used to classify the two types of sensitizers and additionally provide a deeper understanding of the differences in the molecular mechanisms of respiratory and skin sensitization. Specifically, an *in vitro* study using keratinocytes as a cellular model evaluating IL-18 levels following exposure to skin and respiratory sensitizers concluded that the two evaluated respiratory sensitizers did not increase the intracellular cytokine concentration [78]. A panel of inflammatory markers, including, but not limited to, cytokines and chemokines previously identified as promising markers for the identification of respiratory sensitizers, could be incorporated in future studies to evaluate the response of ALIsens® following exposure to respiratory and skin sensitizers as well as non-sensitizer.

5. Conclusions

The correct identification of respiratory sensitizers and differentiating them from skin sensitizers is crucial for the protection of human health, ensuring regulatory compliance, promoting occupational safety, and facilitating informed decision-making in various contexts, including product labeling and medical care.

By conducting this study, our aim was to answer several questions: (a) can the test system correctly identify skin sensitizers and non-sensitizers and potentially differentiate them from respiratory sensitizers? (b) what is/are the specific biomarker(s) that should be measured to correctly classify chemicals as respiratory sensitizers, and (c) what is the threshold for this/these biomarker(s) that provide the best specificity? Based on the obtained results and the calculated accuracy values for the TSLPr cell surface marker, we can conclude that ALIsens® has the potential to identify skin sensitizers and not harmful chemicals as non-respiratory sensitizers. However, since no respiratory sensitizers were included in this study, it is essential to further validate the tentative 150% threshold with a larger dataset comprising data of respiratory sensitizers tested in ALIsens® to ensure robust conclusions.

These findings can fill a knowledge gap in the AOP for respiratory sensitization and provide insights that can inform both the scientific community and the regulatory bodies. The findings of this research have the potential to pave the way towards a better understanding of the mechanisms of respiratory sensitization, ultimately contributing to the identification of respiratory sensitizers in the early development phases of novel chemicals and molecules across various industries, with the main goal of protecting human health.

Supplementary Materials: The following supporting information can be downloaded at: https://www.mdpi.com/article/10.3390/toxics13010029/s1, Table S1. List of materials used in the study.

Author Contributions: Conceptualization, S.B., A.C., T.S., S.C., K.S., E.B., N.S. and A.C.G.; methodology, S.B. and A.C.; investigation, S.B.; formal analysis, S.B.; visualization, S.B. and A.C.G.; writing—original draft preparation, S.B.; writing—review and editing, all authors; supervision, A.C.G. and T.S.; project administration, A.C.G. and T.S.; funding acquisition, A.C.G. and T.S. All authors have read and agreed to the published version of the manuscript.

Funding: This research was funded by the Luxembourg Institute of Science and Technology (LIST) financed internal project DECHIR (2022), the Physicians Committee for Responsible Medicine (PCRM) and the Carroll Petrie Foundation.

Institutional Review Board Statement: Not applicable.

Informed Consent Statement: Not applicable.

Data Availability Statement: Data are available in the article.

Acknowledgments: The authors thank Carmen Socaciu for the supervision, writing assistance and critical review of the manuscript, which enhanced the quality and clarity of the work. The authors also thank Kahina Mehennaoui for their valuable contribution and expertise in statistical analysis.

ChatGPT-3.5 use was limited to language editing to improve readability without changes in the interpretation of scientific data presented in the paper.

Conflicts of Interest: Arno C. Gutleb is the managing director of Invitrolize Sarl. Sabina Burla is an employee of Invitrolize Sarl. The remaining authors declare that the research was conducted in the absence of any commercial or financial relationship that could be considered a potential conflict of interest. The funders had no role in the design of the study, in the collection, analyses, or interpretation of data; in the writing of the manuscript; or in the decision to publish the results.

References

1. Kotz, S.; Pechtold, L.; Jörres, R.A.; Nowak, D.; Chaker, A.M. Occupational rhinitis. *Allergol. Select* **2021**, *5*, 51–56. [CrossRef] [PubMed]
2. Feary, J.; Lindstrom, I.; Huntley, C.C.; de la Hoz, R.E.; Suojalehto, H. Occupational lung disease: When should I think of it and why is it important? *Breathe* **2023**, *19*, 230002. [CrossRef] [PubMed]
3. Kimber, I.; Dearman, R.J. What makes a chemical a respiratory sensitizer? *Curr. Opin. Allergy Clin. Immunol.* **2005**, *5*, 119–124. [CrossRef] [PubMed]
4. Nelson, J.L.; Mowad, C.M. Allergic contact dermatitis: Patch testing beyond the TRUE test. *J. Clin. Aesthet. Dermatol.* **2010**, *3*, 36–41.
5. Zuberbier, T.; Lötvall, J.; Simoens, S.; Subramanian, S.V.; Church, M.K. Economic burden of inadequate management of allergic diseases in the European Union: A GA2LEN review. *Allergy Eur. J. Allergy Clin. Immunol.* **2014**, *69*, 1275–1279. [CrossRef]
6. Nurmagambetov, T.; Kuwahara, R.; Garbe, P. The economic burden of asthma in the United States, 2008–2013. *Ann. Am. Thorac. Soc.* **2018**, *15*, 348–356. [CrossRef]
7. Regulation (EC) No 1272/2008 of the European Parliament and of the Council of 16 December 2008 on Classification, Labelling and Packaging of Substances and Mixtures, Amending and Repealing Directives 67/548/EEC and 1999/45/EC, and Amending Regulation (EC) No 1907/2006. Available online: https://eur-lex.europa.eu/legal-content/EN/TXT/?uri=celex:32008R1272 (accessed on 2 May 2024).
8. Chary, A.; Hennen, J.; Klein, S.G.; Serchi, T.; Gutleb, A.C.; Blömeke, B. Respiratory sensitization: Toxicological point of view on the available assays. *Arch. Toxicol.* **2017**, *92*, 803–822. [CrossRef]
9. Arts, J. How to assess respiratory sensitization of low molecular weight chemicals? *Int. J. Hyg. Environ. Health* **2020**, *225*, 113469. [CrossRef]
10. Sullivan, K.M.; Enoch, S.J.; Ezendam, J.; Sewald, K.; Roggen, E.L.; Cochrane, S. An adverse outcome pathway for sensitization of the respiratory tract by low-molecular-weight chemicals: Building evidence to support the utility of *in vitro* and *in silico* methods in a regulatory context. *Appl. In Vitro Toxicol.* **2017**, *3*, 213–226. [CrossRef]
11. The Adverse Outcome Pathway for Skin Sensitisation Initiated by Covalent Binding to Proteins. Available online: https://www.oecd.org/en/publications/the-adverse-outcome-pathway-for-skin-sensitisation-initiated-by-covalent-binding-to-proteins_9789264221444-en.html (accessed on 25 December 2024).
12. Kimber, I.; Poole, A.; Basketter, D.A. Skin and respiratory chemical allergy: Confluence and divergence in a hybrid adverse outcome pathway. *Toxicol. Res.* **2018**, *7*, 586–605. [CrossRef]
13. Basketter, D.A.; Kimber, I. Phthalic Anhydride: Illustrating a conundrum in chemical allergy. *J. Immunotoxicol.* **2016**, *13*, 767–769. [CrossRef] [PubMed]
14. Dearman, R.J.; Basketter, D.A.; Kimber, I. Inter-relationships between different classes of chemical allergens. *J. Appl. Toxicol.* **2011**, *33*, 558–565. [CrossRef] [PubMed]
15. Vincent, M.J.; Bernstein, J.A.; Basketter, D.; LaKind, J.S.; Dotson, G.S.; Maier, A. Chemical-induced asthma and the role of clinical, toxicological, exposure and epidemiological research in regulatory and hazard characterization approaches. *Regul. Toxicol. Pharmacol.* **2017**, *90*, 126–132. [CrossRef] [PubMed]
16. Identification of Substances as SVHCs Due to Equivalent Level of Concern to CMRs (Article 57(f))—Sensitisers as an Example. Available online: https://echa.europa.eu/documents/10162/17252/svhc_art_57f_sensitisers_en.pdf/a50728cc-6514-486c-9108-193a88b4bc9e?t=1385485688489#:~:text=According%20to%20the%20EU%20Classification,allergic%20response%20following%20skin%20contact (accessed on 2 May 2024).
17. Regulation (EC) No 1907/2006 of the European Parliament and of the Council of 18 December 2006 (REACH Regulation). Available online: https://eur-lex.europa.eu/legal-content/EN/TXT/PDF/?uri=CELEX:02006R1907-20221217 (accessed on 22 October 2024).

18. Candidate List of Substances of Very High Concern for Authorisation. Available online: https://echa.europa.eu/candidate-list-table?p_p_id=disslists_WAR_disslistsportlet&p_p_lifecycle=0&p_p_state=nomal&p_p_mode=view&_disslists_WAR_disslistsportlet_haz_detailed_concern=&_disslists_WAR_disslistsportlet_orderByCol=haz_detailed_concern&_disslists_WAR_disslistsportlet_substance_identifier_field_key=&_disslists_WAR_disslistsportlet_delta=200&_disslists_WAR_disslistsportlet_orderByType=desc&_disslists_WAR_disslistsportlet_dte_inclusionFrom=&_disslists_WAR_disslistsport-let_dte_inclusionTo=&_disslists_WAR_disslistsportlet_doSearch=&_disslists_WAR_disslistsportlet_deltaParamValue=200&_disslists_WAR_disslistsportlet_resetCur=false&_disslists_WAR_disslistsportlet_cur=2 (accessed on 22 October 2024).
19. Guidance on the Application of the CLP Criteria. Guidance to Regulation (EC) No 1272/2008 on Classification, Labelling and Packaging (CLP) of Substances and Mixtures. Available online: https://echa.europa.eu/guidance-documents/guidance-on-clp (accessed on 25 December 2024).
20. Sensitizing Potency of Chemical Respiratory Allergens in Humans. Available online: https://www.rivm.nl/bibliotheek/rapporten/340300004.pdf (accessed on 2 May 2024).
21. Pemberton, M.A.; Arts, J.H.; Kimber, I. Identification of true chemical respiratory allergens: Current Status, limitations and recommendations. *Regul. Toxicol. Pharmacol.* **2024**, *147*, 105568. [CrossRef]
22. North, C.M.; Ezendam, J.; Hotchkiss, J.A.; Maier, C.; Aoyama, K.; Enoch, S.; Goetz, A.; Graham, C.; Kimber, I.; Karjalainen, A.; et al. Developing a framework for assessing chemical respiratory sensitization: A workshop report. *Regul. Toxicol. Pharmacol.* **2016**, *80*, 295–309. [CrossRef]
23. Arts, J.H.E.; de Jong, W.H.; van Triel, J.J.; Schijf, M.A.; de Klerk, A.; van Loveren, H.; Kuper, C.F. The respiratory local lymph node assay as a tool to study respiratory sensitizers. *Toxicol. Sci.* **2008**, *106*, 423–434. [CrossRef]
24. Silva, R.J.; Tamburic, S. A state-of-the-art review on the alternatives to animal testing for the safety assessment of cosmetics. *Cosmetics* **2022**, *9*, 90. [CrossRef]
25. Ban on Animal Testing. Available online: https://single-market-economy.ec.europa.eu/sectors/cosmetics/ban-animal-testing_en (accessed on 2 May 2024).
26. Rogiers, V.; Benfenati, E.; Bernauer, U.; Bodin, L.; Carmichael, P.; Chaudhry, Q.; Coenraads, P.J.; Cronin, M.T.D.; Dent, M.; Dusinska, M.; et al. The way forward for assessing the human health safety of cosmetics in the EU-Workshop proceedings. *Toxicology* **2020**, *436*, 152421. [CrossRef]
27. The SCCS Notes of Guidance for the Testing of Cosmetic Ingredients and Their Safety Evaluation. Available online: https://health.ec.europa.eu/document/download/8d49f487-909c-4498-af89-1f769aaa628c_en (accessed on 2 May 2024).
28. A Strategic Roadmap for Establishing New Approaches to Evaluate the Safety of Chemicals and Medical Products in the United States. Available online: https://ntp.niehs.nih.gov/whatwestudy/niceatm/natl-strategy (accessed on 2 May 2024).
29. Caloni, F.; De Angelis, I.; Hartung, T. Replacement of animal testing by integrated approaches to testing and assessment (IATA): A call for in vivitrosi. *Arch. Toxicol.* **2022**, *96*, 1935–1950. [CrossRef]
30. Hargitai, R.; Parráková, L.; Szatmári, T.; Monfort-Lanzas, P.; Galbiati, V.; Audouze, K.; Jornod, F.; Staal, Y.C.M.; Burla, S.; Chary, A.; et al. Chemical respiratory sensitization—Current status of mechanistic understanding, knowledge gaps and possible identification methods of sensitizers. *Front. Toxicol.* **2024**, *6*, 1331803. [CrossRef]
31. Rothen-Rutishauser, B.; Gibb, M.; He, R.; Petri-Fink, A.; Sayes, C.M. Human lung cell models to study aerosol delivery—Considerations for model design and development. *Eur. J. Pharm. Sci.* **2023**, *180*, 106337. [CrossRef] [PubMed]
32. Posch, W.; Lass-Flörl, C.; Wilflingseder, D. Generation of human monocyte-derived dendritic cells from whole blood. *J. Vis. Exp.* **2016**, *2016*, 54968. [CrossRef]
33. Hölken, J.M.; Teusch, N. The monocytic cell line THP-1 as a validated and robust surrogate model for human dendritic cells. *Int. J. Mol. Sci.* **2023**, *24*, 1452. [CrossRef] [PubMed]
34. dos Santos, G.G.; Reinders, J.; Ouwehand, K.; Rustemeyer, T.; Scheper, R.J.; Gibbs, S. Progress on the development of human *in vitro* dendritic cell based assays for assessment of the sensitizing potential of a compound. *Toxicol. Appl. Pharmacol.* **2009**, *236*, 372–382. [CrossRef] [PubMed]
35. Ashikaga, T.; Yoshida, Y.; Hirota, M.; Yoneyama, K.; Itagaki, H.; Sakaguchi, H.; Miyazawa, M.; Ito, Y.; Suzuki, H.; Toyoda, H. Development of an *in vitro* skin sensitization test using human cell lines: The human cell line activation test (h-CLAT). I. Optimization of the h-CLAT protocol. *Toxicol. Vitro* **2006**, *20*, 767–773. [CrossRef]
36. Sakaguchi, H.; Ashikaga, T.; Miyazawa, M.; Yoshida, Y.; Ito, Y.; Yoneyama, K.; Hirota, M.; Itagaki, H.; Toyoda, H.; Suzuki, H. Development of an *in vitro* skin sensitization test using human cell lines; human cell line activation test (h-CLAT) II. An inter-laboratory study of the h-CLAT. *Toxicol. Vitro* **2006**, *20*, 774–784. [CrossRef]
37. Test No. 442E: *In Vitro* Skin Sensitisation. *In Vitro* Skin Sensitisation Assays Addressing the Key Event on Activation of Dendritic Cells on the Adverse Outcome Pathway for Skin Sensitisation. Available online: https://www.oecd.org/en/publications/test-no-442e-in-vitro-skin-sensitisation_9789264264359-en.html (accessed on 25 December 2024).

38. Martins, J.D.; Maciel, E.A.; Silva, A.; Ferreira, I.; Ricardo, F.; Domingues, P.; Neves, B.M.; Domingues, M.R.M.; Cruz, M.T. Phospholipidomic profile variation on THP-1 cells exposed to skin or respiratory sensitizers and respiratory irritant. *J. Cell Physiol.* **2016**, *231*, 2639–2651. [CrossRef]
39. Ferreira, I.; Silva, A.; Martins, J.D.; Neves, B.M.; Cruz, M.T. Nature and kinetics of redox imbalance triggered by respiratory and skin chemical sensitizers on the human monocytic cell line THP-1. *Redox Biol.* **2018**, *16*, 75–86. [CrossRef]
40. Narita, K.; Okutomi, H.; Kawakami, K.; Sui, H.; Basketter, D.; Ashikaga, T. Behaviour of chemical respiratory sensitizers in *in vitro* methods for skin sensitization. *AATEX* **2021**, *26*, 9–18. [CrossRef]
41. da Silva, A.C.G.; de Morais Carvalho Filho, S.; Valadares, M.C. Biological effects triggered by chemical respiratory sensitizers on THP-1 monocytic cells. *Toxicol. Vitro* **2023**, *90*, 105602. [CrossRef]
42. Forreryd, A.; Johansson, H.; Albrekt, A.S.; Borrebaeck, C.A.K.; Lindstedt, M. Prediction of chemical respiratory sensitizers using GARD, a novel *in vitro* assay based on a genomic biomarker signature. *PLoS ONE* **2015**, *10*, e0118808. [CrossRef] [PubMed]
43. Chary, A.; Serchi, T.; Moschini, E.; Hennen, J.; Cambier, S.; Ezendam, J.; Blömeke, B.; Gutleb, A.C. An *in vitro* coculture system for the detection of sensitization following aerosol exposure. *Altern. Anim. Exp.* **2019**, *36*, 403–418. [CrossRef] [PubMed]
44. da Silva, A.C.G.; de Mendonça, I.C.F.; Valadares, M.C. Characterization and applicability of a novel physiologically relevant 3D-tetraculture bronchial model for *in vitro* assessment of respiratory sensitization. *Toxicology* **2024**, *503*, 153756. [CrossRef] [PubMed]
45. Tanabe, I.; Yoshida, K.; Ishikawa, S.; Ishimori, K.; Hashizume, T.; Yoshimoto, T.; Ashikaga, T. Development of an *in vitro* sensitisation test using a coculture system of human bronchial epithelium and immune cells. *Altern. Lab. Anim.* **2023**, *51*, 387–400. [CrossRef]
46. Roan, F.; Obata-Ninomiya, K.; Ziegler, S.F. Epithelial cell–derived cytokines: More than just signaling the alarm. *J. Clin. Investig.* **2019**, *129*, 1441–1451. [CrossRef]
47. Nakajima, S.; Kabata, H.; Kabashima, K.; Asano, K. Anti-TSLP antibodies: Targeting a master regulator of type 2 immune responses. *Allergol. Int.* **2020**, *69*, 197–203. [CrossRef]
48. Lambrecht, B.N.; Hammad, H. The immunology of asthma. *Nat. Immunol.* **2015**, *16*, 45–56. [CrossRef]
49. Lieber, M.; Todaro, G.; Smith, B.; Szakal, A.; Nelson-Rees, W. A continuous tumor-cell line from a human lung carcinoma with properties of type II alveolar epithelial cells. *Int. J. Cancer* **1976**, *17*, 62–70. [CrossRef]
50. Mizoguchi, I.; Ohashi, M.; Chiba, Y.; Hasegawa, H.; Xu, M.; Owaki, T.; Yoshimoto, T. Prediction of chemical respiratory and contact sensitizers by OX40L expression in dendritic cells using a novel 3D coculture system. *Front. Immunol.* **2017**, *8*, 929. [CrossRef]
51. Edgell, C.J.S.; Mcdonald, C.C.; Graham, J.B. Permanent cell line expressing human factor VIII-related antigen established by hybridization. *Proc. Natl. Acad. Sci. USA* **1983**, *80*, 3734–3737. [CrossRef]
52. Tsuchiya, S.; Yamabe, M.; Yamaguchi, Y.; Kobayashi, Y.; Konno, T.; Tada, K. Establishment and characterization of a human acute monocytic leukemia cell line (THP-1). *Int. J. Cancer* **1980**, *26*, 171–176. [CrossRef] [PubMed]
53. Mizoguchi, I. A novel coculture system for assessing respiratory sensitizing potential by IL-4 in T cells. *Altern. Anim. Exp.* **2022**, *40*, 204–216. [CrossRef] [PubMed]
54. Friedrich, K.; Delgado, I.F.; Santos, L.M.F.; Paumgartten, F.J.R. Assessment of sensitization potential of monoterpenes using the rat popliteal lymph node assay. *Food Chem. Toxicol.* **2007**, *45*, 1516–1522. [CrossRef] [PubMed]
55. Ashikaga, T.; Sakaguchi, H.; Sono, S.; Kosaka, N.; Ishikawa, M.; Nukada, Y.; Miyazawa, M.; Ito, Y.; Nishiyama, N.; Itagaki, H. A comparative evaluation of *in vitro* skin sensitisation tests: The human cell-line activation test (h-CLAT) versus the local lymph node assay (LLNA). *Altern. Lab. Anim.* **2010**, *38*, 275–284. [CrossRef]
56. Gerberick, G.F.; Ryan, C.A.; Kern, P.S.; Schlatter, H.; Dearman, R.J.; Kimber, I.; Patlewicz, G.Y.; Basketter, D.A. Compilation of historical local lymph node data for evaluation of skin sensitization alternative methods. *Dermatitis* **2005**, *16*, 157–202.
57. Test, No. 442C: *In Chemico* Skin Sensitisation. Assays Addressing the Adverse Outcome Pathway Key Event on Covalent Binding to Proteins. Available online: https://www.oecd.org/en/publications/test-no-442c-in-chemico-skin-sensitisation_9789264229709-en.html (accessed on 25 December 2024).
58. Test, No. 442D: *In Vitro* Skin Sensitisation. Assays Addressing the Adverse Outcome Pathway Key Event on Keratinocyte Activation. Available online: https://www.oecd.org/en/publications/test-no-442d-in-vitro-skin-sensitisation_9789264229822-en.html (accessed on 25 December 2024).
59. Avonto, C.; Wang, Z.; Ahn, J.; Verma, R.P.; Sadrieh, N.; Dale, O.; Khan, S.I.; Chittiboyina, A.G.; Khan, I.A. Integrated testing strategy for the safety of botanical ingredients: A case study with german chamomile constituents. *Appl. In Vitro Toxicol.* **2021**, *7*, 129–143. [CrossRef]
60. Klein, S.G.; Serchi, T.; Hoffmann, L.; Blömeke, B.; Gutleb, A.C. An improved 3D tetraculture system mimicking the cellular organisation at the alveolar barrier to study the potential toxic effects of particles on the lung. *Part. Fibre Toxicol.* **2013**, *10*, 31. [CrossRef]

61. Gutleb, A. Figure 1. Schematic representation of the workflow for the build-up of ALIsens® for the detection of respiratory sensitizers. Created in BioRender. 2024. Available online: https://BioRender.com/a97j092 (accessed on 25 December 2024).
62. Guidance Document on Good *In Vitro* Method Practices (GIVIMP). Available online: https://www.oecd.org/en/publications/guidance-document-on-good-in-vitro-method-practices-givimp_9789264304796-en.html (accessed on 16 July 2024).
63. Gloxhuber, C. Toxicological properties of surfactants. *Arch. Toxicol.* **1974**, *32*, 245–270. [CrossRef]
64. Cornells, M.; Dupont, C.; Wepierre, J. *In vitro* cytotoxicity tests on cultured human skin fibroblasts to predict the irritation potential of surfactants. *Altern. Lab. Anim.* **1991**, *19*, 324–336. [CrossRef]
65. AOP: 39. Covalent Binding, Protein, Leading to Increase, Allergic Respiratory Hypersensitivity Response. Available online: https://aopwiki.org/aops/39 (accessed on 7 December 2024).
66. Cochrane, S.A.; Arts, J.H.E.; Ehnes, C.; Hindle, S.; Hollnagel, H.M.; Poole, A.; Suto, H.; Kimber, I. Thresholds in chemical respiratory sensitisation. *Toxicology* **2015**, *333*, 179–194. [CrossRef]
67. Deckers, J.; De Bosscher, K.; Lambrecht, B.N.; Hammad, H. Interplay between barrier epithelial cells and dendritic cells in allergic sensitization through the lung and the skin. *Immunol. Rev.* **2017**, *278*, 131–144. [CrossRef] [PubMed]
68. Tsui, H.C.; Ronsmans, S.; Hoet, P.H.M.; Nemery, B.; Vanoirbeek, J.A.J. Occupational asthma caused by low-molecular-weight chemicals associated with contact dermatitis: A retrospective study. *J. Allergy Clin. Immunol. Pract.* **2022**, *10*, 2346–2354.e4. [CrossRef] [PubMed]
69. Ponder, J.; Rajagopal, R.; Singal, M.; Baker, N.; Patlewicz, G.; Roggen, E.; Cochrane, S.; Sullivan, K. "In litero" screening: Retrospective evaluation of clinical evidence to establish a reference list of human chemical respiratory sensitizers. *Front. Toxicol.* **2022**, *4*, 916370. [CrossRef] [PubMed]
70. Sadekar, N.; Boisleve, F.; Dekant, W.; Fryer, A.D.; Gerberick, G.F.; Griem, P.; Hickey, C.; Krutz, N.L.; Lemke, O.; Mignatelli, C.; et al. Identifying a reference list of respiratory sensitizers for the evaluation of novel approaches to study respiratory sensitization. *Crit. Rev. Toxicol.* **2021**, *51*, 792–804. [CrossRef]
71. Api, A.M.; Belsito, D.; Botelho, D.; Bruze, M.; Burton, G.A.; Cancellieri, M.A.; Chon, H.; Dagli, M.L.; Dekant, W.; Deodhar, C.; et al. Update to RIFM fragrance ingredient safety assessment, Terpineol, CAS Registry Number 8000-41-7. *Food Chem Toxicol* **2024**, *183*, 114147. [CrossRef]
72. Sakaguchi, H.; Ashikaga, T.; Miyazawa, M.; Kosaka, N.; Ito, Y.; Yoneyama, K.; Sono, S.; Itagaki, H.; Toyoda, H.; Suzuki, H. The relationship between CD86/CD54 expression and THP-1 cell viability in an *in vitro* skin sensitization test–human cell line activation test (h-CLAT). *Cell Biol. Toxicol.* **2009**, *25*, 109–126. [CrossRef]
73. Nukada, Y.; Ito, Y.; Miyazawa, M.; Sakaguchi, H.; Nishiyama, N. The relationship between CD86 and CD54 protein expression and cytotoxicity following stimulation with contact allergen in THP-1 cells. *J. Toxicol. Sci.* **2011**, *36*, 313–324. [CrossRef]
74. Yoshida, Y.; Sakaguchi, H.; Ito, Y.; Okuda, M.; Suzuki, H. Evaluation of the skin sensitization potential of chemicals using expression of co-stimulatory molecules, CD54 and CD86, on the naive THP-1 cell line. *Toxicol. Vitro* **2003**, *17*, 221–228. [CrossRef]
75. Huang, S.; Wiszniewski, L.; Constant, S.; Roggen, E. Potential of *in vitro* reconstituted 3D human airway epithelia (MucilAir^{TM}) to assess respiratory sensitizers. *Toxicol. Vitro* **2013**, *27*, 1151–1156. [CrossRef]
76. Hermanns, M.I.; Kasper, J.; Unger, R.E. Assessment of respiratory sensitizers: Cytokine responses in a 3D alveolo-capillary barrier model *in vitro*. *Adv. Biomater. Devices Med.* **2015**, *2*, 1–9.
77. Gibb, M.; Sayes, C. An *in vitro* alveolar model allows for the rapid assessment of chemical respiratory sensitization with modifiable biomarker endpoints. *Chem. Biol. Interact.* **2022**, *368*, 110232. [CrossRef]
78. Corsini, E.; Mitjans, M.; Galbiati, V.; Lucchi, L.; Galli, C.L.; Marinovich, M. Use of IL-18 production in a human keratinocyte cell line to discriminate contact sensitizers from irritants and low molecular weight respiratory allergens. *Toxicol. Vitro* **2009**, *27*, 789–796. [CrossRef]

toxics

MDPI

Article

A Novel Machine Learning Model and a Web Portal for Predicting the Human Skin Sensitization Effects of Chemical Agents

Ricardo Scheufen Tieghi [1,2], José Teófilo Moreira-Filho [1], Holli-Joi Martin [2], James Wellnitz [2], Miguel Canamary Otoch [2], Marielle Rath [2], Alexander Tropsha [2,3,*], Eugene N. Muratov [2,*] and Nicole Kleinstreuer [1,*]

1 National Toxicology Program Interagency Center for Evaluation of Alternative Toxicological Methods (NICEATM), Division of Translational Toxicology, National Institute of Environmental Health Sciences, Research Triangle Park, Durham, NC 27711, USA; ricat@unc.edu (R.S.T.); teofilo.moreirafilho@nih.gov (J.T.M.-F.)

2 UNC Eshelman School of Pharmacy, University of North Carolina, Chapel Hill, NC 27514, USA; holli27@email.unc.edu (H.-J.M.); jwellnitz@unc.edu (J.W.); motoch@email.unc.edu (M.C.O.); mrath@email.unc.edu (M.R.)

3 Predictive LLC, (A.T.), Chapel Hill, NC 27514, USA

* Correspondence: alex_tropsha@unc.edu (A.T.); murik@email.unc.edu (E.N.M.); nicole.kleinstreuer@nih.gov (N.K.); Tel.: +1-(919)-966-2955 (N.K.); Fax: +1-(919)-966-0102 (N.K.)

Abstract: Skin sensitization is a significant concern for chemical safety assessments. Traditional animal assays often fail to predict human responses accurately, and ethical constraints limit the collection of human data, necessitating a need for reliable in silico models of skin sensitization prediction. This study introduces HuSSPred, an in silico tool based on the Human Predictive Patch Test (HPPT). HuSSPred aims to enhance the reliability of predicting human skin sensitization effects for chemical agents to support their regulatory assessment. We have curated an extensive HPPT database and performed chemical space analysis and grouping. Binary and multiclass QSAR models were developed with Bayesian hyperparameter optimization. Model performance was evaluated via five-fold cross-validation. We performed model validation with reference data from the Defined Approaches for Skin Sensitization (DASS) app. HuSSPred models demonstrated strong predictive performance with CCR ranging from 55 to 88%, sensitivity between 48 and 89%, and specificity between 37 and 92%. The positive predictive value (PPV) ranged from 84 to 97%, versus negative predictive value (NPV) from 22 to 65%, and coverage was between 75 and 93%. Our models exhibited comparable or improved performance compared to existing tools, and the external validation showed the high accuracy and sensitivity of the developed models. HuSSPred provides a reliable, open-access, and ethical alternative to traditional testing for skin sensitization. Its high accuracy and reasonable coverage make it a valuable resource for regulatory assessments, aligning with the 3Rs principles. The publicly accessible HuSSPred web tool offers a user-friendly interface for predicting skin sensitization based on chemical structure.

Keywords: skin sensitization; computational toxicology; QSAR; cheminformatics; NAMs

1. Introduction

Skin sensitization testing is employed to determine the potential for a substance to cause allergic contact dermatitis in susceptible individuals [1]. United States and international regulatory authorities require or recommend that chemical manufacturers conduct tests that assess skin sensitization hazards [2]. The most common tests are the human repeat insult patch test, human maximization test, murine local lymph node assay (OECD TG 429 [3]), guinea pig maximization test (OECD TG 406 [4]), and Buehler test (OECD TG 406 [4]), although several non-animal tests now exist (OECD TG 442C; OECD

TG 442D; OECD TG 442E; OECD TG 497) [5–8]. Despite the widespread use of animal tests, it has been shown that animal-based assay outcomes do not always equate with human responses [9] and that animal models are less reproducible than other alternative methods [10].

Human data provide the most accurate assessment of sensitization potential, but ethical considerations preclude the intentional induction of sensitization in human subjects, resulting in limited availability of such data. Despite this sparsity, the human repeat insult patch test (HRIPT) and the human maximization test (HMT) provide sufficient information on the potential of substances to cause allergic contact dermatitis. These assays include results detailing the dose per skin area (DSA) causing induction of sensitization, the number of individuals sensitized, and other parameters reflecting the sensitization potency from test substances. Reliable in silico models leveraging human assay data for skin sensitization predictions could support improved regulatory assessments.

Computational methods have become increasingly vital in toxicology in light of the three Rs (reduce, refine, and replace) of animal testing. In silico approaches are cost-effective, safe, and ethical alternatives to traditional animal and human testing. One computational approach explored here is machine learning (ML), in which algorithms enable systems to learn and improve from experience without being explicitly programmed [11]. Various computational models for skin sensitization already exist (Table 1).

Table 1. Overview of existing skin sensitization prediction software.

Title	Computational Approach	Species/Assay	Access
StopTox [12]	QSAR	Mice/LLNA	Free
PredSkin [13]	QSAR	Mice/LLNA	Free
CASE Ultra [14]	Structural alerts	Mice/LLNA	Commercial
Derex Nexus [15,16]	Structural alerts Expert system	Mice/LLNA	Commercial
OECD QSAR Toolbox [17]	QSAR and read-across	- *	Free
VEGA [18]	Read-across	-	Free
ToxTree [19]	Structural alerts	-	Free
TOPKAT [15]	QSAR	Rat & Other/ Multiple	Commercial

* Cells without input indicate that this information is not publicly available or multiple models are encompassed in the tool, making a single answer not applicable.

The purpose of this work is to present HuSSPred, an in silico human predictive patch test (HPPT)-based tool for Human Skin Sensitization Prediction from the human repeat insult patch test (HRIPT) and the human maximization test (HMT) (Figure 1). This study continues the series of our previous works on skin sensitization [20–22], extending the strategy that we developed earlier [23] in modeling human data. Using an extensively curated HPPT database, we developed QSAR models using random forest (RF), LightGBM, and support vector machine (SVM) algorithms with Bayesian hyperparameter optimization. Our models exhibit comparable or improved performance versus existing tools. We conclude that HuSSPred is one of the first tools built strictly with human data and validated with skin sensitization-defined approaches (DA) [24,25]. This tool provides a reliable, cost-effective, and ethical alternative to traditional testing for skin sensitization. Its accuracy and reasonable coverage make it a valuable resource for regulatory assessments, aligning with the three Rs principles. Furthermore, the HuSSPred web tool (https://husspred.mml.unc.edu/, accessed on 18 October 2024) is publicly accessible and offers a user-friendly interface for predicting skin sensitization based on chemical structure.

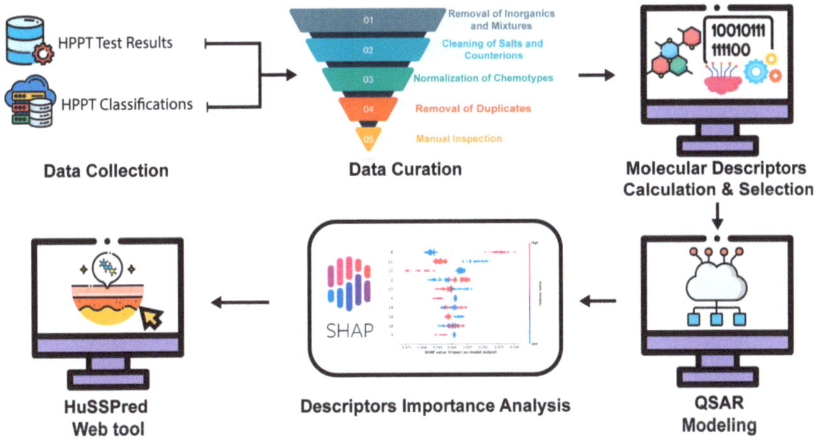

Figure 1. General study design of HuSSPred. Experimental data were collected from the HPPT test results and combined with the HPPT GHS Classifications database. All entries were carefully curated following best practices in the field. Molecular descriptors were calculated and selected to build skin sensitization QSAR models. SHAP analysis was performed to enhance the model's interpretability. The best-performing QSAR models were deployed as a web tool, HuSSPred, available at https://husspred.mml.unc.edu/, accessed on 18 October 2024.

2. Materials and Methods

2.1. Data Collection and Curation

We collected chemical data from the NICEATM human predictive patch test (HPPT) [26,27] and the HPPT Classifications according to the Globally Harmonized System of Classification and Labelling of Chemicals (GHS) [28]. The HPPT database contains 2277 entries for 1366 unique chemicals. Using Chemical Abstracts Service Registry Number (CASRN), the HPPT data were harmonized with the Dose per Skin Area (DSA05) classifications database, resulting in 2014 entries. The data sets were then biologically and chemically curated according to the best practices in the field [29–31]. As for chemical curation, we removed mixtures, inorganics, and large organic compounds, removed counterions, cleaned and neutralized salts, and normalized chemotypes using the ChemAxon Standardizer software [32]. We followed one of the two appropriate procedures for handling duplicates: (i) if the outcomes of all duplicates were concordant (from a hazard perspective), one record was kept with the respective outcome; (ii) if any outcomes disagreed, they were removed. The HPPT data contained four different weight-of-evidence approaches with skin sensitization outcomes: the median location-like parameter (MLLP), median sensitization potency (MSPE), weight of evidence (WoE), and overall weight of evidence (WES). Most curation steps were performed through Konstanz Information Miner (KNIME) [33–35].

2.1.1. Binary Data Curation

After combining the two data sets, we kept only binary labeled outcomes (sensitizers (1) and non-sensitizers (NCs)). Entries with missing outcomes were removed. Then, the dataset-specific curation began for each weight-of-evidence approach, through which we removed outcome entries with the label "unavailable" for MLLP, MSPE, WoE, and WES; this process was performed separately, resulting in 4 independent datasets. Compounds for which the SMILES could not be retrieved were also removed. We labeled all sensitizers as Class 1 and non-sensitizers as Class 0. The resulting data sets were exported as SD files into Python for model development (available in Supplemental File S1).

2.1.2. Multiclass Data Curation

The collected data contained multiclass GHS classifications for skin sensitization potency. The outcome labels were mapped to numbers for model building, where NC (non-sensitizer), 1B (weak sensitizer), and 1A (strong sensitizer) were mapped to classes 0, 1, and 2, respectively. We removed compounds with missing outcomes and SMILES from each weight-of-evidence approach. The four multiclass data sets are available in Supplemental File S1.

2.1.3. Continuous Data Curation

The data set entries also included continuous values for the dose per skin area (DSA), dose per skin area for potential sensitization of 5% of the tested population (DSA05), and dose per skin area for potential sensitization of at least 1 individual (DSA01+). We excluded compounds with missing outcomes, applied our standard chemical curation workflow, and calculated the median for compounds with multiple entries. We applied a logarithmic transformation to the DSA values.

2.2. *Classification QSAR Modeling*

2.2.1. Calculation of Molecular Fingerprints

The RDKit package [36] was used to calculate extended-connectivity fingerprints with a diameter of 4 (ECFP4) and 2048 bits [37]. Molecular ACCess System (MACCS, version 3) Fingerprints [36], Mordred [38], and Saagar descriptors [39] were also used for model development.

2.2.2. Model Development and Performance Assessment

Models were developed with either RF, LightGBM, or SVM algorithms. QSAR models were developed and validated according to the best practices in cheminformatics [40]. In the RF algorithm, trees were decorrelated via bootstrapping with replacement; in LightGBM, a gradient-boosting algorithm optimizes model performance through a leaf-wise tree construction approach. Models using LightGBM [41] and RF [42] were implemented through Scikit-learn, v.1.4.0 [43]. The SVM algorithm finds the optimal hyperplane that maximizes the margin between distinct classes in a multidimensional feature space, thereby enabling robust data point classification. These three ML algorithms were also utilized in similar studies, and they are dominant in the field based on a balance between computational efficiency, predictivity, and interpretability [44].

For binarized models (non-sensitizers vs. sensitizers), the following statistical metrics were used to assess the performance of the classification models (Equations (1)–(5)):

Sensitivity (SE):

$$SE = \frac{N_{TP}}{N_{TP} + N_{FN}}, \tag{1}$$

Specificity (SP):

$$SP = \frac{N_{TN}}{N_{TN} + N_{FP}}, \tag{2}$$

Correct classification rate (CCR):

$$CCR = \frac{SE + SP}{2}, \tag{3}$$

Positive predictive value (PPV):

$$PPV = \frac{N_{TP}}{N_{TP} + N_{FP}}, \tag{4}$$

Negative predictive value (NPV):

$$NPV = \frac{N_{TN}}{N_{TN} + N_{FN}} \quad (5)$$

N represents the number of compounds, N_{TP} and N_{TN} represent the number of true positives and negatives, respectively, and N_{FP} and N_{FN} represent the number of false positives and false negatives, respectively.

For binary classification models, compounds labeled as skin sensitizers were classified as positive (Class 1), and non-sensitizing compounds were classified as negative (Class 0).

For multiclass classification, compounds are labeled as non-sensitizers (Class 0), weak sensitizers (Class 1), or strong sensitizers (Class 2). The one-vs.-rest approach was used when calculating metrics for multiclass models. Metrics (such as SP and NPV) were computed for each class separately and then averaged to obtain an overall score. Classes were weighted equally (macro-average) during calculations for the average

2.2.3. Hyperparameter Optimization

Since the performance of a model can be closely linked to the hyperparameters used during development, the models were optimized using a Bayesian approach implemented through Optuna [45]. Optuna uses a framework to identify ideal hyperparameters for a given set of descriptors and ML algorithms. The best hyper-parameters were then used to fine-tune the models using the entire training set of compounds and tested during the 5-fold cross-validation step.

2.2.4. Dimensionality Reduction and Feature Selection

Dimensionality reduction was performed by applying a low-variance filter with a threshold of 0.01. Descriptors with low variance (threshold of 0.01) were filtered out to exclude non-informative data using Scikit-learn [43]. Subsequently, supervised feature selection was conducted using recursive feature elimination (RFE) to identify the most relevant molecular descriptors, thereby improving model performance and interpretability while reducing overfitting and training time [46].

2.2.5. Normalization of Continuous Descriptors

We normalized the Mordred descriptors using min–max scaling to ensure consistent scale and improve model performance. The normalization was implemented via the "MinMaxScaler" class from the Scikit-Learn library [43], which scales each feature to a range [0–1]. The scaling process adjusts each descriptor's value to be within the specified range, preserving the relationships among features while enhancing model training stability. Scaling was performed within each fold to prevent data leakage, ensuring that the scaling parameters were derived solely from the training data [47].

2.2.6. Data Set Split and 5-Fold Cross-Validation

Five-fold external cross-validation was used [48]. For this, the data set was split into five equal parts, wherein one subset (20%) is used as the test set, and the remaining compounds (80%) compose the training set. This procedure was repeated five times, and each subset was used as the validation set exactly once. Models were built using the training set only, and compounds in the test set had to not be present in the training set. Alternative validation methods, such as 10-fold cross-validation and leave-one-out cross-validation, are available. We selected 5-fold cross-validation due to its established validity within the field, effectively balancing methodological rigor with computational efficiency [44].

2.2.7. Threshold Moving

We tried the threshold-moving calibration of probability estimates to increase prediction confidence without losing data, i.e., without needing to balance the data. Binary QSAR models' probability thresholds were adjusted using this approach, which was incorporated into Python [43,49]. Threshold moving was used to select the binary classification probability threshold for the model that produced the highest geometric mean values on these test sets. The geometric mean was chosen since it better assesses the performance of models when predicting imbalanced data [49–52]. Radar charts to represent model calibration were constructed using the Plotly package [53].

2.2.8. Applicability Domain (AD)

Our group has previously shown the importance of the applicability domain of models when drawing conclusions from predictions derived from QSAR models. The AD must be stated for the given chemical space of predictive models to identify "reliable" and "unreliable" regions for predictions [49,54]. Thus, users should only consider the model's predictions reliable if their predicted compounds fall within the model's AD.

The "Applicability Domain" meta-node was used to assess the AD of our models. The KNIME meta-node uses Euclidean distances to measure the chemical similarity between a compound from the test set and its nearest neighbor in the training set. The prediction may be unreliable if the distance of a compound not present in the test set to its nearest neighbor is higher than an arbitrary parameter (Z = 0.5) that controls the significance level [55].

2.2.9. Model Interpretation

Model interpretation was facilitated using machine learning interpretable packages such as Shapley additive explanations (SHAP) [56,57]. SHAP can be used to identify key features that impact model performance, improving the transparency and understanding of ML models [58].

Contribution maps [59,60] were generated from QSAR models to visualize atoms and fragments contributing to skin sensitization potential. The contribution maps used an approach in which an atom's "weight" was considered a predicted probability difference obtained when bits in the fingerprints corresponding to the atom were removed. Then, the normalized weights were used to color atoms in a topography-like map in which green indicates the contribution to toxicity (i.e., predicted sensitization probability decreases when bits are removed) and red indicates a negative contribution to toxicity (i.e., predicted sensitization probability increases when bits are removed) [60].

2.2.10. Model Implementation—HuSSPred Web Tool

The QSAR models developed in this study were implemented as a web application, HuSSPred, which runs on an Ubuntu server. The HuSSPred application is encoded using Flask [61], uWSGI [62], Nginx [63], Python 3.11.4 [64], RDKit [36], Scikit-learn [43], and JavaScript [65]. HuSSPred also includes the JSME molecule editor [66], written in JavaScript and supported by most popular web browsers. The server takes input chemicals and produces skin sensitization predictions based on chemical structure for the user.

2.3. *Chemical Space Analysis and Grouping*

The MoViz pipeline [67] was employed for chemical space analysis and grouping. The NICEATM team developed the pipeline, which facilitates identifying and organizing structurally similar compounds. Morgan fingerprints with a radius of 2 and 2048 bits were used for molecular descriptor calculation during the analysis. After this, low-variance descriptors were filtered out, and variable selection was performed to enhance the robustness of the analysis. Subsequently, uniform manifold approximation and projection (UMAP) [68] was utilized for dimensionality reduction, enabling the visualization of complex chemical spaces in a simplified manner. UMAP was chosen over linear methods like principal component analysis (PCA) because it is a nonlinear manifold learning technique that can better preserve both local and global structures of high-dimensional data, and it also involves no computational restrictions on embedding dimensions, which is advantageous for handling the high-dimensionality molecular descriptors used [68,69]. The MoViz [67] pipeline allowed for the effective grouping of compounds, supporting the identification of structural similarities and potential activity cliffs.

2.4. Activity Cliff Analysis—Molecular Roughness Calculation

Mordred descriptors were normalized to compute a pairwise distance matrix based on Euclidean distance. The roughness index (ROGI) score was calculated to quantify the ruggedness of the activity landscape [70]. For visualization, we projected the high-dimensional data onto a 2D plane using multidimensional scaling (MDS). The 3D plot was generated by interpolating the activity values across the projected 2D coordinates and plotting the resulting surface. Additionally, a 2D contour plot was created to highlight the activity distribution, utilizing interpolation to create a smooth heatmap overlaying the 2D MDS projection. These visualizations facilitated the identification of activity cliffs, providing insights into structure–activity relationships within the continuous data set. The code for these plots was adapted from https://github.com/coleygroup/rogi-results, accessed on 18 October 2024 [70].

2.5. Case Study: Validation Using Defined Approaches

Our models were validated using compound results from the NIEHS Defined Approaches for Skin Sensitization Web app (DASS app), representing testing strategies that are accepted by international regulatory authorities. The original DASS data set contained 196 compounds. These compounds cover a diverse range of chemical structures and sensitization potencies and have been employed in previous studies to assess the performance of the novel skin sensitization-defined approaches, as recognized by the OECD TG 497 and accepted in 2021 [5]. The data extracted from the DASS app are also available in the National Toxicology Program (NTP) Integrated Chemical Environment (ICE), a highly curated repository of toxicological data for various endpoints [71,72]. We removed compounds with inconclusive outcomes and compounds in our model's training set. Then, we used an automated QSAR-ready workflow to standardize and curate chemical structures [73]. For binary models, we performed validation using (i) the Defined Approaches Integrated Testing Strategy (DA ITS) Call, (ii) 2o3 Defined Approaches Call, and (iii) Basketter Human data Call as references [74,75]. After the curation protocols, the DA ITS call had 138 compounds (38 NC, 100 sensitizers), the 2o3 Call had 142 compounds (61 NC, 81 sensitizers), and the Basketter data set had 63 compounds (21 NC, 42 sensitizers). A consensus heatmap was created using the Seaborn [76] package for the 45 compounds that had predictions for all three validation data sets for comparison.

The "Basketter Potency" and "HPPT Reference Potency" columns were used to validate multiclass models. After the removal of training, unclassified, and automated curation, the data sets comprised 81 compounds (13 NC, 27 weak sensitizers, and 14 strong sensitizers) and 14 compounds (2 NCs, 10 weak sensitizers, and 2 strong sensitizers), respectively.

The Y-randomization test was also used per OECD Guidance Document on the validation of QSAR models [77] to investigate whether the reported accuracy of our models was due to chance correlation. The results of ten rounds of Y-randomization are reported in Table S3.

3. Results and Discussion

3.1. Data Sets Summary

Here, we have collected and curated the HPPT and the HPPT Classifications database into four independent binary classification data sets for modeling and comparison, four multiclass data sets, and three continuous data sets. Table 2 demonstrates the number of compounds in each data set used for binary classification model development. Data curation is essential for the development of QSAR models [29,40]. During curation, we found various inconsistencies in which the same compound would possess the conflicting ranking of both a sensitizer and a non-sensitizer; these entries were removed. We also identified multiple entries of the same compound with concordant outcomes, and these were merged into a single entry. We emphasize the need for rigorous biological curation, as QSAR models developed with duplicate entries will exhibit under-optimistic performance if the outcomes are dissimilar or over-optimistic performance if the outcomes are identical. The number of compounds in the multiclass data set (Table 3) and continuous models (Table 4) are provided below. We note that there was a class imbalance for the binary and multiclass data sets; sensitizers were more abundant than non-sensitizers. Weak sensitizers were roughly three times as abundant for multiclass data as other classes. The majority class was randomly undersampled in the multiclass models to mitigate this imbalance and improve model performance.

Table 2. Number of compounds in each binary data set for MSPE, MLLP, WoE, and WES classifications.

Data Set	Sensitizers	Non-Sensitizers	Total
MLLP	154	41	195
MSPE	177	39	216
WoE	177	41	218
WES	132	37	169

MLLP, median location-like parameter; MSPE, median sensitization potency estimate; WoE, individual weight-of-evidence scores; WES, overall weight-of-evidence scores.

Table 3. Number of compounds in each multiclass data set for MSPE, MLLP, WoE, and WES classifications.

Data Set	Strong Sensitizers	Weak Sensitizers	Non-Sensitizers	Total
MLLP	40	118	41	199
MSPE	44	123	39	206
WoE	42	126	41	209
WES	37	87	34	158

MLLP, Median location-like parameter; MSPE, median sensitization potency estimate; WoE, individual weight-of-evidence scores; WES, overall weight-of-evidence scores.

Table 4. Number of compounds in continuous data sets for DSA, DSA05, and DSA01 classifications.

Data Set	Total
DSA	829
DSA05	170
DSA01	104

DSA, dose per skin area; DSA05, dose per skin area for sensitization of 5% of the tested population; DSA01, dose per skin area for sensitization of at least 1 tested individual.

3.2. *Binary Classification QSAR Models*

Here, we built 36 binary classification models with different combinations of molecular fingerprints and ML methods. The RF, LightGBM, and SVM models were built using Python and validated with 5-fold cross-validation. Table 5 highlights the statistical characteristics of the binary classification models after threshold moving (calibrated). Table S1 highlights the characteristics of the uncalibrated models. Threshold moving was used to increase prediction confidence without losing data (i.e., we attempted a threshold-moving calibration of probability estimates without the need to balance the data). Most cross-validated skin sensitization models showed high-quality predictive accuracy on 5-fold external cross-validation based on metrics such as CCR, SE, SP, PPV, NPV, and coverage.

Table 5 highlights the best-performing model for each data set. Briefly, the performance of these four models showed reasonable CCR (74–88%), SE (63–92%), PPV (90–97%), NPV (41–60%), and AUC (75–92%). The best-performing model, using the WES data set with ECFP4 and SVM, showed the most promise and was deployed as a web tool. The performance of all models in the WES data set was also assessed using the Y-randomization test. The best-performing WES model, with the ECFP4 and SVM, significantly outperformed models built with Y-randomization (Table S3).

Table 5. Calculated metrics for calibrated binary classification QSAR models.

Approach	FP	ML	CCR	SE	SP	PPV	NPV	AUC	Coverage	PT
MLLP	**ECFP4**	**RF**	**0.74**	**0.84**	**0.63**	**0.90**	**0.50**	**0.75**	**83**	**0.77**
	ECFP4	SVM	0.71	0.81	0.61	0.89	0.45	0.77	83	0.78
	ECFP4	LightGBM	0.62	0.55	0.68	0.88	0.27	0.63	83	0.80
	Mordred	RF	0.76	0.69	0.83	0.94	0.40	0.80	93	0.78
	Mordred	SVM	0.62	0.83	0.41	0.85	0.38	0.58	93	0.80
	Mordred	LightGBM	0.74	0.69	0.78	0.93	0.39	0.74	93	0.82
MSPE	ECFP4	RF	0.72	0.85	0.59	0.90	0.47	0.72	88.9	0.80
	ECFP4	SVM	0.68	0.75	0.62	0.90	0.35	0.73	88.9	0.84
	ECFP4	LightGBM	0.55	0.60	0.51	0.85	0.22	0.56	88.9	0.82
	Mordred	**RF**	**0.78**	**0.74**	**0.82**	**0.95**	**0.41**	**0.82**	**94**	**0.77**
	Mordred	SVM	0.70	0.67	0.72	0.91	0.32	0.74	94	0.85
	Mordred	LightGBM	0.73	0.67	0.79	0.94	0.35	0.77	94	0.82
WoE	ECFP4	RF	0.74	0.75	0.73	0.92	0.40	0.75	88.5	0.83
	ECFP4	SVM	0.72	0.88	0.56	0.90	0.52	0.70	88.5	0.78
	ECFP4	LightGBM	0.62	0.49	0.76	0.90	0.25	0.58	88.5	0.81
	Mordred	**RF**	**0.78**	**0.79**	**0.76**	**0.93**	**0.46**	**0.81**	**92**	**0.73**
	Mordred	SVM	0.56	0.64	0.49	0.84	0.24	0.53	92	0.81
	Mordred	LightGBM	0.74	0.72	0.76	0.93	0.39	0.75	92	0.81
WES	ECFP4	RF	0.82	0.89	0.76	0.93	0.65	0.88	90.5	0.75
	ECFP4	**SVM**	**0.88**	**0.83**	**0.92**	**0.97**	**0.60**	**0.92**	**90.5**	**0.84**
	ECFP4	LightGBM	0.64	0.53	0.76	0.89	0.31	0.60	90.5	0.78
	Mordred	RF	0.73	0.67	0.78	0.92	0.4	0.79	91	0.77
	Mordred	SVM	0.6	0.71	0.49	0.83	0.32	0.59	91	0.78
	Mordred	LightGBM	0.74	0.79	0.68	0.9	0.47	0.74	91	0.74

RF, random forest; ECFP4, extended connectivity fingerprints with diameter 4; LightGBM, light gradient-boosting machine; MACCS, molecular access systems keys fingerprint; CCR, correct classification rate; SE, sensitivity; SP, specificity; PPV, positive predictive value; NPV, negative predictive value; coverage, a ratio of the test set or external set compounds within the applicability domain; PT, probability threshold; statistical results all obtained after threshold-moving calibration. Statistical results obtained from the default probability thresholds (uncalibrated) are available in the Supplementary Table S1. Models in bold reflect those with the best performance in the data set.

3.2.1. Threshold Moving

In the developed RF, SVM, and LightGBM models, a continuous value represents the probability of a given compound belonging to a specific class (i.e., sensitizer or non-sensitizer). With the aim of better differentiating between skin sensitization sensitizers and non-sensitizers, despite the imbalance of classes present in the data, the probability threshold was adjusted. Usually, probabilities less than 0.5 are assigned to the non-sensitizing class (Class 0), while values greater than or equal to 0.5 are assigned to the sensitizing class (Class 1). However, when modeling imbalanced data, QSAR models often express lower probability estimates for the minority class [49]. Thus, we tested various probability thresholds ranging from 0 to 1 in order to identify the optimal threshold for model performance. The threshold-moving approach and calibration significantly improve the statistical performance of the QSAR models. Figure 2a–d highlights that, for the best models, the calibrated models outperformed the uncalibrated counterparts or scored similarly, rarely underperforming models without calibration. Table S1 highlights the uncalibrated performance for all binary models.

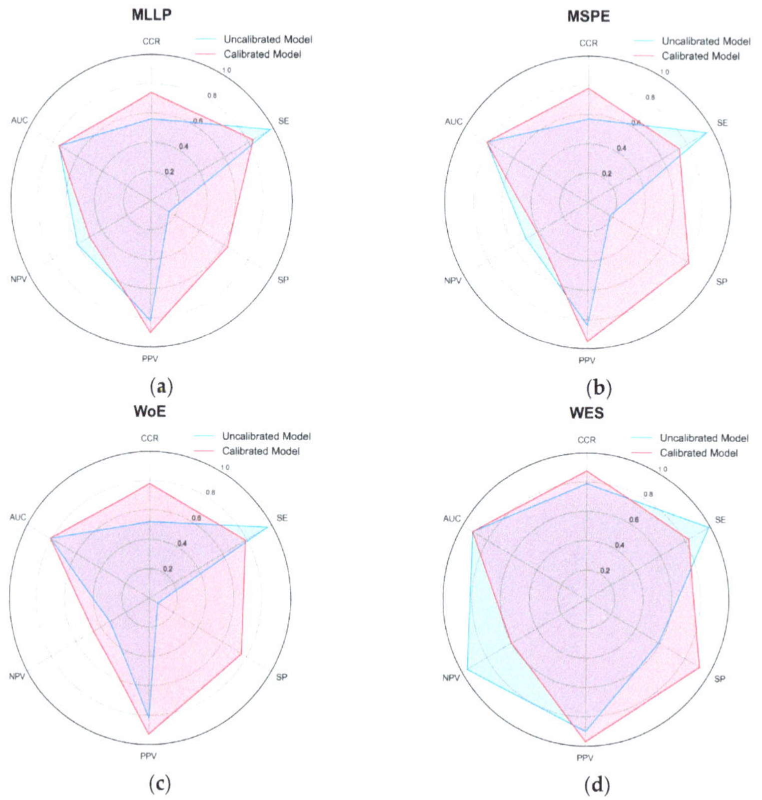

Figure 2. Radar chart comparing metrics before and after calibration for best-performing models. CCR, SE, SP, PPV, NPV, and AUC are included. AUC is a threshold-independent metric. The results for the best-performing (**a**) MLLP, (**b**) MSPE, (**c**) WoE, and (**d**) WES models. The calibrated models outperformed the uncalibrated models or scored similarly, rarely underperforming models without calibration. Generally, a more balanced and symmetrical shape in the radar chart indicates a uniform performance across the metrics, while pronounced peaks and dips highlight potential strengths and weaknesses of the models, respectively.

3.2.2. Binary Model Interpretation with SHAP

SHAP values were calculated to interpret the built binary classification ML models. SHAP calculates the model's feature importance, and it is accessible in the web tool. Figure 3a shows the most relevant features, ranked in order of the impact on the model's predictions. These features exhibited the highest average absolute SHAP values, suggesting that they strongly influenced the model's predictions. Figure 3b demonstrates the bit images of the highest impact features. Bits 7 and 16 shared similar substructures, indicating that this substructure is likely important for the model's predictions. Given the imbalance in the data set, most high-impact bits showed a higher impact on the prediction of sensitizers. It appears that bits 9 and 15 had the highest impact on non-sensitizer prediction. Since a chemical fingerprint of radius 2 was used, many of the substructures highlighted lack robust evidence of chemical significance. However, comparisons can still be drawn for some substructures. For instance, bit 4 resembles the sulfonamide moiety, and drugs with this structural composition have reported common adverse drug reactions such as allergies and hypersensitivity [78–82]. According to the Adverse Outcome Pathway (AOP) for skin sensitization by the OECD [83], step 2 of the AOP involves the substance behaving as a direct-acting electrophile or being converted into a reactive electrophilic metabolite. Given the electrophilic nature of the sulfonamide moiety, as discussed in a recent article evaluating the degradation mechanisms of sulfonamides [84], it is plausible that this structural feature could interact with nucleophilic sites in proteins, potentially contributing to sensitization processes, as outlined in the AOP for skin sensitization. Further, bit 1 resembles a 1,2 alkane diol substructure, and links have been examined between changes in 1,2 alkane diol chain length and skin sensitization potential [85].

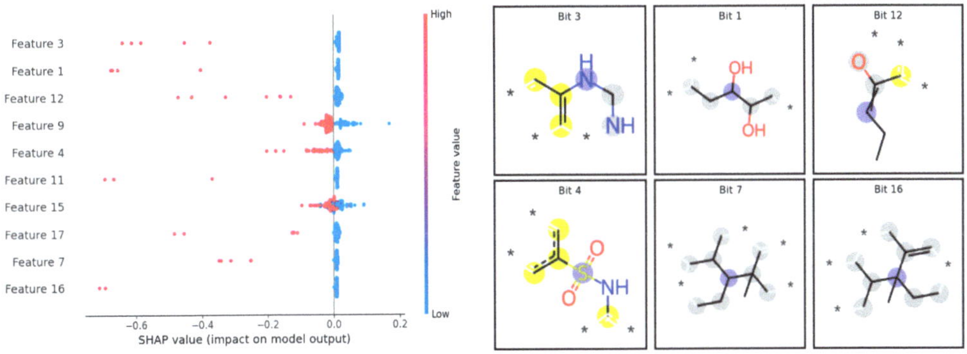

(**a**) (**b**)

Figure 3. SHAP interpretation of the best ECFP4 binary models. (**a**) MLLP model: The x-axis indicates SHAP values and the impact of molecular bits on model output; the y-axis represents compound features (bits). Red denotes a positive effect on model impact; blue represents a negative impact on model prediction. (**b**) The highest impact features on model performance are highlighted. The blue contour atoms are central atoms in the feature; yellow represents aromatic atoms, and gray represents aliphatic ring atoms.

3.2.3. Case Study: Validation of Binary Models with DASS App Defined Approaches

Using the DASS app reference experimental values, we compared our binary model's predictions to the OECD-validated defined approaches for assessing skin sensitization (Table 6), as well as a literature-based human reference set.

Table 6. Case study validation metrics for binary classification QSAR models.

Reference Column	# Compounds (NCs/Sensitizers)	Inside AD	ACC *	SE *	SP *	PPV *	NPV *
DA ITS call	138 (37/101)	115	61%	88%	34%	75%	54%
2o3 DA call	142 (61/81)	117	62%	88%	28%	61%	64%
Basketter human data	63 (21/42)	48	75%	94%	34%	75%	75%

* Metrics reflect predictions inside the model's applicability domain.

The model showed comparable performance across all three validation scenarios, with accuracy ranging from 61 to 75% and SE from 88 to 94%. Notably, the model showed significant strength in predicting sensitizers (high SE and PPV) compared to non-sensitizers (low SP and reasonable NPV). Predictions for the Basketter data set showed promise, which was expected since the Basketter data most closely resemble the model's training data, given that it expresses expert judgment based on human testing, while the DA ITS Call and the 2o3 are based on an in vitro screening of compounds.

Figure S1 highlights the discrepancy between the predictions for each reference, highlighting how the data used for validation can impact the validation metrics. In brief terms, compounds with all reference values concordant across the three data sets (i.e., all three columns are either "1" or "0") showed high correct prediction rates, correctly labeling 30 out of 36 (83%) compounds. When the original outcomes of the external values were divergent, such as in compounds of index 1, 5, 40, and 44, the model usually predicted the Basketter category correctly but incorrectly according to the DA ITS and 2o3 Call. Such discrepancies were anticipated, considering that the model was created using human data, and the Basketter call is also based on human information, whereas the DA ITS Call and the 2o3 Call are based primarily on in vitro experiments (DPRA, h-CLAT, and KeratinoSens assays).

3.3. Chemical Space Analysis

The chemical space analysis performed was investigated for each of the best-performing models. Supervised classification was performed to identify similarities and gaps in the chemical space. Figure 4a highlights that there is not much overlap between classes (i.e., structurally similar compounds but with different outcomes). Despite variable selection, compounds were still aggregating and clustering due to structural similarities. Here, we classified compounds that appeared to cluster together to identify the chemical space, and we found clusters of chlorophenols, sulfanylacetamides, di-amines, and tri-amines to contain mostly skin sensitizers.

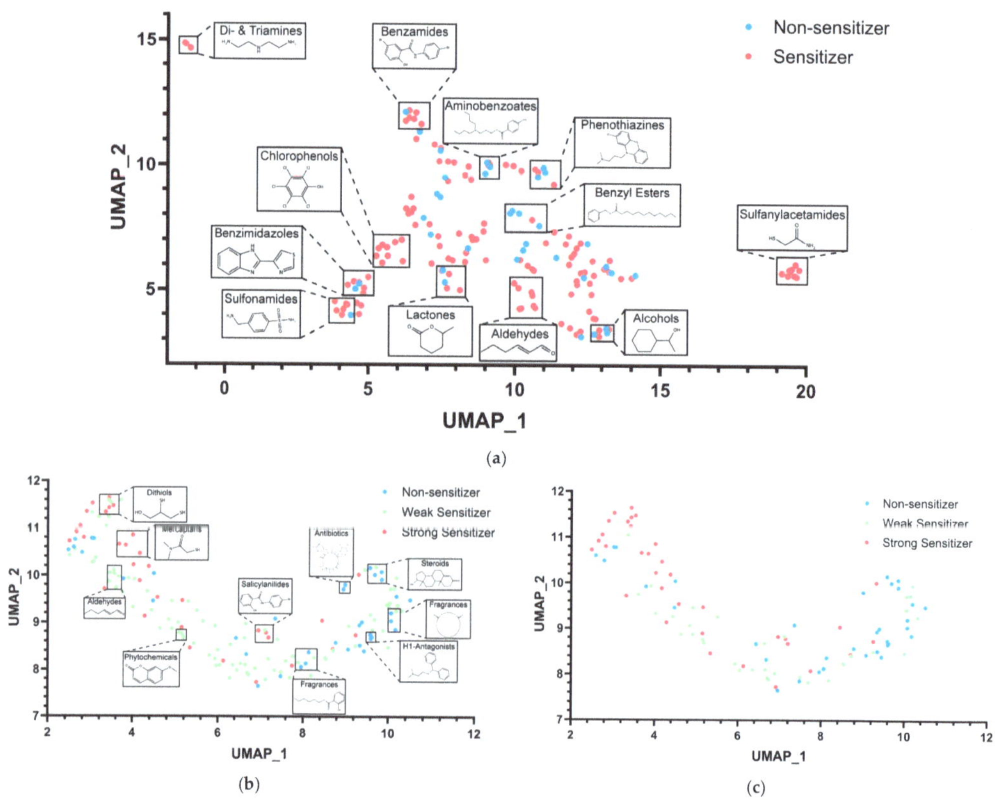

Figure 4. Supervised classification results. Compounds are represented as points on the X and Y axes. The chemical grouping of compounds in the data set was performed after using Morgan descriptors with a radius of 2 and 2048 bits. Low-variance descriptors were filtered, and dimensionality reduction was performed using SVM. Grouping was performed after the calculating was depicted for each of the data sets. Each cluster can be identified by different colors in the chart. If the user downloads the data set and utilizes the pipeline shown in MoViz [67], the option to interact with points in the plot and visualize the chemical structures is available. Here shown are (**a**) clustering for the WES data set, (**b**) clustering for WES multiclass data, and (**c**) clustering for the WES data set after random balancing. Non-sensitizers are in blue, weak sensitizers are in green, and strong sensitizers are in red.

A similar process was conducted for the best-performing multiclass data set (Figure 4b,c). The multiclass also had a highly diverse set of compounds in its chemical space. After random undersampling, Figure 4c highlights that the model still covers most of the chemical space. Random undersampling was beneficial in balancing class distribution by reducing the size of the majority class while mitigating the risk of bias in our ML models. In this situation, random undersampling yielded a reasonable alternative, as most compounds of the majority class (weak sensitizers) were structurally similar and in close proximity to one another (Figure 4b).

3.4. Multiclass Modeling Results

Table 7 demonstrates the metrics for the best-performing multiclass models developed. The performance of all built multiclass models can be found in Table S1. The WES data set showed the most promising results: reasonable accuracy (73%), specificity (82%), and an AUC of 75% with 81% coverage. Compared to the binary models, the multiclass models surprisingly showed better classification of negatives than positives (high SP and NPV). One potential explanation is the balancing of the data sets and the relatively small number of compounds in the data set, which likely contributed to better discrimination between strong sensitizers and non-sensitizers. The best-performing multiclass model also performed better than the models built with Y-randomization (Table S3).

Table 7. Calculated metrics for calibrated multiclass QSAR models with undersampling.

Approach *	FP	ML	CCR	SE	SP	PPV	NPV	AUC	Coverage
WES	ECFP4	RF	0.60	0.47	0.74	0.49	0.73	0.69	49
	ECFP4	SVM	0.58	0.44	0.72	0.44	0.72	0.66	49
	ECFP4	LightGBM	0.56	0.41	0.71	0.42	0.71	0.61	49
	Mordred	**RF**	**0.73**	**0.64**	**0.82**	**0.64**	**0.82**	**0.75**	**81**
	Mordred	SVM	0.50	0.34	0.67	0.23	0.67	0.48	81
	Mordred	LightGBM	0.66	0.55	0.77	0.55	0.78	0.69	81

* Metrics for the remaining approaches are available in Table S1.

Case Study: Validation of Multiclass Models with Defined Approaches

The validation performance of the multiclass models was evaluated using human skin sensitization potency reference values as published by the OECD. Here, we can observe that the model exhibits an average prediction accuracy of (51%), correctly predicting 30 of the 58 compounds. As the confusion matrix highlights (Figure 5), the multiclass model showed good distinctions between the non-sensitizers (NCs) and strong sensitizers (1A), only incorrectly mislabeling one compound as a strong sensitizer. However, as emphasized in Figure 4, the model struggled to correctly distinguish between weak and non-sensitizers due to their structural similarity (Figure 4), so most NCs were incorrectly predicted as weak sensitizers (11 compounds). The non-sensitizing class had the most incorrect labels (17 out of 21 compounds), likely due to the lack of true negatives during model building and structural similarity to weak sensitizers. Table S2 contains each compound's predictions and the validation predictions using HPPT data. For the external set of HPPT compounds in the DASS app, 13 of 14 compounds were inside the applicability domain. The model correctly predicted 9 of the 13 compounds in the HPPT validation set (69%). Overall, when tested with compounds outside the model's training set, the multiclass model showed reasonable promise in predicting human skin sensitization potential, with average class-wise CCR of 63%, SP of 75%, SE of 51%, PPV of 51%, and NPV of 76%.

The validation data set included a diverse array of compounds representing various chemical classes, such as aldehydes, alcohols, esters, acids, amines, phenols, and hydrocarbons (Table S2). The original data in the DASS app are also available in the NTP's Integrated Chemical Environment (ICE) [72,86]. In the ICE toolbox, the data set is accessible via the Chemicals quick list as "OECD Defined Approach to Skin Sensitization: Human" (https://ice.ntp.niehs.nih.gov/ChemicalQuickLists, accessed on 18 October 2024).

The ICE toolbox Chemical Characterization workflow can also be used to enable a better understanding of the chemical composition of the data set.

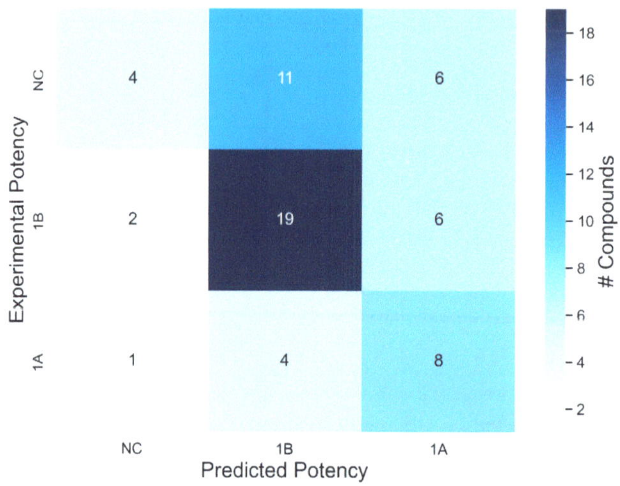

Figure 5. Confusion matrix for multiclass model classification validation results using DA Basketter data. Non-sensitizers were the class with the most incorrect predictions. Color coding represents the number of compounds; the classes are NCs—non-sensitizers; 1B: weak sensitizer; 1A strong sensitizer.

3.5. Visualization of Continuous Data Structure–Property Landscapes

Visualizing the continuous skin sensitization data provides a qualitative representation of molecular roughness. Through the 3D and 2D landscapes, activity cliffs can be easily identified. The data in Figure 6a,b for the DSA data comprised the smoothest landscape of the three, with a ROGI value of 0.19, followed by DSA05 at 0.29 (Figure 6c,d) and then DSA01 at 0.31 (Figure 6e,f). In Figure 6, we can observe that all landscape profiles exhibit a relatively rugged landscape throughout. Further, for the three data sets, we can observe that very similar molecules lead to drastically different outcomes, suggesting that these data sets contain activity cliffs.

The continuous models developed for predicting human skin sensitization did not perform satisfactorily. The box and whisker plots (Figure S2) highlight one of the challenges in modeling human HPPT data due to the high variability and wide range of results for a single chemical that, when expanded to the entire data set, introduces noise and inhibits the model's ability to predict the dose per skin area for compounds correctly. Notably, for the three continuous data sets (Figure S2a–c), most compounds have a wide range and interquartile range (IQR), suggesting that using these compounds for modeling would introduce ambiguity and significant uncertainties into the model's predictions. These results underscore the limitations of using continuous models to capture the underlying patterns in the human predictive patch test data. We recommend that other groups build on our curated continuous data and explore alternative modeling approaches to achieve more reliable predictions.

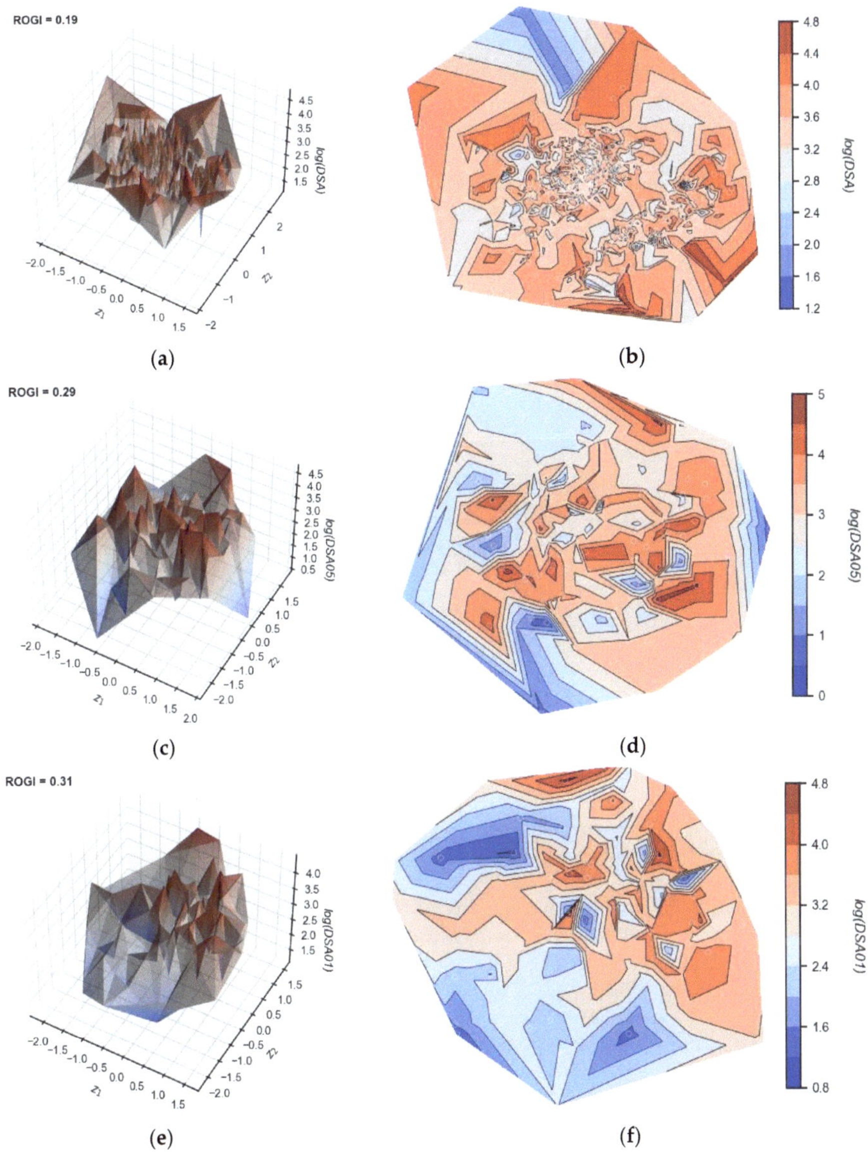

Figure 6. Visualization of molecular landscapes of the different data sets. After Euclidean distance calculation, the three high-dimensionality datasets are projected onto a 2D plane (coordinates z1 and z2). The third dimension corresponds to the activity value of each data set (log values of DSA, DSA05, or DSA01). Mordred descriptors were used to eliminate recursive features. (**a**,**c**,**e**) corresponds to the 3D representation, while (**b**,**d**,**f**) correspond to the two-dimensional contour plots. (**a**,**b**) log(DSA), (**c**,**d**) log(DSA05), and (**e**,**f**) log(DSA01). The ROGI value for each data set is shown. Activity cliffs are visible as areas in proximity with highly different activity values.

3.6. Model Implementation

The most predictive classification model for Human Skin Sensitization Prediction was implemented in the open-access HuSSPred web application (https://husspred.mml.unc.edu, accessed on 18 October 2024). The HuSSPred web tool was designed to possess an intuitive user interface, where users can draw compounds of interest in the "molecular editor" box or directly paste a list of SMILES strings to query chemical compounds (Figure 7). After clicking the "Get Properties" button, the user will be shown the classification outcome of the compounds (sensitizer or non-sensitizer) using the best classification model. Further, the user can opt to include 'Fragment contribution analysis" and "SHAP descriptor importance" in the report. All predictions also contain AD estimates and mechanistic interpretation using color-coded maps of fragment contribution. For the fragment-contribution maps, atoms or fragments promoting positive toxicity are highlighted in green, while those decreasing toxicity are highlighted in purple. The models developed in this study are available within the HuSSPred web application https://husspred.mml.unc.edu/, accessed on 18 October 2024).

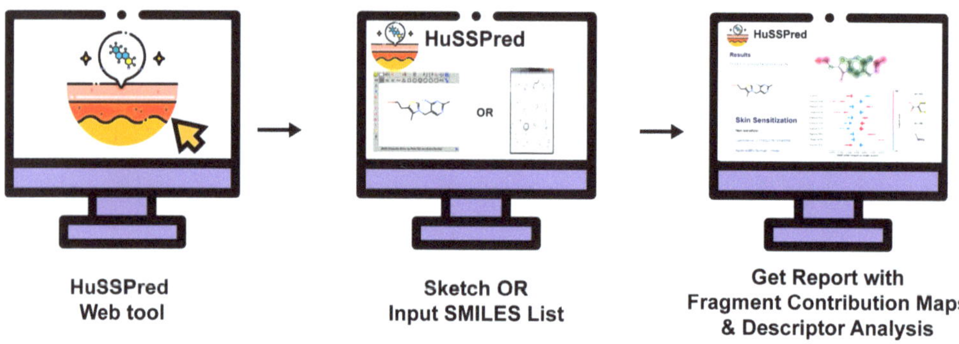

Figure 7. Model implementation and user-friendly pipeline of HuSSPred models.

In the development of NAMs to predict a compound's toxicity and sensitization potential, it is essential for models to accurately predict potential skin sensitizers, rather than non-sensitizers, which can reduce animal testing and the waste of resources by eliminating compounds likely to fail downstream in the development process. Given these safety and resource management considerations, the high sensitivity of the best-performing models (63–92%) and high PPV (90–97%) indicate that sensitizers are frequently correctly labeled. Further, the high PPV and reasonable NPV underscore the model's potential to contribute to advancing the three Rs (reduce, refine, and replace). The model's applicability to binary contexts is further supported by the high sensitivity (94%) and accuracy (75%) obtained during the model's validation with reference human data and other human biology-based DAs for skin sensitization testing.

While the model performance could be drastically improved by including a more comprehensive set of true negatives, we emphasize the difficulty of obtaining such, given the current human skin sensitization testing system. Currently, the Globally Harmonized System of Classification and Labelling of Chemicals (GHS) [74,87] uses a set of frameworks for classifying compounds as skin sensitizers [88]. Therefore, the challenge of obtaining a true negative is commonly caused by not testing substances at a high enough concentration to trigger sensitization. This frequently results in compounds labeled as "inconclusive", which are unsuitable for modeling. To our knowledge, this skin sensitization QSAR model's high sensitivity fares competitively well against other established tools in the field. An evaluation of in silico tools to predict skin sensitization was published in 2017 [89]. From this report, the best-performing and best-known tools were VEGA, Derek Nexus, and the OECD QSAR Toolbox. According to the metrics reported in the assessment, our tool exhibited a good external validation sensitivity score relative to those three, with 94% for the Basketter human data set, compared to 91%, 78%, and 43%, respectively for the other three tools, while also being one of few developed with human data. For other metrics, such as CCR, SP, PPV, and NPV, our models performed similarly or better in the majority of cases.

Here, we suggest that our skin sensitization QSAR models' utility lies in their ability to predict skin-sensitizing substances accurately. The freely available web tool HuSSPred was built on the best-performing models with the most highly curated data set available, making it a valuable tool for toxicity prediction and providing risk assessors with confidence in the model's positive predictions. The overall weight-of-evidence approach used for the web tool (WES) models was the preferred data set due to its robust scoring system and overall performance. The scores assigned to each "extrapolated" classification in the WES database intuitively reflect how a risk assessor might combine multiple HPPT results in a single WoE assessment. Only unambiguous outcomes were reported and assigned a score, so the developed models available in the web tool were built only on the most rigorous HPPT data [27,28].

There is an increasing need to quickly determine a compound's skin sensitization potential from a regulatory standpoint; unfortunately, these studies are time-consuming and expensive, and they raise ethical concerns. On the other hand, our QSAR models can be easily implemented in the early stages of drug development and testing, minimizing resource waste and facilitating the early stages of drug development while reducing animal testing. We hope the QSAR models developed here progress further towards regulatory acceptance.

4. Conclusions

Overall, we have carefully curated the HPPT database to predict skin sensitization, enabling a more accurate assessment of the human response. The models underwent rigorous validation with reference data from the internationally validated Defined Approaches for Skin Sensitization (DASS), resulting in one of the first in silico tools based strictly on human data for skin sensitization, with high accuracy and sensitivity, as validated with human biology-based in vitro outcomes and human data. The rigorously curated data, validation predictions, and models are available on the website (https://husspred.mml.unc.edu/, accessed on 18 October 2024), where users can select the desired models to obtain (i) binary predictions, (ii) multiclass potency predictions for hazard assessments of compounds, and (iii) fragment-contribution maps for each assay prediction.

Supplementary Materials: The following supporting information can be downloaded at https://www.mdpi.com/article/10.3390/toxics12110803/s1. File S1: Curated data sets for binary, multiclass, and continuous model development. Table S1: Uncalibrated model results for binary and WoE approaches for multiclass models. Table S2: Validation exercise data sets and predictions. Table S3: Y-randomization test modeling results. Figure S1: Model predictions for compounds in common across validation datasets. Figure S2: Box and whisker plots for continuous data.

Author Contributions: Conceptualization, R.S.T., M.R., A.T., E.N.M. and N.K; methodology, R.S.T. and J.T.M.-F.; software, J.W., R.S.T., and J.T.M.-F.; data curation, R.S.T. and J.T.M.-F.; writing—original draft preparation, R.S.T., M.C.O., and H.-J.M.; writing—review and editing, R.S.T., H.-J.M.M, J.T.M.-F., N.K., A.T. and E.N.M.; funding acquisition, E.N.M. and A.T. All authors have read and agreed to the published version of the manuscript.

Funding: This study was supported by the National Institutes of Health (NIH) (grants R41ES033589 and R44ES032371).

Data Availability Statement: The data presented in this study are available at https://github.com/molecularmodelinglab/, accessed on 18 October 2024.

Acknowledgments: The authors thank the NIEHS/NICEATM and the NIEHS Scholar's Connect Program (NSCP) for support and funding. This work was supported in part by the NIH Intramural Research Program.

Conflicts of Interest: A.T. and E.N.M. are co-founders of Predictive, LLC., which develops novel alternative methodologies and software for toxicity prediction. All the other authors declare no conflicts.

References

1. Murphy, K.; Weaver, C. *Janeway's Immunobiology*; Garland Science: New York, NY, USA, 2017; p. 840.
2. International Regulatory Requirements for Skin Sensitization Testing—ScienceDirect. Available online: https://www.sciencedirect.com/science/article/pii/S0273230018300667?casa_token=JMU1-S7dZRcAAAAA:ucrEArsu9Pcgw8uJ1MJYfWzIZDiID5pkO4zV8SpOcNFaRxYN4lfnoW5DjfX6KSIzJNcOoJ7G (accessed on 11 September 2024).
3. OECD. *Test No. 429: Skin Sensitisation: Local Lymph Node Assay*; Organisation for Economic Co-Operation and Development: Paris, France, 2010.
4. OECD. *Test No. 406: Skin Sensitisation*; Organisation for Economic Co-Operation and Development: Paris, France, 2022.
5. OECD. *Guideline No. 497: Defined Approaches on Skin Sensitisation*; Organisation for Economic Co-Operation and Development: Paris, France, 2023.
6. OECD. *Test No. 442C: In Chemico Skin Sensitisation: Assays Addressing the Adverse Outcome Pathway Key Event on Covalent Binding to Proteins*; Organisation for Economic Co-Operation and Development: Paris, France, 2024.
7. OECD. *Test No. 442D: In Vitro Skin Sensitisation: Assays Addressing the Adverse Outcome Pathway Key Event on Keratinocyte Activation*; Organisation for Economic Co-Operation and Development: Paris, France, 2024.
8. OECD. *Test No. 442E: In Vitro Skin Sensitisation: In Vitro Skin Sensitisation Assays Addressing the Key Event on Activation of Dendritic Cells on the Adverse Outcome Pathway for Skin Sensitisation*; Organisation for Economic Co-Operation and Development: Paris, France, 2024.
9. Genomic Responses in Mouse Models Poorly Mimic Human Inflammatory Diseases | PNAS. Available online: https://www.pnas.org/doi/abs/10.1073/pnas.1222878110 (accessed on 8 August 2024).
10. Machine Learning of Toxicological Big Data Enables Read-Across Structure Activity Relationships (RASAR) Outperforming Animal Test Reproducibility | Toxicological Sciences | Oxford Academic. Available online: https://academic.oup.com/toxsci/article/165/1/198/5043469 (accessed on 8 August 2024).
11. Mahesh, B. Machine Learning Algorithms—A Review. *Int. J. Sci. Res. (IJSR)* **2019**, *9*, 381–386. [CrossRef]
12. Borba, J.V.B.; Alves, V.M.; Braga, R.C.; Korn, D.R.; Overdahl, K.; Silva, A.C.; Hall, S.U.S.; Overdahl, E.; Kleinstreuer, N.; Strickland, J.; et al. STopTox: An in Silico Alternative to Animal Testing for Acute Systemic and Topical Toxicity. *Environ. Health Perspect.* **2022**, *130*, 027012. [CrossRef] [PubMed]
13. Borba, J.V.B.; Braga, R.C.; Alves, V.M.; Muratov, E.N.; Kleinstreuer, N.; Tropsha, A.; Andrade, C.H. Pred-Skin: A Web Portal for Accurate Prediction of Human Skin Sensitizers. *Chem. Res. Toxicol.* **2021**, *34*, 258–267. [CrossRef] [PubMed]
14. Chakravarti, S.K.; Saiakhov, R.D.; Klopman, G. Optimizing Predictive Performance of CASE Ultra Expert System Models Using the Applicability Domains of Individual Toxicity Alerts. *J. Chem. Inf. Model.* **2012**, *52*, 2609–2618. [CrossRef]
15. Bhhatarai, B.; Wilson, D.M.; Parks, A.K.; Carney, E.W.; Spencer, P.J. Evaluation of TOPKAT, Toxtree, and Derek Nexus in Silico Models for Ocular Irritation and Development of a Knowledge-Based Framework To Improve the Prediction of Severe Irritation. *Chem. Res. Toxicol.* **2016**, *29*, 810–822. [CrossRef]
16. Skin Sensitisation Assessment | Lhasa Limited. Available online: https://www.lhasalimited.org/solutions/skin-sensitisation-assessment/ (accessed on 28 October 2024).
17. Bhatia, S.; Schultz, T.; Roberts, D.; Shen, J.; Kromidas, L.; Marie Api, A. Comparison of Cramer Classification between Toxtree, the OECD QSAR Toolbox and Expert Judgment. *Regul. Toxicol. Pharmacol.* **2015**, *71*, 52–62. [CrossRef]
18. Danieli, A.; Colombo, E.; Raitano, G.; Lombardo, A.; Roncaglioni, A.; Manganaro, A.; Sommovigo, A.; Carnesecchi, E.; Dorne, J.-L.C.M.; Benfenati, E. The VEGA Tool to Check the Applicability Domain Gives Greater Confidence in the Prediction of In Silico Models. *Int. J. Mol. Sci.* **2023**, *24*, 9894. [CrossRef]
19. Patlewicz, G.; Jeliazkova, N.; Safford, R.J.; Worth, A.P.; Aleksiev, B. An Evaluation of the Implementation of the Cramer Classification Scheme in the Toxtree Software. *SAR QSAR Environ. Res.* **2008**, *19*, 495–524. [CrossRef]
20. Alves, V.M.; Muratov, E.; Fourches, D.; Strickland, J.; Kleinstreuer, N.; Andrade, C.H.; Tropsha, A. Predicting Chemically-Induced Skin Reactions. Part I: QSAR Models of Skin Sensitization and Their Application to Identify Potentially Hazardous Compounds. *Toxicol. Appl. Pharmacol.* **2015**, *284*, 262–272. [CrossRef]
21. Alves, V.M.; Capuzzi, S.J.; Muratov, E.N.; Braga, R.C.; Thornton, T.E.; Fourches, D.; Strickland, J.; Kleinstreuer, N.; Andrade, C.H.; Tropsha, A. QSAR Models of Human Data Can Enrich or Replace LLNA Testing for Human Skin Sensitization. *Green Chem.* **2016**, *18*, 6501–6515. [CrossRef]
22. Alves, V.M.; Muratov, E.; Fourches, D.; Strickland, J.; Kleinstreuer, N.; Andrade, C.H.; Tropsha, A. Predicting Chemically-Induced Skin Reactions. Part II: QSAR Models of Skin Permeability and the Relationships between Skin Permeability and Skin Sensitization. *Toxicol. Appl. Pharmacol.* **2015**, *284*, 273–280. [CrossRef] [PubMed]
23. Alves, V.M.; Capuzzi, S.J.; Braga, R.C.; Borba, J.V.B.; Silva, A.C.; Luechtefeld, T.; Hartung, T.; Andrade, C.H.; Muratov, E.N.; Tropsha, A. A Perspective and a New Integrated Computational Strategy for Skin Sensitization Assessment. *ACS Sustain. Chem. Eng.* **2018**, *6*, 2845–2859. [CrossRef]
24. Full Article: Non-Animal Methods to Predict Skin Sensitization (II): An Assessment of Defined Approaches**. Available online: https://www.tandfonline.com/doi/full/10.1080/10408444.2018.1429386 (accessed on 11 September 2024).
25. Computational Application of Internationally Harmonized Defined Approaches to Skin Sensitization: DASS App | BMC Bioinformatics. Available online: https://link.springer.com/article/10.1186/s12859-023-05617-1 (accessed on 11 September 2024).

26. Human Data for Skin Sensitization Method Evaluation. Available online: https://ntp.niehs.nih.gov/whatwestudy/niceatm/test-method-evaluations/skin-sens/hppt (accessed on 9 July 2024).
27. Strickland, J.; Abedini, J.; Allen, D.G.; Gordon, J.; Hull, V.; Kleinstreuer, N.C.; Ko, H.-S.; Matheson, J.; Thierse, H.-J.; Truax, J.; et al. A Database of Human Predictive Patch Test Data for Skin Sensitization. *Arch. Toxicol.* **2023**, *97*, 2825–2837. [CrossRef] [PubMed]
28. Herzler, M.; Abedini, J.; Allen, D.G.; Germolec, D.; Gordon, J.; Ko, H.-S.; Matheson, J.; Reinke, E.; Strickland, J.; Thierse, H.-J.; et al. Use of Human Predictive Patch Test (HPPT) Data for the Classification of Skin Sensitization Hazard and Potency. *Arch. Toxicol.* **2024**, *98*, 1253–1269. [CrossRef]
29. Fourches, D.; Muratov, E.; Tropsha, A. Trust, But Verify: On the Importance of Chemical Structure Curation in Cheminformatics and QSAR Modeling Research. *J. Chem. Inf. Model.* **2010**, *50*, 1189–1204. [CrossRef]
30. Fourches, D.; Muratov, E.; Tropsha, A. Trust, but Verify II: A Practical Guide to Chemogenomics Data Curation. *J. Chem. Inf. Model.* **2016**, *56*, 1243–1252. [CrossRef]
31. Fourches, D.; Muratov, E.; Tropsha, A. Curation of Chemogenomics Data. *Nat. Chem. Biol.* **2015**, *11*, 535. [CrossRef]
32. Chemaxon. Available online: https://chemaxon.com (accessed on 30 December 2023).
33. Open for Innovation | KNIME. Available online: https://www.knime.com/ (accessed on 8 August 2024).
34. Berthold, M.R.; Cebron, N.; Dill, F.; Gabriel, T.R.; Kötter, T.; Meinl, T.; Ohl, P.; Thiel, K.; Wiswedel, B. KNIME-the Konstanz Information Miner: Version 2.0 and Beyond. *AcM SIGKDD Explor. Newsl.* **2009**, *11*, 26–31. [CrossRef]
35. Mazanetz, M.P.; Marmon, R.J.; Reisser, C.B.T.; Morao, I. Drug Discovery Applications for KNIME: An Open Source Data Mining Platform. *Curr. Top. Med. Chem.* **2012**, *12*, 1965–1979. [CrossRef]
36. Landrum, G. Rdkit: Open-Source Cheminformatics Software. 2016. Available online: https://github.com/rdkit/rdkit (accessed on 28 October 2024).
37. Rogers, D.; Hahn, M. Extended-Connectivity Fingerprints. *J. Chem. Inf. Model.* **2010**, *50*, 742–754. [CrossRef]
38. Moriwaki, H.; Tian, Y.S.; Kawashita, N.; Takagi, T. Mordred: A Molecular Descriptor Calculator. *J. Cheminform.* **2018**, *10*, 4. [CrossRef] [PubMed]
39. Sedykh, A.Y.; Shah, R.R.; Kleinstreuer, N.C.; Auerbach, S.S.; Gombar, V.K. Saagar–A New, Extensible Set of Molecular Substructures for QSAR/QSPR and Read-Across Predictions. *Chem. Res. Toxicol.* **2021**, *34*, 634–640. [CrossRef]
40. Tropsha, A. Best Practices for QSAR Model Development, Validation, and Exploitation. *Mol. Inform.* **2010**, *29*, 476–488. [CrossRef]
41. Ke, G.; Meng, Q.; Finely, T.; Wang, T.; Chen, W.; Ma, W.; Ye, Q.; Liu, T.-Y. LightGBM: A Highly Efficient Gradient Boosting Decision Tree. In Proceedings of the Advances in Neural Information Processing Systems 30 (NIP 2017), Long Beach, CA, USA, 4–9 December 2017.
42. Breiman, L. Random Forests. *Mach. Learn.* **2001**, *45*, 5–32. [CrossRef]
43. Pedregosa, F.; Varoquaux, G.; Gramfort, A.; Michel, V.; Thirion, B.; Grisel, O.; Blondel, M.; Prettenhofer, P.; Weiss, R.; Dubourg, V.; et al. Scikit-Learn: Machine Learning in Python. *Mach. Learn. Python* **2011**, *12*, 2825–2830.
44. Wang, M.W.H.; Goodman, J.M.; Allen, T.E.H. Machine Learning in Predictive Toxicology: Recent Applications and Future Directions for Classification Models. *Chem. Res. Toxicol.* **2021**, *34*, 217–239. [CrossRef]
45. Akiba, T.; Sano, S.; Yanase, T.; Ohta, T.; Koyama, M. Optuna: A next-Generation Hyperparameter Optimization Framework. In Proceedings of the 25th ACM SIGKDD International Conference on Knowledge Discovery & Data Mining, Anchorage, AK, USA, 4–8 August 2019; pp. 2623–2631.
46. Darst, B.F.; Malecki, K.C.; Engelman, C.D. Using Recursive Feature Elimination in Random Forest to Account for Correlated Variables in High Dimensional Data. *BMC Genet.* **2018**, *19*, 65. [CrossRef] [PubMed]
47. Bouke, M.A.; Abdullah, A. An Empirical Study of Pattern Leakage Impact during Data Preprocessing on Machine Learning-Based Intrusion Detection Models Reliability. *Expert Syst. Appl.* **2023**, *230*, 120715. [CrossRef]
48. Tropsha, A.; Gramatica, P.; Gombar, V.K. The Importance of Being Earnest: Validation Is the Absolute Essential for Successful Application and Interpretation of QSPR Models. *QSAR Comb. Sci.* **2003**, *22*, 69–77. [CrossRef]
49. Moreira-Filho, J.T.; Braga, R.C.; Lemos, J.M.; Alves, V.M.; Borba, J.V.; Costa, W.S.; Kleinstreuer, N.; Muratov, E.N.; Andrade, C.H.; Neves, B.J. BeeToxAI: An Artificial Intelligence-Based Web App to Assess Acute Toxicity of Chemicals to Honey Bees. *Artif. Intell. Life Sci.* **2021**, *1*, 100013.
50. Zakharov, A.V.; Peach, M.L.; Sitzmann, M.; Nicklaus, M.C. QSAR Modeling of Imbalanced High-Throughput Screening Data in PubChem. *J. Chem. Inf. Model.* **2014**, *54*, 705–712. [CrossRef] [PubMed]
51. Kubat, M.; Matwin, S. *Addressing the Curse of Imbalanced Training Sets: One-Sided Selection*; Citeseer: Hershey, PA, USA, 1997; Volume 97, p. 179.
52. Barandela, R.; Sánchez, J.S.; Garcıa, V.; Rangel, E. Strategies for Learning in Class Imbalance Problems. *Pattern Recognit.* **2003**, *36*, 849–851. [CrossRef]
53. Data Apps for Production | Plotly. Available online: https://plot.ly/ (accessed on 15 September 2024).
54. Trisciuzzi, D.; Alberga, D.; Mansouri, K.; Judson, R.; Novellino, E.; Mangiatordi, G.F.; Nicolotti, O. Predictive Structure-Based Toxicology Approaches to Assess the Androgenic Potential of Chemicals. *J. Chem. Inf. Model.* **2017**, *57*, 2874–2884. [CrossRef] [PubMed]
55. Tropsha, A.; Golbraikh, A. Predictive QSAR Modeling Workflow, Model Applicability Domains, and Virtual Screening. *Curr. Pharm. Des.* **2007**, *13*, 3494–3504. [CrossRef] [PubMed]

56. Welcome to the SHAP Documentation—SHAP Latest Documentation. Available online: https://shap.readthedocs.io/en/latest/index.html (accessed on 15 September 2024).
57. Louhichi, M.; Nesmaoui, R.; Mbarek, M.; Lazaar, M. Shapley Values for Explaining the Black Box Nature of Machine Learning Model Clustering. *Procedia Comput. Sci.* **2023**, *220*, 806–811. [CrossRef]
58. Rodríguez-Pérez, R.; Bajorath, J. Interpretation of Compound Activity Predictions from Complex Machine Learning Models Using Local Approximations and Shapley Values. *J. Med. Chem.* **2020**, *63*, 8761–8777. [CrossRef]
59. Neves, B.J.; Braga, R.C.; Alves, V.M.; Lima, M.N.; Cassiano, G.C.; Muratov, E.N.; Costa, F.T.; Andrade, C.H. Deep Learning-Driven Research for Drug Discovery: Tackling Malaria. *PLoS Comput. Biol.* **2020**, *16*, e1007025. [CrossRef]
60. Riniker, S.; Landrum, G.A. Similarity Maps-a Visualization Strategy for Molecular Fingerprints and Machine-Learning Methods. *J. Cheminform.* **2013**, *5*, 43. [CrossRef]
61. Grinberg, M. *Flask Web Development: Developing Web Applications with Python*; O'Reilly Media, Inc.: Sebastopol, CA, USA, 2018; ISBN 1-4919-9169-0.
62. The uWSGI Project—uWSGI 2.0 Documentation. Available online: https://uwsgi-docs.readthedocs.io/en/latest/ (accessed on 3 March 2024).
63. Nginx News. Available online: https://nginx.org/ (accessed on 3 March 2024).
64. Welcome to Python.Org. Available online: https://www.python.org/ (accessed on 3 March 2024).
65. Home. Available online: https://ecma-international.org/home/ (accessed on 3 March 2024).
66. Bienfait, B.; Ertl, P. JSME: A Free Molecule Editor in JavaScript. *J. Cheminform.* **2013**, *5*, 24. [CrossRef]
67. Moreira-Filho, J.T.; Ranganath, D.; Conway, M.; Schmitt, C.; Kleinstreuer, N.; Mansouri, K. Democratizing Cheminformatics: Interpretable Chemical Grouping Using an Automated KNIME Workflow. *J. Cheminform.* **2024**, *16*, 101. [CrossRef]
68. McInnes, L.; Healy, J.; Saul, N.; Großberger, L. UMAP: Uniform Manifold Approximation and Projection. *J. Open Source Softw.* **2018**, *3*, 861. [CrossRef]
69. Allaoui, M.; Kherfi, M.L.; Cheriet, A. Considerably Improving Clustering Algorithms Using UMAP Dimensionality Reduction Technique: A Comparative Study. In *Image and Signal Processing, Proceedings of the ICISP 2020, Marrakesh, Morocco, 4–6 June 2020*; El Moataz, A., Mammass, D., Mansouri, A., Nouboud, F., Eds.; Springer International Publishing: Cham, Switzerland, 2020; pp. 317–325.
70. Aldeghi, M.; Graff, D.E.; Frey, N.; Morrone, J.A.; Pyzer-Knapp, E.O.; Jordan, K.E.; Coley, C.W. Roughness of Molecular Property Landscapes and Its Impact on Modellability. *J. Chem. Inf. Model.* **2022**, *62*, 4660–4671. [CrossRef] [PubMed]
71. Defined Approaches to Identify Potential Skin Sensitizers. Available online: https://ntp.niehs.nih.gov/whatwestudy/niceatm/test-method-evaluations/skin-sens/da (accessed on 15 October 2024).
72. Bell, S.; Abedini, J.; Ceger, P.; Chang, X.; Cook, B.; Karmaus, A.L.; Lea, I.; Mansouri, K.; Phillips, J.; McAfee, E.; et al. An Integrated Chemical Environment with Tools for Chemical Safety Testing. *Toxicol. Vitr.* **2020**, *67*, 104916. [CrossRef] [PubMed]
73. Mansouri, K.; Moreira-Filho, J.T.; Lowe, C.N.; Charest, N.; Martin, T.; Tkachenko, V.; Judson, R.; Conway, M.; Kleinstreuer, N.C.; Williams, A.J. Free and Open-Source QSAR-Ready Workflow for Automated Standardization of Chemical Structures in Support of QSAR Modeling. *J. Cheminform.* **2024**, *16*, 19. [CrossRef] [PubMed]
74. Basketter, D.A.; Alépée, N.; Ashikaga, T.; Barroso, J.; Gilmour, N.; Goebel, C.; Hibatallah, J.; Hoffmann, S.; Kern, P.; Martinozzi-Teissier, S.; et al. Categorization of Chemicals According to Their Relative Human Skin Sensitizing Potency. *Dermatitis* **2014**, *25*, 11–21. [CrossRef]
75. Basketter, D.A.; Gerberick, G.F. Skin Sensitization Testing: The Ascendancy of Non-Animal Methods. *Cosmetics* **2022**, *9*, 38. [CrossRef]
76. Waskom, M.L. Seaborn: Statistical Data Visualization. *J. Open Source Softw.* **2021**, *6*, 3021. [CrossRef]
77. Guidance Document on the Validation of (Quantitative) Structure-Activity Relationship [(Q)SAR] Models. Available online: https://www.oecd.org/en/publications/guidance-document-on-the-validation-of-quantitative-structure-activity-relationship-q-sar-models_9789264085442-en.html (accessed on 20 October 2024).
78. Sulzberger, M.B.; Kanof, A.; Baer, R.L.; Lowenberg, C. Sensitization by Topical Application of Sulfonamides. *J. Allergy* **1947**, *18*, 92–103. [CrossRef]
79. Chantachaeng, W.; Chularojanamontri, L.; Kulthanan, K.; Jongjarearnprasert, K.; Dhana, N. Cutaneous Adverse Reactions to Sulfonamide Antibiotics. *Asian Pac. J. Allergy Immunol.* **2011**, *29*, 284–289.
80. Chow, T.G.; Khan, D.A. Sulfonamide Hypersensitivity. *Clin. Rev. Allergy Immunol.* **2022**, *62*, 400–412. [CrossRef]
81. Dorn, J.M.; Alpern, M.; McNulty, C.; Volcheck, G.W. Sulfonamide Drug Allergy. *Curr. Allergy Asthma Rep.* **2018**, *18*, 38. [CrossRef] [PubMed]
82. Serrano-Arias, B.; Araya-Zúñiga, A.; Waterhouse-Garbanzo, J.; Rojas-Barrantes, Z.; Arguedas-Chacón, S.; Zavaleta-Monestel, E. A Comprehensive Review of Sulfonamide Hypersensitivity: Implications for Clinical Practice. *Clin. Rev. Allerg Immunol.* **2023**, *65*, 433–442. [CrossRef] [PubMed]
83. The Adverse Outcome Pathway for Skin Sensitisation Initiated by Covalent Binding to Proteins. Available online: https://www.oecd.org/en/publications/the-adverse-outcome-pathway-for-skin-sensitisation-initiated-by-covalent-binding-to-proteins_9789264221444-en.html (accessed on 20 October 2024).
84. Hu, J.; Li, X.; Liu, F.; Fu, W.; Lin, L.; Li, B. Comparison of Chemical and Biological Degradation of Sulfonamides: Solving the Mystery of Sulfonamide Transformation. *J. Hazard. Mater.* **2022**, *424*, 127661. [CrossRef] [PubMed]

85. Lee, E.; An, S.; Cho, S.-A.; Yun, Y.; Han, J.; Hwang, Y.K.; Kim, H.K.; Lee, T.R. The Influence of Alkane Chain Length on the Skin Irritation Potential of 1,2-Alkanediols. *Int. J. Cosmet. Sci.* **2011**, *33*, 421–425. [CrossRef]
86. ICE Search. Available online: https://ice.ntp.niehs.nih.gov/Search (accessed on 20 October 2024).
87. Casati, S.; Asturiol, D.; Browne, P.; Kleinstreuer, N.; Régimbald-Krnel, M.; Therriault, P. Standardisation and International Adoption of Defined Approaches for Skin Sensitisation. *Front. Toxicol.* **2022**, *4*, 943152. [CrossRef]
88. Decision, C.D.; Beckman, E.J.; Beak, P.; Cura, J.J.; Fairbrother, A.; Greene, N.; Henry, C.; Holder, H.; Hutchison, J.E.; Paol, G.M.; et al. Overview of the GHS Classification Scheme in Hazard Classification. In *A Framework to Guide Selection of Chemical Alternatives*; National Academies Press (US): Washington, DC, USA, 2014.
89. Verheyen, G.R.; Braeken, E.; Van Deun, K.; Van Miert, S. Evaluation of in Silico Tools to Predict the Skin Sensitization Potential of Chemicals. *SAR QSAR Environ. Res.* **2017**, *28*, 59–73. [CrossRef]

toxics

MDPI

Article

Increasing Accessibility of Bayesian Network-Based Defined Approaches for Skin Sensitisation Potency Assessment

Tomaz Mohoric [1,†], Anke Wilm [2,†], Stefan Onken [2], Andrii Milovich [1], Artem Logavoch [1], Pascal Ankli [1], Ghada Tagorti [1], Johannes Kirchmair [3], Andreas Schepky [2], Jochen Kühnl [2], Abdulkarim Najjar [2,*,†], Barry Hardy [1,*,†] and Johanna Ebmeyer [2,*,†]

1 Edelweiss Connect GmbH, Hochbergerstrasse 60C, 4057 Basel, Switzerland; tomaz@edelweissconnect.com (T.M.); andrii@edelweissconnect.com (A.M.); artem@edelweissconnect.com (A.L.); pascal@edelweissconnect.com (P.A.); ghada@edelweissconnect.com (G.T.)

2 Beiersdorf AG, Beiersdorfstraße 1-9, 22529 Hamburg, Germany; anke.wilm@beiersdorf.com (A.W.); stefan.onken@beiersdorf.com (S.O.); andreas.schepky@beiersdorf.com (A.S.); jochen.kuehnl@beiersdorf.com (J.K.)

3 Department of Pharmaceutical Sciences, Division of Pharmaceutical Chemistry, Faculty of Life Sciences, University of Vienna, Josef-Holaubek-Platz 2, 1090 Vienna, Austria; johannes.kirchmair@univie.ac.at

* Correspondence: abdulkarim.najjar@beiersdorf.com (A.N.); barry.hardy@edelweissconnect.com (B.H.); johanna.ebmeyer@beiersdorf.com (J.E.)

† These authors contributed equally to this work.

Abstract: Skin sensitisation is a critical adverse effect assessed to ensure the safety of compounds and materials exposed to the skin. Alongside the development of new approach methodologies (NAMs), defined approaches (DAs) have been established to promote skin sensitisation potency assessment by adopting and integrating standardised in vitro, in chemico, and in silico methods with specified data analysis procedures to achieve reliable and reproducible predictions. The incorporation of additional NAMs could help increase accessibility and flexibility. Using superior algorithms may help improve the accuracy of hazard and potency assessment and build confidence in the results. Here, we introduce two new DA models, with the aim to build DAs on freely available software and the newly developed kDPRA for covalent binding of a chemical to skin peptides and proteins. The new DA models are built on an existing Bayesian network (BN) modelling approach and expand on it. The new DA models include kDPRA data as one of the in vitro parameters and utilise in silico inputs from open-source QSAR models. Both approaches perform at least on par with the existing BN DA and show 63% and 68% accuracy when predicting four LLNA potency classes, respectively. We demonstrate the value of the Bayesian network's confidence indications for predictions, as they provide a measure for differentiating between highly accurate and reliable predictions (accuracies up to 87%) in contrast to low-reliability predictions associated with inaccurate predictions.

Keywords: Bayesian network; defined approaches; next-generation risk assessment; NAMs; skin sensitisation

1. Introduction

Skin sensitisation is characterised by four main molecular and physiological Key Events (KEs) [1]. These KEs include (i) the covalent binding of a chemical to skin peptides and proteins (KE1); (ii) the activation of keratinocytes, which constitute the main cell type in skin (KE2); (iii) the activation of dendritic cells (KE3); and (iv) the activation of T-cells (KE4) [2]. Repeated exposure to a compound can result in contact dermatitis. The Next-Generation Risk Assessment (NGRA) of skin sensitisation potential is based on integrating in silico predictions and in vitro test results in defined approaches (DAs). DAs integrate data from various sources, i.e., in vitro, in chemico, and in silico, to support skin sensitisation assessment using a predefined set of resources and data interpretation [3]. Thus, DAs have the potential to align with the principles of the OECD Mutual Acceptance of Data (MAD) agreement. This agreement requires that data generated for regulatory purposes in one member country be recognised and accepted by other OECD member countries [4]. DAs can further be used to support skin sensitisation assessment within Integrated Approaches to Testing and Assessment (IATAs). IATAs use the weight of evidence (WoE) and expert evaluation to support decision-making for regulatory bodies [5].

In 2017, two guidance documents on the reporting of DAs (OECD GD 255 and OECD GD 256) were released [3,6]. In 2018, the U.S. EPA issued a draft science policy document reporting two DAs: two out of three (2o3) and the Key Event 3/1 Sequential Testing Strategy (KE 3/1 STS) [7]. In 2021, the OECD published guideline 497 on defined approaches to skin sensitisation, which incorporated three DAs: the "2o3" DA relies on two concordant tests from three in vitro assays addressing the three first KEs, namely, the Direct Peptide Reactivity Assay (DPRA, KE1), KeratinoSens^{TM} (KE2), and the human Cell Line Activation Test (h-CLAT, KE3) [1]. Furthermore, guideline 497 features two versions of ITS (ITSv1 and ITSv2), which integrate the results of DPRA and h-CLAT with in silico predictions. The primary distinction between these two versions is the use of different in silico tools. ITSv1 uses the Derek Nexus software, whereas ITSv2 incorporates the OECD QSAR Toolbox. In addition, both ITS versions provide a scoring system to predict potency classification based on the United Nations Globally Harmonised System of Classification and Labelling of Chemicals (GHS) [8]. Edelweiss Connect recently developed the SaferSkin approach, a solution package for skin sensitisation risk assessment [9]. The SaferSkin suite offers and applies OECD DAs, IATAs, and new approach methodologies (NAMs) to perform animal-free risk assessment of substances. The solution combines a variety of OECD-required standard in vitro assays with a set of in silico prediction tools based on DAs, machine learning, and Quantitative Structure–Activity Relationship (QSAR) approaches [1,10,11].

In this article, we present a new DA based on the previously reported Bayesian network (BN) ITS-3 model of Jaworska et al. by reflecting the advances in both in vitro experiments and in silico modelling [11,12]. The aim of the study is to increase the transparency and applicability of the available models by integrating a new in vitro assay and replacing commercial in silico tools with freely available ones. A BN is a powerful tool for representing complex relationships between variables. It employs a visual map with nodes depicting factors of interest and arrows indicating their causal or influential connections. Conditional probability tables associated with each node quantify the likelihood of one factor occurring based on the states of others. This allows us to analyse the interplay of variables and make probabilistic inferences within a particular domain. BNs are particularly valuable due to their capacity to elucidate connections among multiple factors, which may not be readily apparent, even to domain experts. In addition, BNs provide a probabilistic assessment of skin sensitisation by integrating diverse data sources to form evidence-based hypotheses and are designed to handle incomplete data easily. Notably, BNs also enable the evaluation of prediction uncertainties based on the quality and completeness of the input data. The primary goals of the present work were to incorporate additional NAMs and to improve the BN model's accessibility. Hence, BN models were developed using public and in-house-generated data and available open-source software. Data used for modelling and evaluation were collected from the literature and carefully curated; new kinetic DPRA (kDPRA) data were also generated to fill data gaps (Tables S1 and S2 in Supplementary Materials). Edelweiss Connect developed a BN model for skin sensitisation assessment based on the previously reported BN model by Jaworska et al. [9,11,12]. The BN model developed by Edelweiss Connect is referred to as "SaferSkin-BN" in this publication. SaferSkin-BN was shown to have similar predictive characteristics to the original model developed by Jaworska [13]. This article describes the changes implemented in the SaferSkin-BN model, which include the incorporation of kDPRA parameters and using inputs from publicly accessible Skin Doctor CP and the open-source software OPERA [14,15]. The original model from Jaworska et al. and SaferSkin-BN use commercial software (TIMES-SS and

ACD/Labs) [16,17]; we investigated the impact of the replacement of the commercial software by Skin Doctor CP and OPERA on model performance. Since potency estimation is one of the main challenges in skin sensitisation assessment, we introduced the kDPRA as a new in vitro assay input parameter. To date, the kDPRA is the only OECD-validated in vitro assay allowing for the classification of skin sensitisers according to the Globally Harmonised System (GHS) classes 1A and 1B/non-classified [18].

In silico toxicology protocols demand the combined use of rule-based and statistics-based models [19]. In order to appreciate this concept, we developed a second model that was able to incorporate additional in silico prediction models, i.e., Derek Nexus. We aimed to either improve the predictive performance and include a specific evaluation of uncertainty and potency or to maintain the original model performance but increase accessibility, which can be used to support risk assessment goals. Thus, we adjusted the SaferSkin-BN model by implementing the following modifications:

1. The use of kDPRA data with the rate constant Kmax and GHS class prediction in addition to DPRA data as input parameters.
2. The replacement of the quinary in silico predictions of skin sensitisation by TIMES-SS with the binary and ternary predictions of Skin Doctor Conformal Predictor (Skin Doctor CP).
3. The replacement of the in silico predictions of physicochemical properties by ACD/Lab with those by OPERA.
4. The extension of the BN model with a rule-based system, i.e., Derek Nexus [20].

The developed BN models were trained by assessing different hyperparametrisation options and tested to determine their predictive performance. In addition, the influencing inputs on the final outcomes were evaluated as follows:

- Using a BN with ternary Skin Doctor CP and kDPRA to assess the predictive performance of the Skin Doctor CP prediction vs. kDPRA;
- Implementing the Skin Doctor CP p-value;
- Using ternary Skin Doctor CP, kDPRA, and Derek Nexus results;
- Assessing the impact of using DPRA and kDPRA.

The current work presents different models based on the proposed modifications. Thus, several models were tested in this approach, and the best-performing ones were incorporated into the SaferSkin software (Edelweiss Connect GmbH, Basel, Switzerland). The performance of the newly developed models was assessed based on a predefined test set (Cosmetics Europe database) that was used previously to evaluate the performance of DA models [21,22]. SaferSkin-BN served as a baseline model for comparison.

2. Materials and Methods

2.1. In Vivo Inputs: LLNA

Regulatory authorities have accepted the Local Lymph Node Assay (LLNA, OECD 429) as a high standard for evaluating skin sensitisation hazards and risks [23–26]. The LLNA assay is based on the induction of lymphocyte proliferation in the lymph nodes by draining the site of the test substance's application. This proliferation is proportional to the dose and potency of the applied allergen and provides a simple means of obtaining a quantitative measurement of sensitisation (OECD 429). This includes the activation and proliferation of antigen-specific T-cells [27]. Skin sensitisation potential is attributed to a chemical if the proliferation rate exceeds a factor of 3 compared with the vehicle-treated control group. The concentration estimated to stimulate a three-fold increase in lymphocyte proliferation (termed "EC3") refers to the threshold concentration, which provides information on the chemical's skin-sensitising potency. Originally, the concentration in LLNA was reported on a "per-weight basis", but for modelling purposes, it was transformed into molar units using the formula pEC3 = log(Mw/250/EC3(%)) [12]. The following pEC3 thresholds were applied to derive four potency classes in molar units: non-sensitiser (<-1.9), weak sensitiser (-1.9, -1.1), moderate sensitiser (-1.1, -0.35), and strong sensitiser (>-0.35) [12]. Thus, the formed variable, the pEC3 class, was the target variable in the BN model. The LLNA data used to develop the model described here were collected from Jaworska et al. [12].

2.2. In Vitro Inputs

2.2.1. DPRA

The DPRA addresses KE1 of the AOP for skin sensitisation, i.e., protein binding, and it determines the reactivity of a test compound towards synthetic cysteine- and lysine-rich peptides [18]. Peptide reactivity is reported as a percentage of peptide depletion. For the DA described by Jaworska et al., the percentage of free peptide remaining in the sample is used for all calculations. For developing and applying the models described in this publication, the percentage depletion was used as the input parameter and calculated with respect to the percentage of the peptide remaining. The DPRA data used to develop the model described here were collected from Jaworska et al. [12].

2.2.2. kDPRA

The kDPRA is currently the only OECD-accepted in vitro/in chemico assay that was reported to enable skin sensitisation potency classification [18]. This assay is a modification of the DPRA. It allows the derivation of a rate constant (Kmax (in 1/sM)) of the depletion of the cysteine-containing peptide upon reaction with the test compound. Depending on the rate constant, a test compound can then be classified as GHS 1A or GHS 1B/not classified. The data were partly derived from studies conducted by Natsch et al. and Natsch and Gerberick [28,29]. For four compounds out of these databases, one study reported an unambiguous result while the other one reported no clear study result. In those cases, the unambiguous result was included in our dataset. No conflicting results were observed between the two data sources. Additionally, new kDPRA data were generated. The kDPRA was performed as described in OECD guideline 442C and in the full method description in the database service on alternative methods (DB-ALM) in protocol 217 [18,30]. The cysteine peptide was provided by RS Synthesis (Louisville, KY, USA). The chemicals were purchased from Sigma-Aldrich (St. Louis, MO, USA) or Alfa Chemistry (Holbrook, NY, USA). The assay was run in 96-well plates. The chemicals were dissolved at 20 mM in acetonitrile (ACN), diluted in acetonitrile, and added to the cysteine peptide solution (0.66 mM in phosphate buffer) to final concentrations of 5, 2.5, 1.25, 0.625, and 0.3125 mM. Monobromobimane (3 mM in ACN) was added after 10, 30, 90, 150, 210, and 1440 min, respectively, and the fluorescence intensity was measured at Ex/Em 390/480 nm. Peptide depletion and Kmax were evaluated using an Excel spreadsheet provided as Supplementary Material in DB-ALM 217 [30]. Not all compounds with missing kDPRA could be tested due to limited availability or solubility.

2.2.3. KeratinoSens^{TM}

The KeratinoSens^{TM} assay (OECD 442D) addresses KE2 of the AOP for skin sensitisation, i.e., the activation of keratinocytes. The assay evaluates the activation of the Keap1-Nrf2-ARE pathway, a key pathway triggered by sensitisers in vivo [31]. The average concentrations (in μM) leading to 1.5-fold and 3-fold induction (EC1.5 and EC3) are reported alongside the concentration leading to 50% cytotoxicity after 24 h (IC50). The data were collected from Jaworska et al. [12].

2.2.4. h-CLAT

The h-CLAT assay addresses KE3 of the AOP for skin sensitisation (OECD guideline 442E), known as dendritic cell activation. The assay quantifies changes in the expression of cell-surface molecules (CD54 and CD86) [18]. Both molecules are essential in the induction of skin sensitisation; CD54 is involved in dendritic cell migration, and CD86 stimulates T-cell activation during antigen presentation to dendritic cells [32]. The average compound concentrations (in μM) inducing 150% of vehicle control expression of CD86 (EC150) and 200% of vehicle control expression of CD54 (EC200) are reported alongside the concentration leading to 75% cell viability after 24 h (CV75). The data were collected from Jaworska et al. [12].

2.3. In Silico Inputs

2.3.1. OPERA

OPERA (v2.9) is a free and open-source suite of QSAR models providing predictions for physicochemical properties, environmental fates, and toxicity endpoints [15,33]. OPERA was used to calculate the octanol–water distribution coefficient of compounds at pH 7 (logD@pH7), the water solubility at pH 7 (Ws@pH7), plasma–protein binding, and the octanol–water partition coefficient (log Kow). The fraction ionised (fion) was calculated using the formulation = ∣1 − 10^logD/10^logKow∣, where "∣ ∣" denotes the absolute value [12].

2.3.2. Skin Doctor CP

Skin Doctor CP is a conformal prediction-based machine learning model for the classification of small organic compounds into two (non-sensitisers and sensitisers) or three (non-sensitisers, weak to moderate sensitisers, and strong to extreme sensitisers) potency classes [34]. The models were trained on publicly available data only. They were transparently published and are accessible via a web service [35]. As a conformal prediction model, Skin Doctor CP will only return a prediction if predefined reliability thresholds are met, and these can be adjusted for the specific use case.

2.3.3. Derek Nexus

Derek Nexus (v6.3) (Lhasa Limited, Leeds, UK) is a rule-based commercial modelling software that can predict the likely toxicity of a given chemical structure based on the absence or presence of certain chemical substructures [20]. It has also been recently included in the DA for skin sensitisation published by the OECD [1]. For skin sensitisation, it is one of only a few models that can predict not just hazards but also EC3 values. These values can then be transferred into five distinct potency classes: non-sensitisers and weak, moderate, strong, and extreme sensitisers. For the modelling, we merged the strong and extreme sensitiser classes into one class.

2.4. Bayesian Network-Based Defined Approach

2.4.1. Dataset

The starting point for model development was the dataset collected by Jaworska et al., which included LLNA, DPRA, KeratinoSens^{TM}, and h-CLAT data, as well as calculated physicochemical properties for 207 compounds [12]. We extended the dataset with kDPRA data from two sources, as well as with newly generated kDPRA data plus sensitiser potency predictions from Skin Doctor CP and Derek Nexus (which replaced TIMES-SS potency predictions in the original dataset) [28,29]. Additionally, the physicochemical parameters originally calculated using commercial software were newly calculated using the open-source OPERA suite. The original split into training (147 compounds) and test sets (60 compounds) was preserved from Jaworska et al. to ensure a fair model comparison [12]. The dataset published by Jaworska et al. included two compounds (Farnesol and Benzyl cinnamate) that are present in training and test set at the same time but each time with different SMILES and a different set of parameters present. Due to comparability we decided to not change the data. The full training and test set used for model building and evaluation within this study can be found in Supplementary Materials Tables S1 and S2, respectively.

2.4.2. Discretisation

BNs require discretised continuous variables. Hence, continuous input variables were discretised into four classes using a supervised clustering algorithm (class–attribute interdependence maximisation, CAIM) implemented in the R package discretisation [36,37].

Discrete variables (potency classes from Skin Doctor CP, Derek Nexus, and kDPRA classification) were not modified.

2.4.3. Building Latent Nodes

In addition to the observed variables (i.e., in vitro and in silico inputs), we used latent variables in the network structure. These variables were not directly observable but simplified the network structure (reduced connections between the nodes) and often improved the model accuracy. Latent variables were designed in a way that they connected similar input variables. These latent nodes were built in the training set by grouping the connected input variables into a predefined number of classes. This grouping was essentially an unsupervised clustering performed through latent class analysis (LCA) using an R package named poLCA: Polytomous Variable Latent Class Analysis [38]. This clustering is essentially a search for a global maximum of the log-likelihood function, so, understandably, its outcome may depend on the initial conditions and number of repetitions. For the test set, the latent nodes were empty, and their values were inferred during the BN prediction.

2.4.4. Bayesian Network

A Bayesian network is a powerful tool to model probabilistic knowledge about certain phenomenon. Nodes in the network represent discrete variables that are connected with arrows. Each arrow represents a conditional probability table (network parameters) between the two connected variables. Bayesian networks allow one to calculate the probability distribution of unobserved discrete variables given the evidence (i.e., values of observed variables).

Building a BN model involves two steps: (i) defining the network structure (i.e., connectivity among the nodes) and (ii) learning the parameters of the network (i.e., conditional probability tables). There were many possible connections among the nodes, and it was not our intention to ascertain the one giving the highest possible performance. Instead, we relied on the BN structure proposed by Jaworska et al., which was developed manually from mechanistic knowledge of skin sensitisation [12]. We tested two different BN structures (Figure 1) that both largely resembled the BN architecture from SaferSkin-BN (or Jaworska et al. [12]) with the following modifications: (i) parameters from the kDPRA were connected in the same way as the DPRACys (i.e., the percentage of cysteine peptide depletion in the DPRA) variable; (ii) the in silico TIMES-SS input was replaced by the Skin Doctor CP (left network) or by the newly formed latent variable INSILICO, which connected the inputs from Skin Doctor CP and Derek Nexus (right network in Figure 1). Once the BN architecture was defined, its parameters (conditional probability tables) were learned from the training set. For the BN modelling, we used the Python library pgmpy and the following settings: estimator = BayesianEstimator, prior_type = K2, equivalent_sample_size = 1 or 5, complete_samples_only = False [39].

The BN model predicted discrete probability distributions of all the unknown variables. For the test set, the unknowns were the target variable, latent nodes, and possibly some input nodes. However, due to the non-uniform class distribution of target variables in the training set, the resulting probability was biassed towards the more populated classes. To correct this bias, we followed the approach presented by Jaworska et al. and calculated Bayes factors [12]. Then, the predicted class was the one with the highest Bayes factor. The size of the Bayes factor was also directly related to the confidence in a prediction [40]: with Bayes factor < 3.2 indicating weak confidence, 3.2 ≤ Bayes factor < 10 indicating substantial confidence, 10 ≤ Bayes factor < 32 indicating strong confidence, and Bayes factor > 32 indicating very strong confidence. With latent class analysis, we introduced several hyperparameters that required optimisation: (i) each latent node had a defined number of classes; we tested three to five classes for each node; (ii) the number of repetitions of latent class analysis could also influence the BN model performance; we tested three values, i.e., 10, 20, and 50. We performed an extensive grid search in the space of hyperparameters and selected the set of hyperparameters that resulted in the highest balanced accuracy on the test set. Balanced accuracy was calculated using the method *sklearn.metrics.balanced_accuracy_score* from the *scikit-learn* Python library version 1.5.1. The method calculates balanced accuracy as an average recall over the four potency classes [41].

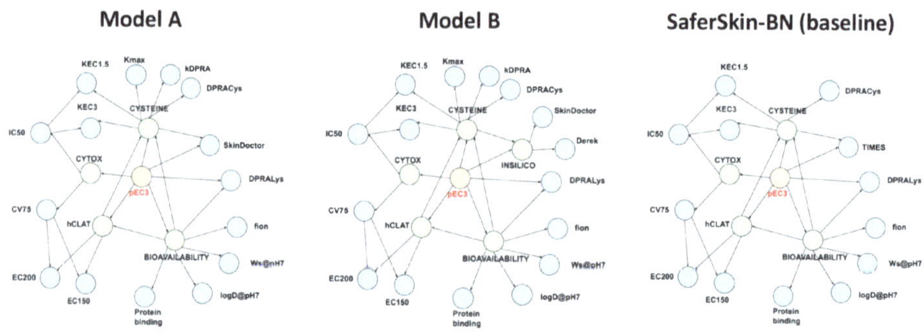

Figure 1. Architecture of the three Bayesian networks (BNs): model A (**left**), model B (**middle**), and the baseline SaferSkin-BN model (**right**). Blue nodes denote input parameters; green nodes represent latent variables; red nodes designate the target variable (pEC3 class).

During hyperparameter optimisation, we selected a set of hyperparameters for each network structure, leading to the highest balanced accuracy on the test set. However, this outcome could require further investigation to avoid overfitting the test set by the hyperparameters. Hence, we estimated the accuracy of the two best models using 10-fold cross-validation. The full dataset (training set + test set) was randomly split into 10 folds. Nine folds were used to train the discretisation scheme, run latent class analysis (just with the set of hyperparameters determined during hyperparameter optimisation in Section 3.1), and train the BN parameters. The trained model was used to generate predictions for the 10th fold.

2.4.5. Evaluation Dataset

The performance of model A and model B in comparison with the baseline SaferSkin-BN model was evaluated using the Cosmetics Europe Database, a dataset compiled by the Cosmetics Europe consortium [21,22]. It was previously used by Kleinstreuer et al. to evaluate the performance of different DA models [22]. The dataset covers, e.g., in vitro data (DPRA, KeratinoSens^{TM}, h-CLAT) and LLNA data for 128 substances. The dataset was prepared and curated to include additional required in silico and in vitro data for the current modelling approach; physicochemical parameters were generated using OPERA. Derek Nexus and Skin Doctor CP were used to generate predictions for the INSILICO nodes in model A and model B. kDPRA data were collected from Natsch and Gerberick [29]. TIMES-SS data for comparison with the SaferSkin-BN were collected from Hofmann et al. [42]. Of the 128 substances in the original dataset, 6 were of natural origin with unknown exact compositions. We had to omit those 6 compounds as converting mass concentrations into molar concentrations or deriving a defined chemical structure for in silico predictions was impossible. The full evaluation set is summarised in Supplementary Materials Table S3.

3. Results and Discussion

The current work presents the development of novel BN models prioritising transparency and reproducibility to address regulatory concerns and support future regulatory acceptance. The starting point for model development was SaferSkin-BN, based on the ITS-3 BN approach developed by Jaworska et al. [12]. The Jaworska ITS-3 model uses two commercial software packages, TIMES-SS (for predicting skin sensitisation potency) and ACD/Labs (for predicting physical–chemical properties), as well as in vitro input parameters (DPRA, KeratinoSens^{TM}, and h-CLAT). SaferSkin-BN was built using OPERA, TIMES-SS, DPRA, KeratinoSens^{TM}, and h-CLAT.

Within the present study, we successfully added a new in chemico assay, kDPRA, which was also recently included in the updated OECD guideline 422C on assays addressing KE1, i.e., covalent binding to proteins. In addition, the successful replacement of commercial in silico tools by freely available ones could be shown.

We developed two slightly different models, models A and B, incorporating these two steps. The first step involved adding two input nodes to the BN representing data from the kDPRA, i.e., Kmax and kDPRA classification according to GHS classes. Both nodes were considered equivalent to the DPRACys node representing the cysteine depletion value determined in the DPRA and were therefore connected to the CYSTEINE latent node just like DPRACys. This pattern was the same in both model A and model B.

These changes resulted in model A, which is based solely on freely accessible in silico tools.

Another variant was developed to complement the statistics-based Skin Doctor CP model with a rule-based in silico model and sensitisation potency prediction. To this end, we integrated Derek Nexus as an additional input. While we strived to use open-access tools, the inclusion of Derek Nexus was favourable for considering the recommendations for in silico toxicology protocols due to its recent adoption in the OECD guideline 497 on defined approaches on skin sensitisation and to improve the performance with respect to that of model A [1,19]. To reduce connectivity in the BN and optimise its performance, we created a new latent node, INSILICO, that replaced the TIMES node in SaferSkin-BN. The new INSILICO node was connected to the two in silico inputs from Skin Doctor CP and Derek Nexus. The resulting architecture represented model B.

3.1. Hyperparameter Optimisation

After determining the final network structures, we ran an extensive hyperparameter search to maximise the balanced accuracy on the test set. Hyperparameters refer to the latent class analysis during data preparation. In the latent class analysis, we assumed the number of classes (N_{CYTOX}, $N_{CYSTEINE}$, N_{HCLAT}, N_{BIOAV}, $N_{INSILICO}$) for each latent node (CYTOX, CYSTEINE, HCLAT, BIOAV, INSILICO). Furthermore, we assumed the number of times the global optimum is estimated using different initial conditions, NREP. The space of these six hyperparameters was extensively searched, and Table 1 summarises the optimal set of parameters found for models A and B. Further results reported in this section are based on the analysis of those optimised versions of models A and B in comparison with SaferSkin-BN.

Table 1. Optimal parameters and balanced accuracies of BN models A and B predicting four potency classes on the test set.

Model Architecture	Optimal Hyperparameters						Balanced Accuracy on Test Set (%)
	N_{CYTOX}	$N_{CYSTEINE}$	N_{HCLAT}	N_{BIOAV}	$N_{INSILICO}$	N_{REP}	
Baseline (SaferSkin-BN)	3	4	4	3	N/A	20	66
Model A	3	3	4	3	N/A	20	66
Model B	3	3	3	3	4	20	70

3.2. Incorporation of kDPRA into In Vitro Parameters

The OECD recently updated test guideline 422C on assays addressing KE1, i.e., covalent binding to proteins, with a new kDPRA. The kDPRA shows promising predictive capacity for risk assessment, distinguishing between GHS 1A-classified sensitisers and GHS 1B/non-sensitisers with good accuracy [18]. More kDPRA data might become available in the public domain in the future or might be generated during a specific compound's risk assessment. Therefore, in the present work, we aimed to broaden the scope and source of the in vitro parameters that could be considered for DA.

The main parameter obtained in the kDPRA was Kmax, a rate constant for the binding of the compound to the peptide. The assay also defines a threshold in Kmax that best distinguishes between GHS 1A sensitisers and GHS 1B/non-sensitisers. Therefore, this assay provided us with two input parameters for the BN model: Kmax (a continuous variable that we discretised as described in the Methods section) and kDPRA (a binary variable representing the classification into GHS 1A or GHS 1B/non-sensitiser). Table 2 compares BN models with and without kDPRA input parameters. Balanced accuracies for the test set were calculated using four model variants (i.e., models A and B, each with and without kDPRA nodes). The individual predictions for all compounds from the test set by the baseline model, as well as model A and model B, can be found in Supplementary Materials Table S4.

Table 2. Balanced accuracies of BN models A and B predicting four potency classes on the test set with and without kinetic DPRA (kDPRA) data.

Model	Balanced Accuracy (%)
Baseline (SaferSkin-BN)	66
Model A	66
Model A without kDPRA nodes	64
Model B	70
Model B without kDPRA nodes	68

Including the kDPRA data resulted in the comparable or even slightly improved predictive capacity of the BN model, thus enabling the use of the model with kDPRA data, DPRA data, or both, depending on their availability. In the field of skin sensitisation prediction, where relatively sparse data are available in the public domain, this flexibility will offer an advantage, especially if we expect the kDPRA to be conducted more frequently in the future due to its easier and faster application and its ability to predict a measure of potency.

3.3. Implementation of In Silico Skin Sensitisation Prediction Models

One critical parameter of the Jaworska ITS-3 model and SaferSkin-BN is the in silico prediction model of skin sensitisation potency, i.e., the TIMES-SS (TImes MEtabolism Simulator) V.2.27.13 commercial software [43]. TIMES-SS is a hybrid expert system that encodes structure–activity and structure–metabolism relationships and is therefore capable of predicting the sensitising potency of a parent compound, as well as that of its metabolites. In their analysis, Jaworska et al. demonstrated that this parameter had the highest predictive power of all the input parameters [11].

Within the present study, TIMES-SS inputs were replaced with predictions from the Skin Doctor CP model, which was transparently published in the literature, was trained on public LLNA data, and is available via a web service [35]. Table 3 compares the accuracies of BN models incorporating TIMES-SS or Skin Doctor CP inputs. In model A, we directly replaced the TIMES-SS input node with the Skin Doctor CP node. However, in model B, we added one more in silico parameter, i.e., the prediction of skin sensitisation potency from the Derek Nexus software. Hence, in model B, we actually created a new latent node, INSILICO, that connected both in silico inputs and replaced the TIMES-SS node.

The statistical analysis showed that replacing TIMES-SS with Skin Doctor CP maintained the model's prediction accuracy. Addition of Derek Nexus predictions (model B) slightly increased the balanced accuracy. TIMES-SS predicted the skin sensitisation potential for the parent compound and predicted metabolites. The most conservative input was provided to the DA. Our newly generated models A and B were based on Skin Doctor CP and Derek Nexus. Both tools predicted the skin sensitisation potential based on the parent compound. Metabolites were not predicted. The impact of a potential sensitising metabolite was only indirectly considered, as these models were trained on LLNA in vivo data, which covered the effect of native, although not human, skin metabolism and, thus, captured the effects of both the parent compound and the metabolites formed. The impact of using in silico predictions for the metabolites on the overall performance of the model still needed proper evaluation. The superior predictive performance of model B over model A demonstrated a high impact of the in silico tool Derek Nexus on improving the model's accuracy. This was aligned with the observation by Jaworska et al. that the in silico input (TIMES-SS in their case) was the input parameter with the largest impact on the final prediction [12].

Table 3. Balanced accuracies of BN models A and B predicting four potency classes with different in silico inputs.

Model	Balanced Accuracy (%)
Baseline (SaferSkin-BN)	66
Variation of model A with TIMES-SS replacing the Skin Doctor CP node	66
Model A (with Skin Doctor CP node)	66
Model B (with Skin Doctor CP and Derek Nexus nodes)	70

3.4. Replacement of Physicochemical Parameter Prediction Models

In the Jaworska ITS-3 model, the physicochemical parameters (octanol–water partition coefficient, water solubility, fraction of compound ionised, protein binding) were calculated with the commercial QSAR models from ACD/Labs [12,17]. To increase the accessibility and usability of the DA, we calculated these parameters using the open-source OPERA suite. Table 4 summarises a head-to-head comparison of the newly developed models trained with physicochemical properties coming from ACD/Labs and OPERA while keeping the rest of the parameters unchanged (the network architecture was as depicted in Figure 1). Note that we determined the best set of hyperparameters for each of the two variants of model A (trained with ACD/Labs and trained with OPERA phys-chem). Incidentally, the same set of hyperparameters gave the highest test set accuracy for both model variants.

Table 4. Balanced accuracies of the BN models A and B predicting four potency classes on the test set using physicochemical properties from ACD/Labs and OPERA.

Model	Balanced Accuracy (%)
Baseline (SaferSkin-BN)	66
Model A trained with ACD/Labs phys-chem.	66
Model A trained with OPERA phys-chem.	66
Model B trained with ACD/Labs phys-chem.	69
Model B trained with OPERA phys-chem.	70

Replacing the physicochemical parameters from the commercial QSAR models with those from the open-source suite did not decrease the model's accuracy.

3.5. Model Cross-Validation

After the optimal hyperparameters were determined, we ran a 10-fold cross-validation on the entire dataset (i.e., merged the training and test sets) to obtain a more realistic estimate of the model's accuracy. Table 5 shows the cross-validated balanced accuracy (full dataset) of the two models, and these were clearly comparable to the balanced accuracies on the test set (Table 1). Furthermore, we calculated the minimum, median, and maximum balanced accuracy per fold to illustrate the variability in balanced accuracy on small datasets (a typical fold contained 20 compounds). We can see that the balanced accuracy was difficult to estimate with a relatively small set of compounds.

Table 5. Cross-validated and fold-balanced accuracies of BN models A and B in predicting four potency classes.

Model Architecture	Cross-Validated	Fold-Balanced Accuracy (%)		
		Minimum	Median	Maximum
Model A	66	55	65	83
Model B	68	52	65	90

To illustrate the robustness of the choice of hyperparameters, we ran a 10-fold cross-validation for every set of hyperparameters and collected the minimum, average, and maximum balanced accuracy obtained. The results are presented in Table 6. We can conclude that the hyperparameters selected based on the balanced accuracy of the test set gave close to the best cross-validated balanced accuracy, too. Another observation was that we could expect about 10 to 15% variation in balanced accuracy between the best-performing and the worst-performing sets of hyperparameters. Expectedly, this variation was substantially lower than the variation in balanced accuracy among the folds.

Table 6. Cross-validated balanced accuracies of BN models A and B in predicting four potency classes with different sets of hyperparameters.

Model Architecture	Cross-Validated Balanced Accuracy (%)		
	Minimum	Median	Maximum
Model A	59	64	68
Model B	61	68	75

3.6. Relation between the Prediction Confidence and Accuracy

An advantage of a BN is its ability to predict sensitiser potency alongside an indication of prediction confidence. More concordant evidence supports a particular prediction with a higher confidence level than non-concordant evidence [12]. To illustrate the influence of the confidence level on prediction accuracy, the accuracy of predictions with different confidence levels was calculated (Table 7). Importantly, accuracy improved substantially as confidence levels progressed from weak to very strong.

Table 7. Accuracy of BN models A and B on the test set when predicting four potency classes, calculated separately for each confidence class. The number of compounds for each class, N, is provided in parentheses.

Model Architecture	Accuracy (%)			
	Weak Confidence	Substantial Confidence	Strong Confidence	Very Strong Confidence
Model A	48 (N = 33)	55 (N = 69)	55 (N = 31)	88 (N = 69)
Model B	40 (N = 25)	56 (N = 67)	67 (N = 33)	90 (N = 69)

3.7. Prediction of the LLNA EC3 Value for Use in Risk Assessment

Although the proposed BN models provide discrete probability distributions, it is possible to derive a predicted continuous LLNA EC3 value from those distributions. This is of substantial value for any quantitative risk assessment. To the authors' current knowledge, only a few approaches (i.e., the BN, the neural network, and the multiple linear regression approaches) predict EC3 values [44,45]. The discrete probability distribution over potency classes could be turned into a distribution over pEC3 values (by applying the pEC3 thresholds to the class limits). The most likely pEC3 value is the value at the 50th percentile of that distribution. However, it reports more restrictive values, e.g., at the 70th or the 90th percentile, which might be considered. The currently developed BN reports the probability distribution of pEC3, unlike the existing two DAs, which simply deliver a pEC3 value without a confidence interval.

The results in Figure 2 display the predicted pEC3 values at the 50th (left), 70th (middle), and 90th percentiles (right) versus the true pEC3 values. The red lines indicate a perfect fit. The pEC3 values derived from the 50th percentile are scattered around the red line. Deriving pEC3 values from higher percentiles resulted in overestimating the potency (predicting lower EC3 values than the true LLNA results). Within expectations, the highest R^2-values can be found for the 50th percentile. With values of 0.47 and 0.54, they are in line with the R^2-values evaluated for Natsch et al., which are 0.53, 0.52, and 0.54 for EQ1, EQ4 and EQ5, respectively. Small R^2-values for the 70th and 90th percentile, of course, result from the intention to avoid underprediction which is not suited for R^2 optimisation.

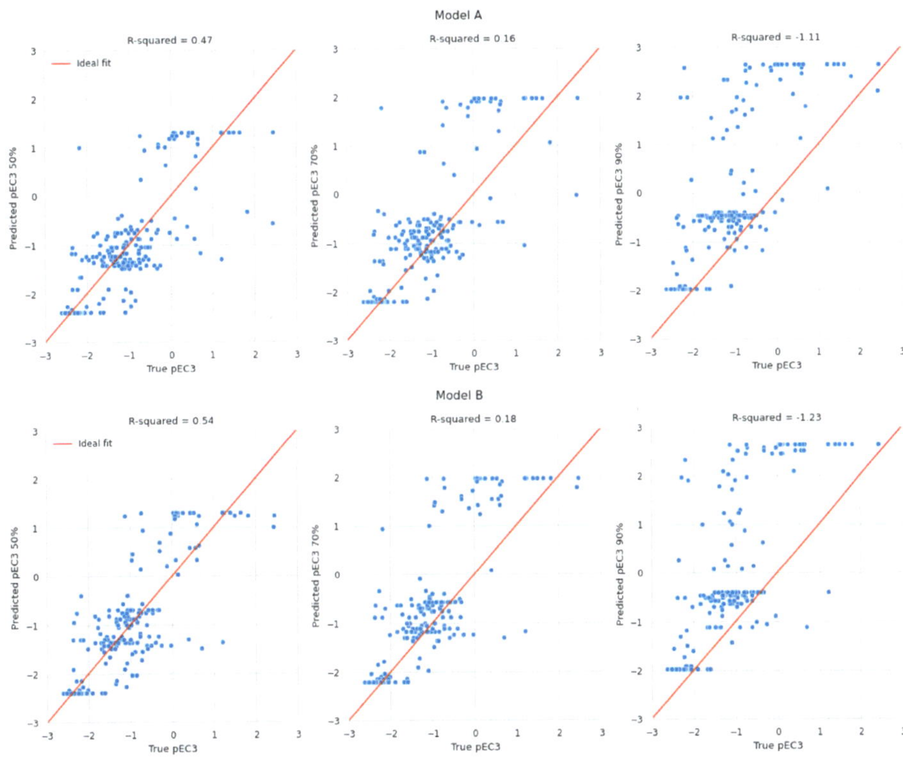

Figure 2. Predicted pEC3 taken at the 50th (**left**), 70th (**middle**), and 90th percentiles (**right**) versus the true pEC3 values for models A and B.

3.8. Model Evaluation Using the Cosmetics Europe Database

Recently, a dataset was compiled by the Cosmetics Europe consortium to enable comparisons of skin sensitisation DA models [21,22]. The dataset covers in vitro data (DPRA, KeratinoSens™, h-CLAT), LLNA data, and human potency data for 128 substances. To enable a consistent comparison with other DA models, we decided to use the same dataset to evaluate the newly developed DAs. The data were used within the freshly generated DAs of model A and model B, as well as SaferSkin-BN, for comparison. The outcomes are summarised in Tables 8 and 9. The individual predictions for all compounds from the Cosmetics Europe dataset by the baseline model, as well as model A and model B, can be found in Supplementary Materials Table S5.

Table 8. Accuracy of the baseline and BN models A and B in predicting four LLNA potency classes on the Cosmetics Europe dataset calculated separately for each confidence class. The number of compounds for each class, N, is provided in parentheses.

Model Architecture	Accuracy (%)				
	Complete Dataset	Weak Confidence	Substantial Confidence	Strong Confidence	Very Strong Confidence
Baseline (SaferSkin-BN)	61 (N = 122)	46 (N = 28)	62 (N = 39)	44 (N = 25)	87 (N = 30)
Model A	63 (N = 122)	33 (N = 30)	65 (N = 54)	83 (N = 12)	85 (N = 26)
Model B	68 (N = 122)	43 (N = 37)	72 (N = 43)	86 (N = 22)	85 (N = 20)

Table 9. Accuracy of the baseline and BN models A and B in predicting three human potency classes on the Cosmetics Europe dataset calculated separately for each confidence class. The number of compounds for each class, N, is provided in parentheses.

Model Architecture	Accuracy (%)				
	Complete Dataset	Weak Confidence	Substantial Confidence	Strong Confidence	Very Strong Confidence
Baseline (SaferSkin-BN)	57 (N = 122)	54 (N = 28)	44 (N = 29)	64 (N = 25)	70 (N = 30)
Model A	55 (N = 122)	43 (N = 30)	52 (N = 54)	58 (N = 12)	73 (N = 26)
Model B	61 (N = 122)	49 (N = 37)	63 (N = 43)	64 (N = 22)	75 (N = 20)

Considering the overall predictions of the complete set (N = 122), model B's accuracy was superior (68%) to both model A's performance and that of the baseline model, SaferSkin-BN (63% and 61% accuracy, respectively). The database was originally implemented to evaluate the performance of several DA models by Kleinstreuer et al.: Kao Sequential Testing Strategy, Kao Integrated Testing Strategy, Shiseido Artificial Neural Network (DPRA/h-CLAT or DPRA/h-CLAT/KeratinoSens^{TM}), and P&G BN ITS-3 [22]. The resulting accuracies for predicting the LLNA sensitising potency ranged from 65 to 70%. Thus, the accuracies of the newly developed models were of the same order of magnitude as those of the existing DA models. Additional benefits of the newly developed DAs are improved accessibility, transparency, and a more flexible spectrum of in vitro input parameters. Beyond that, our newly developed models A and B provide four-class potency prediction, while the DAs evaluated by Kleinstreuer et al. [22] provide up to three potency classes. This difference needs to be considered when comparing the derived accuracy values. A higher number of potency classes is beneficial for safety assessment since it decreases the range of possible potencies for compounds assigned to a certain potency class.

To predict the three human potency classes, the four LLNA classes were combined into three classes (strong, weak = moderate + weak, non-sensitising), and accuracies were evaluated on the complete set (N = 122), as well as for different confidence levels (Table 9). The predictive capacity for LLNA was superior to the prediction of human potency. This can be explained by the fact that all three models were trained on the LLNA data. LLNA data, despite their variability, are experimental values obtained in reproducible experiments, whereas human potency data are expert gradings that also incorporate the exposure of the population [25]. For the human endpoint, we again observed that the accuracy of model A was comparable to that of the baseline model. In contrast, model B offered a slight improvement over the baseline model. To compare the newly developed models with other DAs, we again turned to the work by Kleinstreuer et al. [22], where DAs were evaluated on the Cosmetics Europe dataset. To briefly summarise the results from [22] on the accuracy of predicting three human potency classes, all DAs performed comparable or better than the LLNA assay (59% accuracy). The worst DA in [22] was the Bayesian network approach with 55% accuracy, the same as the newly developed model A. The best performing DAs were Kao STS and Kao ITS with 64 and 69% accuracy. With model B, we reach the same performance as the neural network approaches by Shiseido ANN (D_hC) and ANN (D_hC_KS) with 61 and 63% accuracy.

An advantage of the presented BN models is their capacity to provide a confidence estimation for a single prediction. For each sensitiser potency class, the prediction probability is calculated based on the evidence from in vitro assays and in silico tools. Consistent results from different in silico and in vitro tools increase confidence, whereas conflicting results decrease confidence. A good correlation between prediction accuracy and the estimated confidence level was observed (Tables 8 and 9). Predictions with very strong confidence reached accuracies of up to 88%, whereas the predictions with weak confidence were associated with accuracies below 50%. Therefore, predictions with weak confidence should be scrutinised thoroughly and may require an additional safety assessment factor.

3.9. Applicability and Limitations of Proposed Approach

Among the strengths of the proposed approach are (i) its ability to handle incomplete input parameters; (ii) its provision of the confidence level of predictions, which are strongly correlated to the model's accuracy; (iii) and its provision of the EC3 value that is require for risk assessment.

However, the approach also has limitations. One is the discretization of continuous input variables, which will necessarily result in potentially mispredicted compounds that have continuous input parameters close to their discretization thresholds. Such cases should be inspected carefully, and predictions could be generated with slight variation of such input parameters.

Another limitation is that the approach requires a known chemical structure to convert in vitro parameters into molar units. So, the issue of how to handle cases of natural substances with unknown chemical structures is not very well defined.

4. Conclusions

In this article, we describe the development of two new DA models, model A and model B, with the aim to increase the applicability and flexibility of current approaches by allowing input from a new in vitro method as well as from freely available in silico tools. Both models developed are based on previously published Bayesian network approaches, ITS-3 and SaferSkin-BN [9,12]. The existing DAs were upgraded by expanding the set of in vitro input parameters with integrated kDPRA data and by increasing accessibility and transparency by replacing the commercial in silico software TIMES-SS and ACD/Labs with the publicly accessible software Skin Doctor CP and OPERA. The most accurate model, model B, includes predictions from the commercial in silico tool Derek Nexus to comply with recommendations for in silico toxicology protocols. The combination of rule-based and statistical in silico prediction tools increased the performance of model B compared with model A and the baseline model SaferSkin-BN, respectively. Using the Cosmetics Europe dataset to evaluate the newly developed models enables direct comparison with other published DA models evaluated on the same dataset [22]. Models A and B both showed comparable or even superior accuracy compared to the baseline model, SaferSkin-BN. A comparative evaluation of published DAs showed similar accuracy at 63% to 68%. The models' accuracy increased by considering a higher confidence level of up to 85%. However, the newly developed models enable a more granular potency assessment because they predict four instead of three classes. The incorporation of additional LLNA data would further improve the predictive capacity, as well as the confidence in predictions. While the models are trained and optimised for LLNA prediction, we could also show their capability to predict skin sensitisation in humans. In addition, we could show that the incorporated measure of reliability can indicate the model's applicability to an individual molecule of interest. This would allow us to distinguish between highly and less reliable predictions. Our analysis shows a general trend in potency overestimation for the predicted pEC3 derived from higher percentiles. This overestimation results in a more conservative and, consequently, more protective calculation of safe concentrations using NGRA approaches. Ultimately, this work supports the goal of reaching regulatory acceptance of newly developed DA models for NGRA approaches to skin sensitisation.

Supplementary Materials: The following supporting information can be downloaded at: https://www.mdpi.com/article/10.3390/toxics12090666/s1, Table S1: Data set used for training of the BN models; Table S2: Data set used for evaluation of the BN models; Table S3: Cosmetics Europe data set enriched with results from kDPRA, Derek, Skin Doctor CP and OPERA as used for further evaluation of the BN models; Table S4a: Predictions on the test set returned by the baseline model; Table S4b: Predictions on the test set returned by model A; Table S4c: Predictions on the test set returned by model B; Table S5a: Predictions on the Cosemtics Europe data set returned by the baseline model; Table S5b: Predictions on the Cosemtics Europe data set returned by model A; Table S5c: Predictions on the Cosemtics Europe data set returned by model B.

Author Contributions: Conceptualisation, T.M., A.W., J.K. (Jochen Kühnl), A.N., B.H. and J.E.; methodology, T.M., A.W., S.O., J.K. (Johannes Kirchmair) and A.N.; software, T.M., A.M. and A.L.; validation, T.M.; formal analysis, T.M. and A.W.; investigation, T.M., A.W., S.O., P.A., J.K. (Johannes Kirchmair), A.N., B.H. and J.E.; resources, A.S. and B.H.; data curation, T.M., A.W. and S.O.; writing—original draft preparation, T.M., A.W., A.N., B.H. and J.E.; writing—review and editing, T.M., A.W., S.O., G.T., J.K. (Johannes Kirchmair), A.S., J.K. (Jochen Kühnl), A.N., B.H. and J.E.; visualisation, T.M.; supervision, J.K. (Jochen Kühnl), A.N., B.H. and J.E.; project administration, A.N., B.H. and J.E. All authors have read and agreed to the published version of the manuscript.

Funding: This research received no external funding.

Institutional Review Board Statement: Not applicable.

Informed Consent Statement: Not applicable.

Data Availability Statement: All relevant data are within the manuscript and its Supplementary Materials.

Conflicts of Interest: Authors: Tomaz Mohoric; Andrii Milovich; Artem Logavoch; Pascal Ankli; Ghada Tagorti and Barry Hardy were employed by the Edelweiss Connect GmbH, and authors:Anke Wilm; Stefan Onken; Andreas Schepky; Jochen Kühnl; Abdulkarim Najjar and Johanna Ebmeyer were employed by Beiersdorf AG. The remaining authors declare that the research was conducted in the absence of any commercial or financial relationships that could be construed as a potential conflict of interest.

References

1. OECD. OECD Guideline No. 497: Defined Approaches on Skin Sensitisation. In *OECD Guidelines for the Testing of Chemicals, Section 4*; OECD: Paris, France, 2023; ISBN 978-92-64-90300-5.
2. Piipponen, M.; Li, D.; Landén, N.X. The Immune Functions of Keratinocytes in Skin Wound Healing. *Int. J. Mol. Sci.* **2020**, *21*, 8790. [CrossRef] [PubMed]
3. OECD. OECD Guidance Document on the Reporting of Defined Approaches to Be Used Within Integrated Approaches to Testing and Assessment. In *OECD Series on Testing and Assessment*; OECD: Paris, France, 2017; ISBN 978-92-64-27482-2.
4. OECD. *Decision of the Council Concerning the Mutual Acceptance of Data in the Assessment of Chemicals, OECD/LEGAL/0194 (Revised in 1997)*; OECD Publishing: Paris, France, 1981.
5. OECD. OECD Guidance Document for the Use of Adverse Outcome Pathways in Developing Integrated Approaches to Testing and Assessment (IATA). In *OECD Series on Testing and Assessment*; OECD: Paris, France, 2017; ISBN 978-92-64-89104-3.
6. OECD. OECD Guidance Document on the Reporting of Defined Approaches and Individual Information Sources to Be Used within Integrated Approaches to Testing and Assessment (IATA) for Skin Sensitisation. In *OECD Series on Testing and Assessment*; OECD: Paris, France, 2017; ISBN 978-92-64-27928-5.
7. US EPA. *Draft Interim Science Policy: Use of Alternative Approaches for Skin Sensitization as a Replacement for Laboratory Animal Testing*; US EPA: Washington, DC, USA, 2018; OCSPP, OPP, OPPT. Available online: https://downloads.regulations.gov/EPA-HQ-OPP-2016-0093-0090/content.pdf (accessed on 5 June 2024).
8. United Nations Economic Commission for Europe. *Globally Harmonized System of Classification and Labelling of Chemicals (GHS Rev. 9)*; United Nations: New York, NY, USA; Geneva, Switzerland, 2021. Available online: https://unece.org/transport/documents/2021/09/standards/ghs-rev9 (accessed on 13 July 2024).
9. SaferWorldbyDesign, Edelweiss Connect GmbH. SaferSkin™: Integrative Skin Sensitisation Assessment Tool. Available online: https://saferworldbydesign.com/saferskin/in-silico/skin-sensitization-app/app/ (accessed on 15 July 2024).

10. Berry, W. Multiple Regression—An Overview. Encyclopedia of Social Measurement 2005. Available online: https://www.sciencedirect.com/topics/social-sciences/multiple-regression (accessed on 9 May 2023).
11. Jaworska, J.; Dancik, Y.; Kern, P.; Gerberick, F.; Natsch, A. Bayesian Integrated Testing Strategy to Assess Skin Sensitization Potency: From Theory to Practice. *J. Appl. Toxicol.* **2013**, *33*, 1353–1364. [CrossRef] [PubMed]
12. Jaworska, J.S.; Natsch, A.; Ryan, C.; Strickland, J.; Ashikaga, T.; Miyazawa, M. Bayesian Integrated Testing Strategy (ITS) for Skin Sensitization Potency Assessment: A Decision Support System for Quantitative Weight of Evidence and Adaptive Testing Strategy. *Arch. Toxicol.* **2015**, *89*, 2355–2383. [CrossRef] [PubMed]
13. Exner, T.; Abdelaziz, A.; Hardy, B. An Integrated Decision Strategy Development for Skin Sensitization Assessment. In Proceedings of the 56th Annual Meeting and ToxExpo, Baltimore, MD, USA, 12–16 March 2017; Poster No. 2948. p. 459. Available online: https://www.toxicology.org/pubs/docs/tox/2017tox.pdf (accessed on 16 July 2024).
14. Wilm, A.; Stork, C.; Bauer, C.; Schepky, A.; Kühnl, J.; Kirchmair, J. Skin Doctor: Machine Learning Models for Skin Sensitization Prediction That Provide Estimates and Indicators of Prediction Reliability. *Int. J. Mol. Sci.* **2019**, *20*, 4833. [CrossRef] [PubMed]
15. Mansouri, K.; Grulke, C.M.; Judson, R.S.; Williams, A.J. OPERA Models for Predicting Physicochemical Properties and Environmental Fate Endpoints. *J. Cheminform.* **2018**, *10*, 10. [CrossRef]
16. Patlewicz, G.; Kuseva, C.; Mehmed, A.; Popova, Y.; Dimitrova, G.; Ellis, G.; Hunziker, R.; Kern, P.; Low, L.; Ringeissen, S.; et al. TIMES-SS—Recent Refinements Resulting from an Industrial Skin Sensitisation Consortium. *SAR QSAR Environ. Res.* **2014**, *25*, 367–391. [CrossRef]
17. ACDLabs. Advanced Chemistry Development. In *Chemistry Software for Analytical and Chemical Knowledge Management*; ACDLabs: Toronto, ON, Canada, 2024. Available online: https://www.acdlabs.com/ (accessed on 27 May 2024).
18. OECD. OECD Test No. 442C: In Chemico Skin Sensitisation: Direct Peptide Reactivity Assay (DPRA). In *OECD Guidelines for the Testing of Chemicals, Section 4*; OECD: Paris, France, 2024; ISBN 978-92-64-22970-9.
19. Johnson, C.; Ahlberg, E.; Anger, L.T.; Beilke, L.; Benigni, R.; Bercu, J.; Bobst, S.; Bower, D.; Brigo, A.; Campbell, S.; et al. Skin Sensitization in Silico Protocol. *Regul. Toxicol. Pharmacol.* **2020**, *116*, 104688. [CrossRef]
20. Lhasa Limited. Derek Nexus [Software]. Available online: https://www.lhasalimited.org/solutions/skin-sensitisation-assessment/ (accessed on 2 May 2023).
21. Hoffmann, S.; Kleinstreuer, N.; Alépée, N.; Allen, D.; Api, A.M.; Ashikaga, T.; Clouet, E.; Cluzel, M.; Desprez, B.; Gellatly, N.; et al. Non-Animal Methods to Predict Skin Sensitization (I): The Cosmetics Europe Database. *Crit. Rev. Toxicol.* **2018**, *48*, 344–358. [CrossRef]
22. Kleinstreuer, N.C.; Hoffmann, S.; Alépée, N.; Allen, D.; Ashikaga, T.; Casey, W.; Clouet, E.; Cluzel, M.; Desprez, B.; Gellatly, N.; et al. Non-Animal Methods to Predict Skin Sensitization (II): An Assessment of Defined Approaches. *Crit. Rev. Toxicol.* **2018**, *48*, 359–374. [CrossRef]
23. Frank Gerberick, G.; Ryan, C.A.; Dearman, R.J.; Kimber, I. Local Lymph Node Assay (LLNA) for Detection of Sensitization Capacity of Chemicals. *Methods* **2007**, *41*, 54–60. [CrossRef]
24. OECD. OECD Test No. 429: Skin Sensitisation: Local Lymph Node Assay. In *OECD Guidelines for the Testing of Chemicals, Section 4*; OECD: Paris, France, 2010; ISBN 978-92-64-07110-0.
25. Anderson, S.E.; Siegel, P.D.; Meade, B.J. The LLNA: A Brief Review of Recent Advances and Limitations. *J. Allergy* **2011**, *2011*, 1–10. [CrossRef]
26. Roberts, D.W.; Schultz, T.W.; Api, A.M. Chemical Applicability Domain of the Local Lymph Node Assay (LLNA) for Skin Sensitisation Potency. Part 3. Apparent Discrepancies between LLNA and GPMT Sensitisation Potential: False Positives or Differences in Sensitivity? *Regul. Toxicol. Pharmacol.* **2016**, *80*, 260–267. [CrossRef] [PubMed]
27. OECD. OECD The Adverse Outcome Pathway for Skin Sensitisation Initiated by Covalent Binding to Proteins. In *OECD Series on Testing and Assessment*; OECD: Paris, France, 2014; ISBN 978-92-64-22144-4.
28. Natsch, A.; Haupt, T.; Wareing, B.; Landsiedel, R.; Kolle, S.N. Predictivity of the Kinetic Direct Peptide Reactivity Assay (kDPRA) for Sensitizer Potency Assessment and GHS Subclassification. *ALTEX-Altern. Anim. Exp.* **2020**, *37*, 652–664. [CrossRef]
29. Natsch, A.; Gerberick, G.F. Integrated Skin Sensitization Assessment Based on OECD Methods (I): Deriving a Point of Departure for Risk Assessment. *ALTEX-Altern. Anim. Exp.* **2022**, *39*, 636–646. [CrossRef] [PubMed]
30. DB-ALM. DB-ALM Protocol 217: The Kinetic Direct Peptide Reactivity Assay (kDPRA). 2020. Available online: https://jeodpp.jrc.ec.europa.eu/ftp/jrc-opendata/EURL-ECVAM/datasets/DBALM/LATEST/online/DBALM_docs/217_P_kDPRA_final_27Oct20.pdf (accessed on 13 July 2024).
31. OECD. OECD Test No. 442D: In Vitro Skin Sensitisation: ARE-Nrf2 Luciferase Test Method. In *OECD Guidelines for the Testing of Chemicals, Section 4*; OECD: Paris, France, 2024; ISBN 978-92-64-22982-2.
32. Ashikaga, T.; Yoshida, Y.; Hirota, M.; Yoneyama, K.; Itagaki, H.; Sakaguchi, H.; Miyazawa, M.; Ito, Y.; Suzuki, H.; Toyoda, H. Development of an in Vitro Skin Sensitization Test Using Human Cell Lines: The Human Cell Line Activation Test (h-CLAT). *Toxicol. Vitr.* **2006**, *20*, 767–773. [CrossRef]
33. Mansouri, K. OPERA. 2022. Available online: https://github.com/kmansouri/OPERA (accessed on 27 May 2024).
34. Wilm, A.; Norinder, U.; Agea, M.I.; De Bruyn Kops, C.; Stork, C.; Kühnl, J.; Kirchmair, J. Skin Doctor CP: Conformal Prediction of the Skin Sensitization Potential of Small Organic Molecules. *Chem. Res. Toxicol.* **2021**, *34*, 330–344. [CrossRef]
35. SkinDoctor. 2024. Available online: https://nerdd.univie.ac.at/skinDoctorII/ (accessed on 13 July 2024).
36. Kurgan, L.A.; Cios, K.J. CAIM Discretization Algorithm. *IEEE Trans. Knowl. Data Eng.* **2004**, *16*, 145–153. [CrossRef]

37. Kim, H. Discretization: Data Preprocessing, Discretization for Classification. 2022. Available online: https://cran.r-project.org/web/packages/discretization/index.html (accessed on 21 November 2023).
38. Linzer, D.; Lewis, J. poLCA: Polytomous Variable Latent Class Analysis. 2022. Available online: https://cran.r-project.org/web/packages/poLCA/index.html (accessed on 21 November 2023).
39. Ankur. Supported Data Types—Pgmpy 0.1.23 Documentation. 2023. Available online: https://pgmpy.org/ (accessed on 21 November 2023).
40. Goodman, S.N. Toward Evidence-Based Medical Statistics. 1: The P Value Fallacy. *Ann. Intern. Med.* **1999**, *130*, 995. [CrossRef]
41. Scikit-Learn Developers. Sklearn Metrics Balanced Accuracy Score. In Scikit-Learn: Machine Learning in Python. Available online: https://scikit-learn.org/stable/modules/generated/sklearn.metrics.balanced_accuracy_score.html (accessed on 4 September 2024).
42. Hoffmann, S.; Alépée, N.; Gilmour, N.; Kern, P.S.; Van Vliet, E.; Boislève, F.; Bury, D.; Cloudet, E.; Klaric, M.; Kühnl, J.; et al. Expansion of the Cosmetics Europe Skin Sensitisation Database with New Substances and PPRA Data. *Regul. Toxicol. Pharmacol.* **2022**, *131*, 105169. [CrossRef]
43. Patlewicz, G.; Dimitrov, S.D.; Low, L.K.; Kern, P.S.; Dimitrova, G.D.; Comber, M.I.H.; Aptula, A.O.; Phillips, R.D.; Niemelä, J.; Madsen, C.; et al. TIMES-SS—A Promising Tool for the Assessment of Skin Sensitization Hazard. A Characterization with Respect to the OECD Validation Principles for (Q)SARs and an External Evaluation for Predictivity. *Regul. Toxicol. Pharmacol.* **2007**, *48*, 225–239. [CrossRef]
44. Hirota, M.; Fukui, S.; Okamoto, K.; Kurotani, S.; Imai, N.; Fujishiro, M.; Kyotani, D.; Kato, Y.; Kasahara, T.; Fujita, M.; et al. Evaluation of Combinations of in Vitro Sensitization Test Descriptors for the Artificial Neural Network-based Risk Assessment Model of Skin Sensitization. *J. Appl. Toxicol.* **2015**, *35*, 1333–1347. [CrossRef] [PubMed]
45. Natsch, A.; Emter, R.; Gfeller, H.; Haupt, T.; Ellis, G. Predicting Skin Sensitizer Potency Based on In Vitro Data from KeratinoSens and Kinetic Peptide Binding: Global Versus Domain-Based Assessment. *Toxicol. Sci.* **2015**, *143*, 319–332. [CrossRef] [PubMed]

toxics

MDPI

Article

In Vitro Prediction of Skin-Sensitizing Potency Using the GARDskin Dose–Response Assay: A Simple Regression Approach

Robin Gradin [1,*], Fleur Tourneix [2], Ulrika Mattson [1], Johan Andersson [1], Frédéric Amaral [2], Andy Forreryd [1], Nathalie Alépée [2] and Henrik Johansson [1]

1 Senzagen AB, 22381 Lund, Sweden; ulrika.mattson@senzagen.com (U.M.); johan.andersson@senzagen.com (J.A.); andy.forreryd@senzagen.com (A.F.); henrik.johansson@senzagen.com (H.J.)

2 L'Oréal, Research & Innovation, 93600 Aulnay-sous-Bois, France; fleur.tourneix@loreal.com (F.T.); frederic.amaral@loreal.com (F.A.); nathalie.alepee@loreal.com (N.A.)

* Correspondence: robin.gradin@senzagen.com

Abstract: Toxicological assessments of skin sensitizers have progressed towards a higher reliance on non-animal methods. Current technological trends aim to extend the utility of non-animal methods to accurately characterize skin-sensitizing potency. The GARDskin Dose–Response assay has previously been described; it was shown that its main readout, cDV_0 concentration, is associated with skin-sensitizing potency. The ability to predict potency from cDV_0 in the form of NESILs derived from LLNAs or human NOELs was evaluated. The assessment of a dataset of 30 chemicals showed that the cDV_0 values still correlated strongly and significantly with both LLNA EC3 and human NOEL values (ρ = 0.645–0.787 [$p < 1 \times 10^{-3}$]). A composite potency value that combined LLNA and human potency data was defined, which aided the performance of the proposed model for the prediction of NESILs. The potency model accurately predicted sensitizing potency, with cross-validation errors of 2.75 and 3.22 fold changes compared with NESILs from LLNAs and humans, respectively. In conclusion, the results suggest that the GARDskin Dose–Response assay may be used to derive an accurate quantitative continuous potency estimate of skin sensitizers.

Keywords: NAM; GARDskin Dose–Response; sensitizing potency; quantitative risk assessment; point of departure

1. Introduction

Chemicals possess a wide range of physiochemical properties. While many are beneficial, making them ubiquitous in industries and society, some can induce hazardous effects in humans. Therefore, it is essential to investigate the potential of chemicals to induce adverse effects. One type of hazard that is routinely screened for is the capacity to induce skin sensitization [1,2]. Skin sensitization is a hypersensitivity reaction that follows a topical exposure to an inducing chemical, leading to the establishment of a specific immunological memory [3,4]. The underlying mechanisms of this reaction have been extensively studied and explained, and the major key events (KEs) are recognized and have been summarized in an Adverse Outcome Pathway (AOP) [5]. As the sensitization reaction results in an immunological memory, a sensitized individual will suffer from adverse effects upon repeated exposure to that specific chemical, making the initial avoidance of hazardous exposure levels imperative.

The induction of skin sensitization is a dose-dependent event, and the likelihood that sensitization occurs typically increases with dose [6]. However, chemicals' efficiencies to induce sensitization vary greatly [7]. For example, a strong skin sensitizer may cause sensitization from a few micrograms of topical exposure, while weak sensitizers would require milligrams to reach a similar probability of sensitization on the same surface area. Therefore, it is not only relevant to evaluate the ability of a chemical to induce skin sensitization, but also to characterize an identified skin sensitizer's potency, i.e., to derive an estimate of an exposure level that is not expected to induce skin sensitization in the general population.

Traditionally, the assessment of skin sensitizers has relied on information from in vivo assays. Guinea pig models were among the earliest standardized methods, which included the Buehler assay [8–10] and the Guinea Pig Maximization Test (GPMT) [11]. These were largely superseded by the more recent murine Local Lymph Node Assay (LLNA) [12–14], which brought several improvements over the guinea pig assays. These include aspects of animal welfare and an improved scientific rationale, as the assay provides an objective and quantitative readout [15]. Further, while potency information could be gathered from the guinea pig models [6], the LLNA incorporates dose–response measurements, and the derived effective concentration for a stimulation index of 3 (EC3) serves as an immediate indicator of sensitizing potency, which has been shown to correspond well with human-derived potency data [16–19]. Human potency data are often described in terms of No Observed Effect Levels (NOELs) or Lowest Observed Effect Levels (LOELs) from Human Repeat Insult Patch Tests (HRIPTs). However, testing on humans is problematic [20], especially given the potential severity and enduring effects of skin sensitization, and is typically avoided unless used in confirmatory scenarios where a lack of response has been demonstrated [21]. Nevertheless, when available, data from both humans and LLNAs may be used as information sources to derive No Expected Sensitization Induction Levels (NESILs), which may be used as a Point of Departure (PoD) in risk assessment strategies [22,23].

More recent advancements in the field of skin-sensitization assessment, however, are being made toward the development of non-animal methods, or New Approach Methodologies (NAMs), aiming to replace animal tests, not only for hazard identification but also for hazard characterization and potency evaluation. An initial milestone, which has to a significant degree already been realized, is the establishment of NAMs capable of accurately and reliably identifying skin-sensitizing hazards. Today, many different NAMs can be used to generate such hazard information, and validated and regulatory approved methods are described in various OECD test guidelines [24–26], each of which is associated with a specific KE of the AOP. However, none of the methods described in the TGs have so far been approved for stand-alone use. Therefore, endeavors have also been made to integrate results across NAMs to increase confidence in their joint classification outcomes, using so-called Defined Approaches for Skin Sensitization (DASS) [27–29]. In addition to these NAMs, skin-sensitizing hazard data can also be obtained using in silico methods [30–33], which may contribute to weight-of-evidence assessments.

A second milestone in the replacement of animal tests is the development of methods capable of informing about chemicals' skin-sensitizing potencies. Indeed, some potency information may already be derived in a discrete manner, as described in the current OECD test guideline 497, where integrated testing strategies incorporating multiple information sources (including in chemico, in vitro, and in silico) are used to predict sensitizing potency per the globally harmonized system for classification and labelling (UN GHS) [29]. However, the apogee of NAM development for the purpose of potency assessment is, nevertheless, likely a readout provided as a quantitative continuous potency value. Some progress has already been made towards this objective. For example, probabilistic approaches based on Bayesian or Artificial Neuronal Network (ANN) frameworks that incorporate multiple information sources, including several NAMs, have been described for the prediction of PoDs [34–37]. Similarly, several alternative regression models based on the output of established NAMs have been proposed for the prediction of LLNA EC3 values [38]. Besides methods incorporating data from individual NAMs, novel test methods have also been proposed for directly generating potency information [24,39].

While the development of NAMs for the assessment of skin sensitizers progresses, frameworks for interpreting their results and managing risks are also being established. For example, Next-Generation Risk Assessment (NGRA) strategies that incorporate information from several NAMs have recently been proposed [40,41].

We have previously described an extended protocol of the GARDskin assay that incorporates dose–response measurements to derive a concentration estimate that correlates strongly and significantly with both LLNA EC3 and human NOELs [42]. The classical GARDskin assay is an OECD-approved assay described in test guideline 442E [26], together with other methods addressing the key event of dendritic cell activation. The GARDskin assay, and the dose–response adaptation, relies on the monitoring of genetic changes in a biomarker signature following chemical exposure. The biomarker signature includes several genes relevant to immune activation, including cd86, hmox1, nqo1, and nlrp12 [43]. The dose–response adaptation studies the summarized gene response over multiple concentrations to identify the lowest concentration capable of inducing a positive classification in the GARDskin assay. As such, the main methodology is a typical toxicological approach, which is also applied in the LLNA, where a PoD is determined from a dose–response curve and used as a measure of potency.

In this work, we further characterize the GARDskin Dose–Response assay and provide results that show that its readout, the cDV_0 value, correlates significantly with established potency metrics on a larger dataset and can be used to accurately predict potency. Notably, the redundancy and potential ambiguity of having repeated NESILs derived from the LLNA and human NOELs for model fitting was rectified using a composite potency value describing a latent potency signal. This approach improved the predictive performance compared with methods relying on individual references, generating NESIL predictions relevant for risk assessment frameworks such as NGRA [40,41].

2. Materials and Methods

2.1. The GARDskin Dose–Response Assay

The GARDskin Dose–Response assay is based on the protocol of the GARDskin assay, publicly available in the Tracking System for Alternative methods towards Regulatory acceptance (TSAR) [44]. The assay measures the expression levels of genes in the GARDskin Genomic Prediction Signature (GPS) following treatment with a test chemical to the cell system [26,43,45], i.e., the Senzacell cell line (ATCC depository PTA-123875). The interpretation of gene response is made with a support vector machine (SVM) (explicitly described in supporting documents to the OECD TG 442E [46]), which outputs a decision value (DV) on which the subsequent classification is based. A test chemical is classified as a skin sensitizer if the generated mean DV is greater than zero ($DV \geq 0$) and as a non-sensitizer if the mean DV is less than zero ($DV < 0$).

The GARDskin Dose–Response assay extends the standard protocol by evaluating the test chemical at several concentrations [42]. DVs are calculated for every concentration and the lowest concentration expected to induce a positive classification is determined, which corresponds to the lowest concentration that predicts a DV of 0. Practically, this concentration can be estimated using linear interpolation between the two closest concentrations that generate predictions on the opposite sides of the GARDskin classification threshold. The estimated concentration is the main readout of the GARDskin Dose–Response assay, termed cDV_0 [42].

In more detail, a test item assessment starts with an examination of the cytotoxic properties of the chemical, using a protocol based on propidium iodide staining and flow cytometry analysis. A concentration inducing low-to-non-toxic conditions, as determined by the relative viability of the cell system following exposure, is determined. This concentration is termed the GARD input concentration and constitutes the sole assessment concentration in the standard GARDskin assay and the highest evaluated concentration in the GARDskin Dose–Response assay. Subsequent concentrations in the dose–response assay are typically calculated using a dilution series with a dilution factor of approximately 0.5. The actual design of the concentration curve can be modified for particular purposes, i.e., a higher number of replicates with smaller steps between concentrations may result in reduced uncertainty of an estimated cDV_0 value but at the cost of resources. As this study comprises chemicals from several separate experiments, the designs of the concentration curves are not identical. For example, in Gradin et al.'s study [42], chemicals were analyzed by evaluating 12 different concentrations with a dilution factor of 0.6 and a single replicate per concentration. Since then, data have been acquired using more standardized dilution schemes optimized from accumulated observations, typically including 6 concentrations with two or three replicates. The actual dilution schemes used for each chemical in this study are available in Supplementary Tables S2 and S3.

Cells are treated with the test chemical at the determined concentrations for 24 h, following which cells are harvested and total RNA is isolated and purified. Gene expression levels are quantified using the NanoString nCounter system using a custom GARDskin codeset, as described previously [26,45]. The data analysis pipeline for transforming raw gene expression levels into decision values has been publicly described, including all parameter values [46]. For the dose–response analysis, the decision values are examined against the treatment concentrations to verify the presence of a dose–response relationship A cDV_0 concentration is estimated when a dose-dependent response is observed, using linear interpolation between the two closest points (in terms of concentration) that are on adjacent sides of the classification threshold.

2.2. Dataset

This work was based on a dataset of 30 chemicals, as listed in Table 1. More complete details regarding the chemicals are presented in Supplementary Table S1. GARDskin Dose–Response data for 18 of the chemicals were previously described in Gradin et al.'s work [42]. In that work, chemicals were selected to span a wide range of expected potency values while preferably having both human and LLNA reference data. In addition, they were selected to cover the potency span relatively evenly to allow for an efficient correlation analysis between cDV_0 and potency references. That dataset of 18 chemicals was extended here with 12 new chemicals. The new chemicals raised skin-sensitization alerts and had LLNA and human NOEL data. The chemicals were butyl resorcinol (18979-61-8); citral (5392-40-5); chlorpromazine (50-53-3); (R)-(+)-Limonene (5989-27-5); alpha-iso-Methylionone (127-51-5); phenylacetaldehyde (122-78-1); ethyl acrylate (140-88-5); cinnamic alcohol (104-54-1); 3-Propylidenephthalide (17369-59-4); 5-Methyl-2,3-hexanedione (13706-86-0); and carvone (6485-40-1). A majority of weak and moderate, i.e., UN GHS Cat. 1B, skin sensitizers were selected to reduce the uncertainty in the estimated relationship between cDV_0 and the sensitization potency for this category of ingredients, as precision in this potency range may be of particular relevance for the cosmetics industry.

The analysis included repeated runs for 10 of the chemicals. Repeated cDV_0 values were merged by their geometric mean values before incorporation in downstream calculations (including correlation evaluations and regression analyses). Individual cDV_0 values and repeated runs' geometric means are described in Table 1 and in Supplementary Table S1. The geometric mean was used as it consistently maintains the relative differences between measurements. For example, the geometric mean of 0.5 µg/mL and 2 µg/mL is 1 µg/mL, which is 2-fold from either of the original values.

LLNA EC3, human NOELs, and human Lowest Observed Effect Levels (LOELs) were collected from the published literature [47–53] (see Supplementary Table S1 for details). LLNA EC3 values were converted into NESIL estimates in µg/cm^2 by multiplying the percentage values by 250 [54].

Table 1. Description of the chemical dataset. The cPV column describes the computed composite potency values. Chemicals that have data from repeated tests show results from the individual studies within parentheses and the summarized values (the geometric mean), which are used for subsequent calculations, outside the parentheses. Entries with dashes ("-") represent reference data points that could not be located or composite potency values that could not be defined due to at least one missing reference value.

				NESIL µg/cm²			
Chemical	**CAS**	**MW (g/mol)**	**cDV0 (µg/mL)**	**LLNA**	**HRIPT NOEL**	**Composite Potency Value (cPV)**	**Human LOEL (µg/cm²)**
Benzalkonium chloride	8001-54-5	424.15	0.350	25.0	-	-	-
2,4-Dinitrochlorobenzene	97-00-7	202.55	0.443	13.5	8.8	9.80	8.8
Cinnamic aldehyde	104-55-2	132.16	0.524	250	591	378	775
Citral	5392-40-5	152.23	1.11 (0.737, 1.67)	1450	1417	1440	3876
Diethyl maleate	141-05-9	172.18	1.03 (0.754, 1.40)	525	1600	921	-
Dimethyl fumarate	624-49-7	144.13	0.874	87.5	88	82.8	-
Methylisothiazolinone	2682-20-4	115.15	0.904	325	15	63.4	-
Benzyl Alcohol	100-51-6	108.14	1.37	NS	5905	-	8858
Chlorpromazine	50-53-3	318.9	1.38	35.0	1150	200	17,241
alpha-Isomethylionone	127-51-5	206.32	3.10 (1.48, 6.51)	5450	70,860	21,400	-
Iodopropynyl butylcarbamate	55406-53-6	281.09	1.61	225	-	-	-
Isoeugenol	97-54-1	164.21	1.70	325	250	275	775
p-Mentha-1,8-dien-7-al	2111-75-3	150.22	2.41 (1.73, 3.38)	1010	709	835	2760
Phenylacetaldehyde	122-78-1	120.15	2.68	750	591	654	1181
Carvone	6485-40-1	150.22	5.13 (3.58, 7.35)	3250	2657	2980	18,898
7-Hydroxycitronellal	107-75-5	172.26	5.70	5275	4960	5260	5814
5-Methyl-2,3-hexanedione	13706-86-0	128.169	8.97	6500	3448	4830	3450
Eugenol	97-53-0	164.21	9.29	2900	5906	4270	-
Ethyl acrylate	140-88-5	100.12	9.47	8188	1600	3630	4000
Cinnamic alcohol	104-54-1	134.17	10.3	5775	2953	4200	4724
Butyl resorcinol	18979-61-8	166.22	8.83 (10.3, 7.55)	950	-	-	-
Farnesol	4602-84-0	222.37	11.8 (11.5, 12.1)	1200	2755	1850	6897
Geraniol	106-24-1	154.25	15.4 (12.7, 18.7)	4025	11,811	7220	-
Imidazolidinyl urea	39236-46-9	388.29	14.9	6000	2000	3490	2000
3-Propylidenephthalide	17369-59-4	174.2	18.5 (18.8, 18.2)	925	945	928	2760
Pentachlorophenol	87-86-5	266.34	20.1	5000	2155	3310	6897
(R)-(+)-Limonene	5989-27-5	136.23	14.8 (20.2, 10.8)	13,125	10,000	12,000	-
3-Dimethylaminopropylamine	109-55-7	102.18	25.7	875	-	-	-
Benzyl salicylate	118-58-1	228.25	37.4	725	17,715	3790	-
Linalool	78-70-6	154.25	43.0	8875	14,998	12,100	-

2.3. Creation of a Composite Potency Value

The reference composite potency value was created by fitting a robust errors-in-variables model between NESIL values derived from LLNA EC3 and human NOELs, and by orthogonally projecting the datapoints onto the fitted line, creating the composite potency values. More explicitly, a Passing–Bablok regression model [55] was fitted between the log-transformed NESILs from LLNA EC3 and human NOEL values (see Table 1 for a description of the data). The 25 chemicals with continuous LLNA and human data were included in the fit. The Passing–Bablok fit was created in R (version 4.2.0) [56] with the R-package Deming (version 1.4) [57]. Confidence intervals for the fitted coefficients were estimated using bootstrap [58], taking 10,000 bootstrap replicates. The composite potency values were calculated as follows.

Given the linear PB fit as line l in the form $y = ax + b$, and data point p with coordinates (x_i, y_i), the coordinates of the projection of point p onto l was calculated by finding the intersect between line l and the line that is orthogonal to l and passes through coordinates (x_i, y_i), as described by Equation (1). The point's projected coordinates $(\hat{x}_i, \hat{y}_i)$ are thus given by Equations (2) and (3).

$$y = -\frac{x}{a} + y_i + \frac{x_i}{a}, \quad (1)$$

$$\hat{x}_i = \frac{y_i + \frac{x_i}{a} + b}{a + \frac{1}{a}}, \quad (2)$$

$$\hat{y}_i = a \times \hat{x}_i + b, \tag{3}$$

The relative distances between the projected points were calculated with the Pythagorean theorem, using the data point with the smallest composite potency value as a reference (note that $\min(\hat{x})$ and $\min(\hat{y})$ refer to the same data point). As the relative distances between the projected points may be inflated in this step, the data were scaled by the expected inflation rate per original potency unit. The relative distances were calculated as described in Equation (4).

$$d_i = \sqrt{(\hat{x}_i - \min(\hat{x}))^2 + (\hat{y}_i - \min(\hat{y}))^2} \times \frac{1}{\sqrt{1+a^2}}, \tag{4}$$

where d_i is the distance between chemical i and the chemical with the smallest composite potency value on the composite potency scale, $\hat{x}_i$ is the projected x-coordinate for chemical i, $\hat{y}_i$ is the projected y-coordinate for chemical i, $\min(\hat{x})$ and $\min(\hat{y})$ correspond to the minimum values of the projected x- and y-coordinates for the set of chemicals, and a is the slope of the fitted PB model onto which the projections are made. Finally, the location of the composite potency value was shifted so that the average differences between the composite scale and the projected x- and y-coordinates were minimized (see Equation (5)).

$$cPV = d + \frac{Avg(\hat{x} - d) + Avg(\hat{y} - d)}{2}, \tag{5}$$

where cPV is the composite potency value, d is the vector of distance calculated in Equation (4), $\hat{x}$ and $\hat{y}$ are the projected x- and y-coordinates, and Avg is a function for the calculation of the arithmetic mean.

2.4. Fitting of Potency Prediction Models

Two main types of models were considered for predicting potency values from cDV_0. The first one was standard linear regression, where overall prediction errors are minimized. The other type of model was a robust implementation of linear regression, where the fit was achieved using iterative re-weighted least squares, incorporating Huber loss to down-weight samples with relatively outlying values [59]. The constant k in the Huber weighting was kept at the default value of 1.345. The standard regression models were fitted using R (version 4.2.0) and base package stats (version 4.2.0) [56]. The robust regression models were fitted in R (version 4.2.0) with the package MASS (version 7.3-56) [60].

The performances of the fits were evaluated using repeated cross-validation, with 50 repeats and 10 folds. Classification accuracy was based on the absolute geometric mean fold change, defined as described in Equation (6). The cross-validation performance was summarized within repeats and then across the repeats.

$$Absolute\ geometric\ mean\ fold\ change = e^{\frac{\sum_{i=1}^{n} |\log(\frac{Prediction_i}{Reference_i})|}{n}} \tag{6}$$

where $Prediction_i$ is the potency prediction of chemical i, $Reference_i$ is the reference potency value for chemical i, and n is the number of chemicals. Note that $Prediction_i$ and $Reference_i$ are not log-transformed in Equation (6).

2.5. General Statistical Calculations and Visualizations

All calculations were performed in R (version 4.2.0) [56]. Correlation metrics and regression models were calculated and fitted on log10-transformed concentrations or dose values, unless explicitly stated otherwise. Linear correlations were calculated with the Pearson correlation coefficient and rank correlations with the Spearman correlation coefficient. Calculations of prediction errors were based on relative differences, expressed as fold changes. Aggregated performance figures summarized over a set of chemicals were calculated as absolute geometric mean fold changes (see Equation (6)).

Figures 1–3 were created using ggplot2 (version 3.4.0) [61].

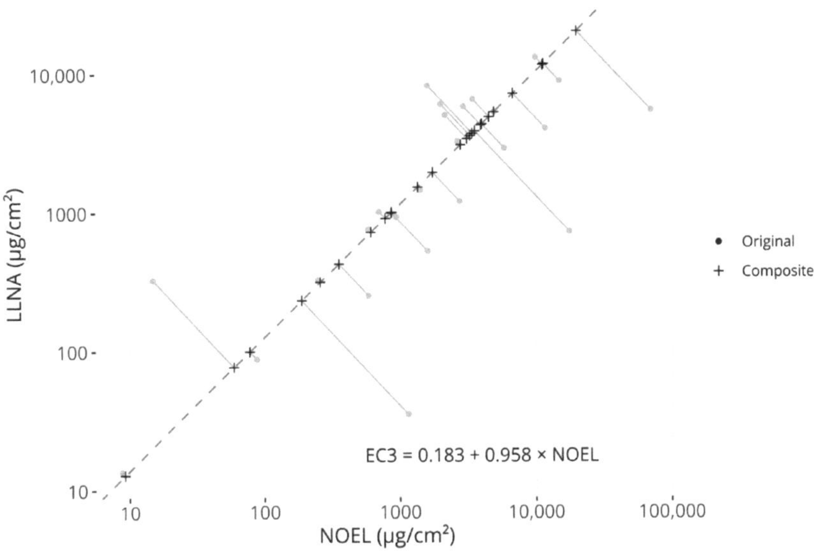

Figure 1. Visualization of the creation of the composite potency values. The grey circular points describe the original reference values (i.e., LLNA EC3 values and human NOEL values). The dashed line describes the Passing–Bablok regression model fitted to the potency references (its equation is described in the figure). Plus signs describe the projection of the original reference values onto the model. The grey linear segments visualize the projections. Axes are log-transformed.

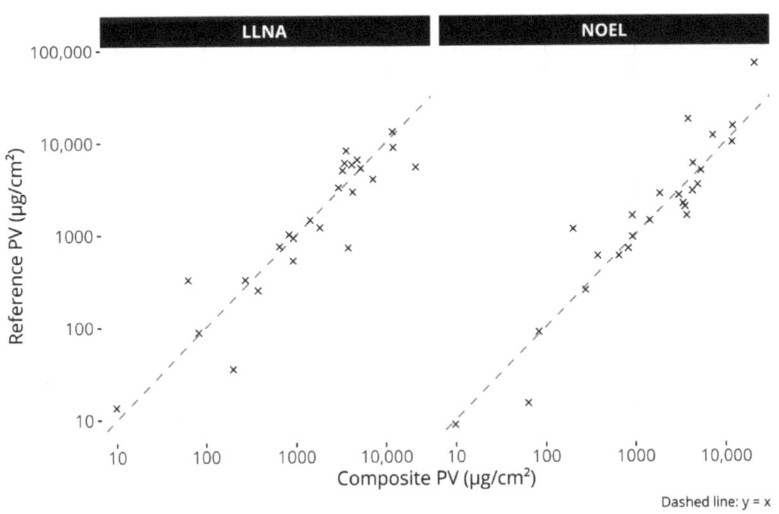

Figure 2. Scatter plots comparing the composite potency values with the individual references. The left plot compares the composite potency values with LLNA EC3 values and the right plot with human NOEL values. The dashed lines describe the identity lines, i.e., y = x. All axes are log-transformed.

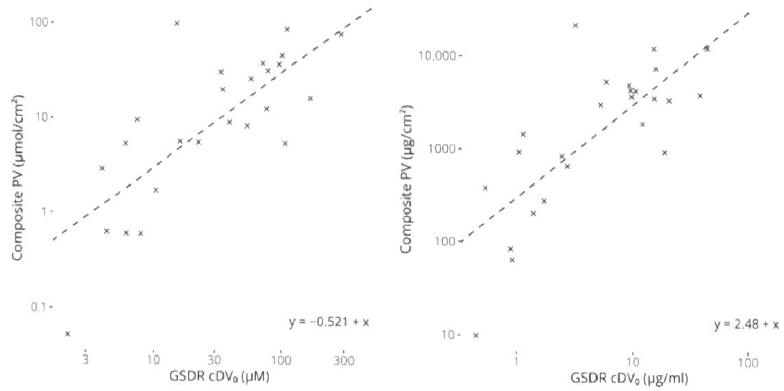

Figure 3. Scatter plots comparing GARDskin Dose–Response cDV_0 values with composite potency values. Concentration units are expressed in molar-based concentrations on the left and mass-based concentrations on the right. The dashed lines describe the regression models for predicting potency from cDV_0 values. The equations of the lines are described in the figures. All axes are log-transformed.

3. Results

3.1. Potency Information in cDV_0 Values

The GARDskin Dose–Response assay is a method for the quantitative assessment of skin-sensitizing potency. cDV_0 has been shown to be significantly associated with skin-sensitizing potency, defined from either LLNA EC3 values or human NOEL values. To provide a more detailed understanding of these associations, and to examine methods capable of predicting such potency values from cDV_0, data on 12 chemicals were generated and added to a dataset of 18 previously described chemicals [42]. The results for the complete dataset of 30 chemicals are described in Table 1.

The correlations between cDV_0 values and LLNA EC3 and human NOELs were first assessed (see Table 2). As can be seen, both linear and rank correlations are significant and range from 0.645 to 0.787. Generally, the cDV_0 values appear more strongly correlated with LLNA EC3 than with human NOELs. Moreover, the magnitudes of the correlation scores are similar to those calculated between the two references.

Table 2. Linear (Pearson) and rank (Spearman) correlation coefficients between GARDskin Dose–Response cDV_0, LLNA EC3, and human NOEL values.

	On Molar-Based Concentrations		On Mass-Based Concentrations	
	Linear Correlation	**Rank Correlation**	**Linear Correlation**	**Rank Correlation**
cDV_0 vs. LLNA EC3	0.787 ($p = 4.20 \times 10^{-7}$)	0.709 ($p = 2.74 \times 10^{-5}$)	0.743 ($p = 3.92 \times 10^{-6}$)	0.669 ($p = 7.18 \times 10^{-5}$)
cDV_0 vs. human NOEL	0.645 ($p = 3.70 \times 10^{-4}$)	0.664 ($p = 3.08 \times 10^{-4}$)	0.652 ($p = 3.11 \times 10^{-4}$)	0.656 ($p = 2.74 \times 10^{-4}$)
LLNA EC3 vs. human NOEL	0.738 ($p = 2.54 \times 10^{-5}$)	0.773 ($p = 1.02 \times 10^{-5}$)	0.736 ($p = 2.72 \times 10^{-5}$)	0.709 ($p = 7.20 \times 10^{-5}$)

3.2. Creation of Reference Composite Potency Values

As shown, the cDV_0 values correlate significantly with both LLNA EC3 and human NOELs, which suggests that the ranking of chemicals obtained from the GARDskin Dose–Response assay is, in itself, informative of potency. However, for the metric to be useful in NGRA, the cDV_0 value is advantageously translated into a more appropriate unit such as the more readily interpretable $\mu g/cm^2$.

However, two distinct sets of references are available for constructing a potential prediction model, i.e., NESILs derived from either LLNA EC3 values or human NOEL values. While it may be argued that the human endpoint is the more relevant one (as we are attempting to predict sensitization in humans), LLNA EC3 values are a more consistent estimator of a PoD, as human NOELs do not necessarily represent the highest possible non-sensitizing concentration (only the highest observed non-sensitizing concentration).

As both metrices have potential advantages and disadvantages, while simultaneously describing an underlying potency signal, it was hypothesized that one could leverage both information sources against each other to create a composite potency value to which a model could be fitted. Therefore, a robust errors-in-variables model, a Passing–Bablok (PB) regression model, was fitted between the 25 chemicals with NESILs from both LLNA EC3 and human NOEL values. The fit between the measures is displayed in Figure 1. The fitted coefficients of the PB model are described in Table 3. The estimated coefficients were close to, and not significantly different from, the identity line (slope = 1, intercept = 0), indicating that no systematic differences between the references could be detected on the examined dataset.

Table 3. Regression coefficients of the PB models fitted between the reference NESILs. The values within the parentheses describe the estimated 95% confidence intervals of the coefficients.

	Intercept	Slope
Molar	0.0435 (−0.459, 0.331)	0.988 (0.700, 1.30)
Mass	0.183 (−0.778, 1.22)	0.958 (0.624, 1.22)

The composite potency values were created by projecting the individual data points onto the fitted line. The individual reference values and the composite scores are described in Table 1. Figure 2 compares the composite scores with the original potency references. The figure suggests that the composite values capture the overall potency information from either reference well.

3.3. Prediction of Potency Values from cDV_0

Figure 3 shows a scatter plot comparing the cDV_0 values with the derived composite potency values. As expected, based on the underlying reference values, they correlate strongly and significantly. The linear correlation for molar-based concentrations was estimated to be 0.770 ($p = 6.68 \times 10^{-6}$) and the rank correlation to be 0.710 ($p = 1.07 \times 10^{-4}$). The corresponding values for the mass-based concentrations were 0.762 ($p = 9.56 \times 10^{-6}$) and 0.709 ($p = 1.10 \times 10^{-4}$), for linear and rank correlation respectively. Interestingly, for the 25 chemicals that had references from LLNA EC3 and human NOEL values, the correlation scores between the composite potency values and cDV_0 were higher compared with any correlation score calculated between cDV_0 and either of the individual references.

Regression models were fitted with the aim of predicting the composite potency values from the cDV_0 values. Two main variants of regression models were considered: a standard linear regression model and a robust regression model that reduced the influence of deviating observations. The model types were fitted both with and without a parameter for a slope. When the slope was not estimated, a constant value of 1 was assumed. The performances of all four types of models were examined using repeated cross-validation, which showed that the prediction errors were consistently smaller for the models that only estimated an intercept. The two regression techniques produced relatively similar results, but the smallest errors were observed for the robust regression. Therefore, the robust regression model that only included an intercept was selected as the most appropriate for potency predictions. From the cross-validation, the prediction errors for both molar- and mass-based models were estimated at 2.75 and 3.22 fold changes compared with the NESIL values from the LLNA EC3 and human NOELs, respectively.

For comparative reasons, and to allow for the assessment of the utility of the composite potency values for the fitting of the models, the cross-validation procedure was repeated using the individual potency references for fitting and evaluating the regression models (e.g., a model was trained and evaluated on LLNA data). The result from this cross-validation is described in Table 4. As can be seen, the prediction errors of these models were generally higher compared with the errors obtained for the models fitted to the composite potency values, supporting the hypothesis that it may be beneficial to leverage potency references against each other.

Table 4. Results from repeated cross-validation using individual references for fitting the models. The fold changes describe the absolute geometric mean fold changes from the cross-validation.

Potency Reference	Regression Method	Concentration Unit	Fold-Change Error—Intercept and Slope	Fold-Change Error—Intercept Only
LLNA	Linear regression	Mass	2.97	2.84
LLNA	Linear regression	Molar	2.96	2.84
NOEL	Linear regression	Mass	3.45	3.24
NOEL	Linear regression	Molar	3.45	3.24
LLNA	Robust regression	Mass	2.98	2.82
LLNA	Robust regression	Molar	2.98	2.82
NOEL	Robust regression	Mass	3.5	3.24
NOEL	Robust regression	Molar	3.48	3.24

Based on available data and the presented results, the robust regression model that only estimated an intercept and that was fitted on the composite potency values was found to be the overall best-performing potency prediction model. Table 5 shows the fitted coefficients. Given the simplicity of the final models and the fact that they were fitted on log-transformed values, predictions can in practice be obtained by multiplying a cDV_0 value by a constant, which equals the value of the intercept expressed in base 10 (see Equations (7) and (8); coefficients are rounded to three significant figures).

$$\text{Predicted potency in } \mu\text{mol/cm}^2 = cDV_0 \text{ in } \mu\text{M} \times 0.301, \quad (7)$$

$$\text{Predicted potency in } \mu\text{g/cm}^2 = cDV_0 \text{ in } \mu\text{g/mL} \times 304, \quad (8)$$

Table 5. Coefficients of the fitted potency models. Values within the parentheses describe the estimated 95% confidence intervals for the coefficients.

Model	Estimated Intercept
Molar-based robust regression	−0.521 (−0.689, −0.353)
Mass-based robust regression	2.48 (2.32, 2.65)

The fitted values of the proposed models and their predictions for the samples not included in the training set (i.e., the five chemicals without complete references; data are detailed in Supplementary Table S1) were further examined. The geometric mean prediction errors expressed in fold change were estimated to be 2.69 and 3.12 compared with NESILs from LLNA EC3 and human NOELs, respectively (for both mass- and molar-based concentrations). For comparison, the prediction error of LLNA EC3 compared with human NOELs was 2.63 fold changes. On the same set of 25 chemicals, the GARDskin Dose–Response's prediction error of human NOELs was 2.94 fold changes. While the LLNA predicted human potency with an overall lower error on this set of chemicals, the difference in performance compared with the proposed prediction model was not significant ($p = 0.508$ [Wilcoxon signed-rank test]).

Figure 4 displays the relative errors between the proposed model's predictions and the individual references' potency values for the chemicals, including LLNA EC3 (black bars), human NOELs (grey bars), and the composite reference values (white bars). Fourteen chemicals were consistently predicted within 3 fold changes compared with any of the references (i.e., the largest fold change for any reference was below 3). Three chemicals were predicted with a greater than 5-fold difference compared with any of the references (i.e., the smallest fold change for any reference was greater than 5). These chemicals were 2,4-dinitrochlorobenzene and 3-propylenephthalide (underpredicted), and alpha-isomethylionone (overpredicted).

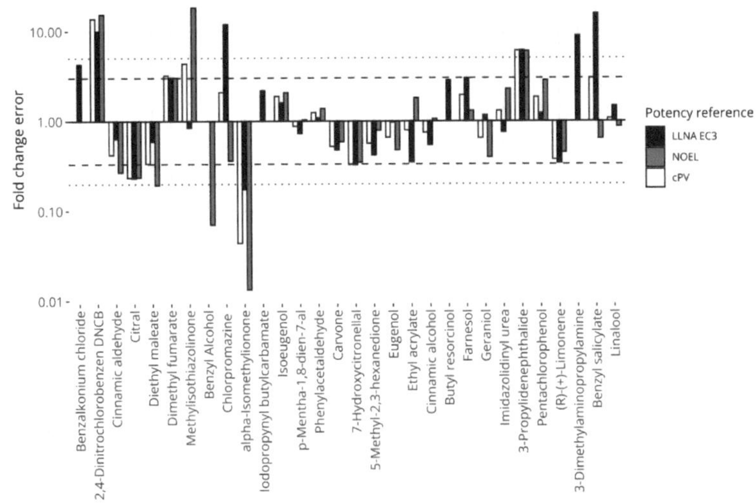

Figure 4. Prediction errors expressed in fold changes for the model defined on mass-based concentrations, comparing predictions of the proposed potency model with reference values. The *y*-axis is log-transformed. The dashed horizontal lines mark the 3-fold thresholds and the dotted horizontal lines describe the 5-fold thresholds.

Finally, considering the comparison between potency predictions and the reference values, due to the simplicity of the proposed potency prediction model, the correlation coefficients between predictions and references are identical to those described in Table 2 (ranging between 0.645 and 0.787 when compared with LLNA EC3 and human NOELs).

4. Discussion

The development of NAMs able to inform on chemicals' skin-sensitizing hazards has progressed relatively rapidly and far, and several methods have gained regulatory approval for identifying hazards, following the demonstration of their reliabilities and performances [24–26,29]. In addition, more recent advancements have seen integrated testing strategies approved for categorical potency assessment in line with the subcategorization defined by the UN GHS [29]. While all of these progressions reduce and replace the necessity of animal methods for the assessment of skin sensitizers, there are still areas where NAMs can be further improved. One such area is the assessment of skin-sensitizing potency, particularly considering the derivation of quantitative continuous potency metrices that may act as replacements for, e.g., the well-established LLNA EC3 value. This is an active field of research and several interesting methods have been described [34,36–39].

In this work, we have expanded on previously published results, describing how the GARDskin assay can be used to derive a continuous concentration metric, the cDV_0 value, associated with skin-sensitizing potency. Specifically, the GARDskin Dose–Response assay extends the validated GARDskin assay's protocol by incorporating measurements at multiple concentrations. Using the derived dose–response data, the lowest concentration capable of inducing a positive classification is identified. Here, we describe such cDV_0 concentrations for a set of 30 chemicals and further characterize their association with both human and LLNA data. It was confirmed that cDV_0 values maintained strong and significant correlations with both LLNA EC3 and human NOELs on this extended set of chemicals. Generally, the correlation coefficients were higher for LLNA EC3 compared to human NOELs. This could potentially be attributed to the fact that human NOEL values do not necessarily describe the highest possible non-sensitizing concentration. However, the uncertainty in the correlation estimates was too high to conclude that either metric correlated significantly higher with cDV_0 than the other. This is, however, a trend that could be considered in future works.

Having shown that cDV_0 values are associated with potency, a model capable of predicting NESILs was defined. Several approaches could be considered for creating prediction models that take cDV_0 as input, including the creation of separate models for the classification of NESILs derived from LLNA EC3 and human NOELs (i.e., distinct models trained on LLNA EC3 and on human NOELs). However, this may not be ideal, as the same cDV_0 value could give rise to two distinct potency values, which could impede interpretation. Therefore, it was explored if a single potency model could be defined from a composite potency reference constructed using information from both LLNA EC3 and human NOELs. A similar endeavor has been described in the construction of the reference chemical potency list (RCPL), where a consensus potency value was defined by evaluating both human and animal data [62]. While that scale, in part, relies on expert judgment for potency value assignment, the approach taken here was solely data- and computationally driven, combining NESILs from LLNA EC3 and human NOELs. The main assumption for the creation of the potency scale was that both LLNA EC3 and human NOELs describe the same underlying phenomenon, i.e., sensitizing potency, and that both metrics are associated with errors. The composite potency value was then designed to attempt to describe the common underlying potency signal, which was accomplished using a Passing–Bablok regression model [55] that robustly captured the association between the two references. Importantly, models fitted to the composite potency values received lower errors when predicting the original reference potency values, as compared with models fitted to the individual references. Moreover, the correlation coefficients were also marginally higher for the composite potency values compared with correlations to the individual references when considering the same set of 25 chemicals. While the reason for these results is not completely understood, it is possible that composite potency values inflict some degree of moderation or regularization to the original references, which may have a stabilizing effect during model fitting. This does not seem unreasonable, as a latent signal may be better described via the aggregation of several noisy measurements.

During initial model exploration, it was observed that the fitted parameters for the slopes were very close to 1, suggesting that they may not need to be estimated from the model. Indeed, models without an explicit estimate for the slope achieved lower values of Akaike's Information Criterion (AIC) [63]. Because of this, for the actual evaluation of the models' performances in cross-validation, they were fitted both with and without a parameter for the slope. In addition, the biological implication of a fixed slope of 1 could further indicate that the system used for the evaluation of potency is relevant, as the relative difference of skin-sensitizing potency is sustained. The consistently best-performing models corresponded to those without slope estimates. This is also convenient, as it leads to very simple prediction models. As the final models were fitted to log-transformed potency values and only contained estimates for the intercepts, in practice, they can be applied by multiplying an untransformed cDV_0 value with a constant, as specified by equations 7 and 8. Moreover, as noted, an appealing property of these models is that they suggest that an n-factor difference in cDV_0 values corresponds to an n-factor difference in the predicted NESIL. Finally, as models only contain a single estimated parameter, the uncertainties of the fits become relatively small and constant over the prediction range on the log scale.

The errors for the prediction model were estimated at 2.75 and 3.22 fold changes compared with LLNA EC3 and human NOELs, respectively, evaluated using a cross-validation approach. These figures were also close to the errors observed for the fitted values of the final model (i.e., as predicted by the current fits without cross-validation), which were 2.69 fold changes and 3.12 fold changes for LLNA EC3 and human NOELs, respectively. In line with observations for the correlation coefficients, LLNA EC3 was predicted with a lower error, though not significantly. As previously noted, human NOEL does not necessarily represent the highest non-sensitizing concentration of a chemical, which could potentially explain prediction discrepancies. Despite this, the numbers of over/under-predicted chemicals were quite evenly balanced. Nevertheless, of the 12 chemicals whose potency was underpredicted compared with human NOELs, LOEL values could be located for 10. Interestingly, only three of the underpredicted potency instances produced predicted NESILs greater than the LOEL values (i.e., 7/10 predictions were between NOEL and LOEL).

To put the performance figures into greater context, the prediction performances of other potency sources were also considered. First, the error of the LLNA assay when predicting human NOELs in this dataset was calculated, and was found to be 2.63 fold changes. The error of the GARDskin Dose–Response model was 2.94 fold changes, compared with human NOELs on the same set of 25 chemicals. While the LLNA achieved an overall lower error, the prediction errors were not found to be significantly different. Other sources of potency predictions include, e.g., the relatively recent publication where several regression models incorporating data from established NAMs, including KeratinoSens, DPRA, h-CLAT, and the kDPRA, for predicting EC3 values were described. While their prediction metrices were based on a larger dataset (n = 188), they reported geometric mean absolute fold changes that ranged between 3.1 and 3.5 (median values ranged between 2.3 and 2.7). This appears to be comparable with the errors described in this work. One aspect of consideration is, however, that the prediction model proposed in this work is very simple and only requires a single information source, i.e., the cDV_0 value, for potency predictions.

Considering the largest prediction errors obtained with this dataset, GARDskin Dose–Response predicted three chemicals with a fold change error greater than 5 compared with both human and LLNA references. These were alpha-isomethylionone, 2,4-dinitrochlorobenzene, and 3-propylidenephthalide. The potency of 3-propylidenephthalide was underpredicted by approximately 6-fold. It had two similar cDV_0 values, suggesting that the assay outcomes were reproducible for the test item. Moreover, the NESILs from the LLNA and human NOEL were also very similar to each other. Currently, no explanation has been recognized for this prediction discrepancy. The potency of alpha-isomethylionone was overpredicted compared with both LLNA and human NOELs. The chemical had two runs in the GARDskin Dose–Response assay. However, these differed more than 4-fold, indicating some uncertainty in the chemical's actual cDV_0. Even so, a prediction generated with the larger of the two cDV_0 estimates still overpredicted the potency of the chemical. The relative error for the run with the higher cDV_0 value compared with LLNA EC3 was approximately 2.5 fold changes, which is, nevertheless, still a significant overprediction of potency compared with the human NOEL (>25 fold). It is relevant to note that the LLNA also overpredicts the potency of alpha-isomethylione. A tentative hypothesis for LLNA's overprediction was previously provided where it was suggested that the autoxidation of the chemical, which is more likely to occur in the LLNA than under the occluded patches in HRIPTs, could explain the higher potency in the LLNA, as the oxidation products are typically strong sensitizers [64]. Although only a hypothesis, it is possible that this effect is also relevant for the overprediction of its potency in the GARDskin Dose–Response assay, which could potentially be explored in follow-up studies. Lastly, 2,4-dinitrochlorobenzene displayed a clear and steep response and the cDV_0 value was considered reliable, despite only consisting of data from a single experiment. Interestingly, the underprediction of its potency does not appear to be a unique observation in the GARDskin Dose–Response assay, as similar tendencies were also described by Natsch et al. [65]. A clear explanation for this is, however, not available. Finally, it is important to consider that the GARDskin Dose–Response assay represents a relatively simple model of biological processes and ignores potential components that may contribute to the observed in vivo potency values. Additional data generated from chemical assessments in the GARDskin Dose–Response assay will likely provide interesting avenues for future investigations. These will help to further establish the predictive performance of the assay and other important properties, including the reproducibility of the numerical readout. Indeed, data for these purposes are being generated and some have already been published [41].

To conclude, we have described the GARDskin Dose–Response assay and demonstrated a significant association between its readout and available measures of skin-sensitizing potency on a set of 30 chemicals. A simple model for the prediction of sensitizing potency in the form of a quantitative continuous PoD was developed that only consisted of a single parameter. The performance of the prediction model was evaluated using cross-validation, and errors were estimated at 2.75 fold changes compared with NESILs derived from the LLNA, and 3.22 fold changes compared with NESILs based on human NOELs. The GARDskin Dose–Response method is therefore considered a useful tool for deriving skin-sensitizing potency information, which may be used as a PoD in subsequent risk assessments.

Supplementary Materials: The following supporting information can be downloaded at: https://www.mdpi.com/article/10.3390/toxics12090626/s1, Table S1: Chemical information; Tables S2 and S3: Dilution schemes for chemicals assayed in GARDskin Dose-Response.

Author Contributions: Conceptualization, R.G., F.T., F.A., N.A. and H.J.; methodology, R.G., F.T., U.M., F.A., N.A. and H.J.; software, R.G.; validation, R.G., F.T., U.M., J.A., F.A., N.A. and H.J.; formal analysis, R.G., F.T. and H.J.; investigation, R.G., F.T., U.M., J.A., F.A., A.F., N.A. and H.J.; resources, J.A. and U.M.; data curation, R.G., F.T., F.A., U.M. and J.A.; writing—original draft preparation, R.G., F.T., A.F., F.A. and H.J.; writing—review and editing, R.G., F.T., U.M., J.A., F.A., A.F., N.A. and H.J.; visualization, R.G.; supervision, F.A., N.A. and H.J.; project administration, F.T., F.A., N.A., U.M. and H.J.; funding acquisition, F.A. and H.J. All authors have read and agreed to the published version of the manuscript.

Funding: This research received no external funding.

Institutional Review Board Statement: Not applicable.

Informed Consent Statement: Not applicable.

Data Availability Statement: Data are contained within the article and Supplementary Materials.

Conflicts of Interest: R.G., U.M., J.A., A.F. and H.J. are employees of Senzagen AB, which offers the GARD™ platform assays as commercial services. F.T., F.A. and N.A. are employees of L'Oréal and have no financial interest regarding the GARD™ platform.

References

1. Daniel, A.B.; Strickland, J.; Allen, D.; Casati, S.; Zuang, V.; Barroso, J.; Whelan, M.; Régimbald-Krnel, M.J.; Kojima, H.; Nishikawa, A.; et al. International Regulatory Requirements for Skin Sensitization Testing. *Regul. Toxicol. Pharmacol.* **2018**, *95*, 52–65. [CrossRef]
2. EFSA Panel on Additives and Products or Substances used in Animal Feed (FEEDAP); Bampidis, V.; Azimonti, G.; Bastos, M.d.L.; Christensen, H.; Durjava, M.; Dusemund, B.; Kouba, M.; López-Alonso, M.; López Puente, S.; et al. Guidance on the Assessment of the Safety of Feed Additives for the Users. *EFSA J.* **2023**, *21*, e8469. [CrossRef] [PubMed]
3. Kimber, I.; Basketter, D.A.; Gerberick, G.F.; Ryan, C.A.; Dearman, R.J. Chemical Allergy: Translating Biology into Hazard Characterization. *Toxicol. Sci.* **2011**, *120*, S238–S268. [CrossRef]
4. Esser, P.R.; Martin, S.F. Pathomechanisms of Contact Sensitization. *Curr. Allergy Asthma Rep.* **2017**, *17*, 83. [CrossRef] [PubMed]
5. OECD. *The Adverse Outcome Pathway for Skin Sensitisation Initiated by Covalent Binding to Proteins*; OECD Series on Testing and Assessment; OECD: Paris, France, 2014; ISBN 978-92-64-22144-4.
6. Kimber, I.; Basketter, D.A.; Berthold, K.; Butler, M.; Garrigue, J.-L.; Lea, L.; Newsome, C.; Roggeband, R.; Steiling, W.; Stropp, G.; et al. Skin Sensitization Testing in Potency and Risk Assessment. *Toxicol. Sci.* **2001**, *59*, 198–208. [CrossRef] [PubMed]
7. Basketter, D.A.; Gerberick, F.; Kimber, I. The Local Lymph Node Assay and the Assessment of Relative Potency: Status of Validation. *Contact Dermat.* **2007**, *57*, 70–75. [CrossRef]
8. Buehler, E.V. Delayed Contact Hypersensitivity in the Guinea Pig. *Arch. Dermatol.* **1965**, *91*, 171–175. [CrossRef]
9. Ritz, H.; Buehler, E. Planning, Conduct and Interpretation of Guinea Pig Sensitization Patch Tests. In *Current Concepts in Cutaneous Toxicity*; Academic Press: New York, NY, USA, 1980; pp. 25–40.
10. Robinson, M.K.; Nusair, T.L.; Robert Fletcher, E.; Ritz, H.L. A Review of the Buehler Guinea Pig Skin Sensitization Test and Its Use in a Risk Assessment Process for Human Skin Sensitization. *Toxicology* **1990**, *61*, 91–107. [CrossRef]
11. Magnusson, B.; Kligman, A.M. The Identification of Contact Allergens by Animal Assay. The Guinea Pig Maximization Test. *J. Investig. Dermatol.* **1969**, *52*, 268–276. [CrossRef]
12. Kimber, I.; Weisenberger, C. A Murine Local Lymph Node Assay for the Identification of Contact Allergens. *Arch. Toxicol.* **1989**, *63*, 274–282. [CrossRef]

13. Kimber, I.; Dearman, R.J.; Basketter, D.A.; Ryan, C.A.; Gerberick, G.F. The Local Lymph Node Assay: Past, Present and Future. *Contact Dermat.* **2002**, *47*, 315–328. [CrossRef] [PubMed]
14. OECD. *Test No. 429: Skin Sensitisation*; OECD: Paris, France, 2010.
15. Anderson, S.E.; Siegel, P.D.; Meade, B.J. The LLNA: A Brief Review of Recent Advances and Limitations. *J. Allergy* **2011**, *2011*, 1–10. [CrossRef]
16. Gerberick, G.F.; Robinson, M.K.; Ryan, C.A.; Dearman, R.J.; Kimber, I.; Basketter, D.A.; Wright, Z.; Marks, J.G. Contact Allergenic Potency: Correlation of Human and Local Lymph Node Assay Data. *Am. J. Contact Dermat.* **2001**, *12*, 156–161. [CrossRef] [PubMed]
17. Basketter, D.A.; Clapp, C.; Jefferies, D.; Safford, B.; Ryan, C.A.; Gerberick, F.; Dearman, R.J.; Kimber, I. Predictive Identification of Human Skin Sensitization Thresholds. *Contact Dermat.* **2005**, *53*, 260–267. [CrossRef]
18. Api, A.M.; Basketter, D.; Lalko, J. Correlation between Experimental Human and Murine Skin Sensitization Induction Thresholds. *Cutan. Ocul. Toxicol.* **2015**, *34*, 298–302. [CrossRef] [PubMed]
19. Bil, W.; Schuur, A.G.; Ezendam, J.; Bokkers, B.G.H. Probabilistic Derivation of the Interspecies Assessment Factor for Skin Sensitization. *Regul. Toxicol. Pharmacol.* **2017**, *88*, 34–44. [CrossRef]
20. Basketter, D.A. The Human Repeated Insult Patch Test in the 21st Century: A Commentary. *Cutan. Ocul. Toxicol.* **2009**, *28*, 49–53. [CrossRef]
21. Na, M.; Ritacco, G.; O'Brien, D.; Lavelle, M.; Api, A.M.; Basketter, D. Fragrance Skin Sensitization Evaluation and Human Testing: 30-Year Experience. *Dermatitis* **2021**, *32*, 339–352. [CrossRef]
22. Lee, I.; Na, M.; Lavelle, M.; Api, A.M. Derivation of the No Expected Sensitization Induction Level for Dermal Quantitative Risk Assessment of Fragrance Ingredients Using a Weight of Evidence Approach. *Food Chem. Toxicol.* **2022**, *159*, 112705. [CrossRef]
23. Api, A.M.; Basketter, D.; Bridges, J.; Cadby, P.; Ellis, G.; Gilmour, N.; Greim, H.; Griem, P.; Kern, P.; Khaiat, A.; et al. Updating Exposure Assessment for Skin Sensitization Quantitative Risk Assessment for Fragrance Materials. *Regul. Toxicol. Pharmacol.* **2020**, *118*, 104805. [CrossRef]
24. OECD. *Test No. 442C: In Chemico Skin Sensitisation*; OECD: Paris, France, 2023.
25. OECD. *Test No. 442D: In Vitro Skin Sensitisation*; OECD: Paris, France, 2022.
26. OECD. *Test No. 442E: In Vitro Skin Sensitisation*; OECD: Paris, France, 2023.
27. Kleinstreuer, N.C.; Hoffmann, S.; Alépée, N.; Allen, D.; Ashikaga, T.; Casey, W.; Clouet, E.; Cluzel, M.; Desprez, B.; Gellatly, N.; et al. Non-Animal Methods to Predict Skin Sensitization (II): An Assessment of Defined Approaches. *Crit. Rev. Toxicol.* **2018**, *48*, 359–374. [CrossRef] [PubMed]
28. Tourneix, F.; Alépée, N.; Detroyer, A.; Eilstein, J.; Martinozzi Teissier, S.; Nardelli, L.; Noçairi, H.; Pauloin, T.; Piroird, C.; Del Bufalo, A. Assessment of a Defined Approach Based on a Stacking Prediction Model to Identify Skin Sensitization Hazard. *Toxicol. In Vitro* **2019**, *60*, 134–143. [CrossRef] [PubMed]
29. OECD. *Guideline No. 497: Defined Approaches on Skin Sensitisation*; OECD: Paris, France, 2023.
30. Russo, G.; Crispino, E.; Corsini, E.; Iulini, M.; Paini, A.; Worth, A.; Pappalardo, F. Computational Modelling and Simulation for Immunotoxicity Prediction Induced by Skin Sensitisers. *Comput. Struct. Biotechnol. J.* **2022**, *20*, 6172–6181. [CrossRef] [PubMed]
31. Iulini, M.; Russo, G.; Crispino, E.; Paini, A.; Fragki, S.; Corsini, E.; Pappalardo, F. Advancing PFAS Risk Assessment: Integrative Approaches Using Agent-Based Modelling and Physiologically-Based Kinetic for Environmental and Health Safety. *Comput. Struct. Biotechnol. J.* **2024**, *23*, 2763–2778. [CrossRef]
32. Golden, E.; Macmillan, D.S.; Dameron, G.; Kern, P.; Hartung, T.; Maertens, A. Evaluation of the Global Performance of Eight in Silico Skin Sensitization Models Using Human Data. *ALTEX* **2021**, *38*, 33–48. [CrossRef]
33. Johnson, C.; Ahlberg, E.; Anger, L.T.; Beilke, L.; Benigni, R.; Bercu, J.; Bobst, S.; Bower, D.; Brigo, A.; Campbell, S.; et al. Skin Sensitization in Silico Protocol. *Regul. Toxicol. Pharmacol.* **2020**, *116*, 104688. [CrossRef] [PubMed]
34. Reynolds, J.; MacKay, C.; Gilmour, N.; Miguel-Vilumbrales, D.; Maxwell, G. Probabilistic Prediction of Human Skin Sensitiser Potency for Use in next Generation Risk Assessment. *Comput. Toxicol.* **2019**, *9*, 36–49. [CrossRef]
35. Reynolds, J.; Gilmour, N.; Baltazar, M.T.; Reynolds, G.; Windebank, S.; Maxwell, G. Decision Making in next Generation Risk Assessment for Skin Allergy: Using Historical Clinical Experience to Benchmark Risk. *Regul. Toxicol. Pharmacol.* **2022**, *134*, 105219. [CrossRef] [PubMed]
36. Jaworska, J.S.; Natsch, A.; Ryan, C.; Strickland, J.; Ashikaga, T.; Miyazawa, M. Bayesian Integrated Testing Strategy (ITS) for Skin Sensitization Potency Assessment: A Decision Support System for Quantitative Weight of Evidence and Adaptive Testing Strategy. *Arch. Toxicol.* **2015**, *89*, 2355–2383. [CrossRef]
37. Hirota, M.; Fukui, S.; Okamoto, K.; Kurotani, S.; Imai, N.; Fujishiro, M.; Kyotani, D.; Kato, Y.; Kasahara, T.; Fujita, M.; et al. Evaluation of Combinations of in Vitro Sensitization Test Descriptors for the Artificial Neural Network-based Risk Assessment Model of Skin Sensitization. *J. Appl. Toxicol.* **2015**, *35*, 1333–1347. [CrossRef]
38. Natsch, A.; Gerberick, G.F. Integrated Skin Sensitization Assessment Based on OECD Methods (I): Deriving a Point of Departure for Risk Assessment. *ALTEX* **2022**, *39*, 636–646. [CrossRef]
39. Natsch, A.; Haupt, T.; Wareing, B.; Landsiedel, R.; Kolle, S.N. Predictivity of the Kinetic Direct Peptide Reactivity Assay (kDPRA) for Sensitizer Potency Assessment and GHS Subclassification. *ALTEX* **2020**, *37*, 652–664. [CrossRef]

40. Gilmour, N.; Alépée, N.; Hoffmann, S.; Kern, P.; Van Vliet, E.; Bury, D.; Miyazawa, M.; Nishida, H.; Cosmetics Europe. Applying a next Generation Risk Assessment Framework for Skin Sensitisation to Inconsistent New Approach Methodology Information. *ALTEX* **2023**, *40*, 439–451. [CrossRef] [PubMed]
41. Donthamsetty, S.; Forreryd, A.; Sterchele, P.; Huang, X.; Gradin, R.; Johansson, H.; Mattson, U.; Lee, I.; Api, A.M.; Ladics, G. GARDskin Dose-Response Assay and Its Application in Conducting Quantitative Risk Assessment (QRA) for Fragrance Materials Using a Next Generation Risk Assessment (NGRA) Framework. *Regul. Toxicol. Pharmacol.* **2024**, *149*, 105597. [CrossRef]
42. Gradin, R.; Forreryd, A.; Mattson, U.; Jerre, A.; Johansson, H. Quantitative Assessment of Sensitizing Potency Using a Dose–Response Adaptation of GARDskin. *Sci. Rep.* **2021**, *11*, 18904. [CrossRef] [PubMed]
43. Johansson, H.; Lindstedt, M.; Albrekt, A.-S.; Borrebaeck, C.A.K. A Genomic Biomarker Signature Can Predict Skin Sensitizers Using a Cell-Based in Vitro Alternative to Animal Tests. *BMC Genom.* **2011**, *12*, 399. [CrossRef] [PubMed]
44. EURL ECVAM. *GARDskin Assay Protocol*; EURL ECVAM: Ispra, Italy, 2021.
45. Forreryd, A.; Zeller, K.S.; Lindberg, T.; Johansson, H.; Lindstedt, M. From Genome-Wide Arrays to Tailor-Made Biomarker Readout—Progress towards Routine Analysis of Skin Sensitizing Chemicals with GARD. *Toxicol. In Vitro* **2016**, *37*, 178–188. [CrossRef]
46. OECD. *Supporting Document to the Genomic Allergen Rapid Detection Test Method for Skin Sensitisation (GARDTMskin), Described in Test Guideline 442E*; OECD: Paris, France, 2023.
47. Hoffmann, S.; Kleinstreuer, N.; Alépée, N.; Allen, D.; Api, A.M.; Ashikaga, T.; Clouet, E.; Cluzel, M.; Desprez, B.; Gellatly, N.; et al. Non-Animal Methods to Predict Skin Sensitization (I): The Cosmetics Europe Database. *Crit. Rev. Toxicol.* **2018**, *48*, 344–358. [CrossRef]
48. OECD. *Supporting Document to the OECD Guideline 497 on Defined Approaches for Skin Sensitisation*; OECD: Paris, France, 2023.
49. Basketter, D.A.; Alépée, N.; Ashikaga, T.; Barroso, J.; Gilmour, N.; Goebel, C.; Hibatallah, J.; Hoffmann, S.; Kern, P.; Martinozzi-Teissier, S.; et al. Categorization of Chemicals According to Their Relative Human Skin Sensitizing Potency. *Dermatitis* **2014**, *25*, 11–21. [CrossRef] [PubMed]
50. Na, M.; O'Brien, D.; Lavelle, M.; Lee, I.; Gerberick, G.F.; Api, A.M. Weight of Evidence Approach for Skin Sensitization Potency Categorization of Fragrance Ingredients. *Dermatitis* **2022**, *33*, 161–175. [CrossRef]
51. Gautier, F.; Tourneix, F.; Assaf Vandecasteele, H.; Van Vliet, E.; Bury, D.; Alépée, N. Read-across Can Increase Confidence in the Next Generation Risk Assessment for Skin Sensitisation: A Case Study with Resorcinol. *Regul. Toxicol. Pharmacol.* **2020**, *117*, 104755. [CrossRef]
52. Api, A.M.; Parakhia, R.; O'Brien, D.; Basketter, D.A. Fragrances Categorized According to Relative Human Skin Sensitization Potency. *Dermatitis* **2017**, *28*, 299–307. [CrossRef]
53. Na, M.; O'Brien, D.; Gerberick, G.F.; Kern, P.S.; Lavelle, M.; Lee, I.; Parakhia, R.; Ryan, C.; Api, A.M. Benchmarking Performance of SENS-IS Assay against Weight of Evidence Skin Sensitization Potency Categories. *Regul. Toxicol. Pharmacol.* **2022**, *130*, 105128. [CrossRef]
54. Griem, P.; Goebel, C.; Scheffler, H. Proposal for a Risk Assessment Methodology for Skin Sensitization Based on Sensitization Potency Data. *Regul. Toxicol. Pharmacol.* **2003**, *38*, 269–290. [CrossRef]
55. Bablok, W.; Passing, H.; Bender, R.; Schneider, B. A General Regression Procedure for Method Transformation. Application of Linear Regression Procedures for Method Comparison Studies in Clinical Chemistry, Part III. *J. Clin. Chem. Clin. Biochem.* **1988**, *26*, 783–790. [CrossRef]
56. R Core Team. *R: A Language and Environment for Statistical Computing*; R Foundation for Statistical Computing: Vienna, Austria, 2022.
57. Therneau, T. *Deming: Deming, Theil-Sen, Passing-Bablock and Total Least Squares Regression*; CRAN: Vienna, Austria, 2018.
58. Efron, B.; Tibshirani, R. *An Introduction to the Bootstrap*; Monographs on Statistics and Applied Probability; Chapman & Hall: New York, NY, USA, 1993; ISBN 978-0-412-04231-7.
59. Huber, P.J. *Robust Statistics*, 1st ed.; Wiley Series in Probability and Statistics; Wiley: Hoboken, NJ, USA, 1981; ISBN 978-0-471-41805-4.
60. Venables, W.N.; Ripley, B.D. *Modern Applied Statistics with S*, 4th ed.; Springer: New York, NY, USA, 2002; ISBN 0-387-95457-0.
61. Wickham, H. *Ggplot2*, 2nd ed.; Use R! Springer International Publishing: Basel, Switzerland, 2016; ISBN 978-3-319-24277-4.
62. Irizar, A.; Bender, H.; Griem, P.; Natsch, A.; Vey, M.; Kimber, I. Reference Chemical Potency List (RCPL): A New Tool for Evaluating the Accuracy of Skin Sensitisation Potency Measurements by New Approach Methodologies (NAMs). *Regul. Toxicol. Pharmacol.* **2022**, *134*, 105244. [CrossRef]
63. Akaike, H. A New Look at the Statistical Model Identification. *IEEE Trans. Autom. Control* **1974**, *19*, 716–723. [CrossRef]
64. Roberts, D.W.; Api, A.M. Chemical Applicability Domain of the Local Lymph Node Assay (LLNA) for Skin Sensitisation Potency. Part 4. Quantitative Correlation of LLNA Potency with Human Potency. *Regul. Toxicol. Pharmacol.* **2018**, *96*, 76–84. [CrossRef]
65. Natsch, A. Integrated Skin Sensitization Assessment Based on OECD Methods (III): Adding Human Data to the Assessment. *ALTEX* **2023**, *40*, 571–583. [CrossRef]

toxics

MDPI

Article

Limitations and Modifications of Skin Sensitization NAMs for Testing Inorganic Nanomaterials

Britta Wareing [1,†], Ayse Aktalay Hippchen [1,†], Susanne N. Kolle [1], Barbara Birk [2], Dorothee Funk-Weyer [1] and Robert Landsiedel [1,3,*]

1 BASF SE, Experimental Toxicology and Ecology, 67057 Ludwigshafen, Germany; britta.wareing@basf.com (B.W.); ayse.aktalay-hippchen@basf.com (A.A.H.); susanne.kolle@basf.com (S.N.K.); dorothee.funk-weyer@basf.com (D.F.-W.)
2 BASF SE, Agriculture Solutions, 67117 Limburgerhof, Germany; barbara.birk@basf.com
3 Pharmacy, Pharmacology and Toxicology, Free University of Berlin, 14195 Berlin, Germany
* Correspondence: robert.landsiedel@fu-berlin.de or robert.landsiedel@basf.com
† These authors contributed equally to this work.

Abstract: Since 2020, the REACh regulation requires toxicological data on nanoforms of materials, including the assessment of their skin-sensitizing properties. Small molecules' skin sensitization potential can be assessed by new approach methodologies (NAMs) addressing three key events (KE: protein interaction, activation of dendritic cells, and activation of keratinocytes) combined in a defined approach (DA) described in the OECD guideline 497. In the present study, the applicability of three NAMs (DPRA, LuSens, and h-CLAT) to nine materials (eight inorganic nanomaterials (NM) consisting of CeO_2, $BaSO_4$, TiO_2 or SiO_2, and quartz) was evaluated. The NAMs were technically applicable to NM using a specific sample preparation (NANOGENOTOX dispersion protocol) and method modifications to reduce interaction of NM with the photometric and flowcytometric read-outs. The results of the three assays were combined according to the defined approach described in the OECD guideline No. 497; two of the inorganic NM were identified as skin sensitizers. However, data from animal studies (for ZnO, also human data) indicate no skin sensitization potential. The remaining seven test substances were assessed as "inconclusive" because all inorganic NM were outside the domain of the DPRA, and the achievable test concentrations were not sufficiently high according to the current test guidelines of all three NAMs. The use of these NAMs for (inorganic) NM and the relevance of the results in general are challenged in three ways: (i) NAMs need modification to be applicable to insoluble, inorganic matter; (ii) current test guidelines lack adequate concentration metrics and top concentrations achievable for NM; and (iii) NM may not cause skin sensitization by the same molecular and cellular key events as small organic molecules do; in fact, T-cell-mediated hypersensitivity may not be the most relevant reaction of the immune system to NM. We conclude that the NAMs adopted by OECD test guidelines are currently not a good fit for testing inorganic NM.

Keywords: inorganic nanomaterials; skin sensitization; OECD test guidelines; new approach methodologies; applicability; h-CLAT; LuSens; DPRA

1. Introduction

The applications of nanomaterials (NM) range through various fields, including coatings, cosmetics, medicine, and electronics [1]. As the application fields grew, extensive investigation of NM with respect to their potential effects on human health and the environment became necessary [2–4]. During the last decades, numerous publicly funded projects have generated methods and know–how on the characterization, toxicological testing, and grouping of NMs [5]. Since 2020, NM falls under the REACH regulations; thus, information on NM skin sensitization potential is required, as listed in Annex VII. Non-animal methods (or new approach methodologies, NAMs) are a priority for data generation under

REACh, for small organic molecules and NM alike. Thereupon, an in vivo test shall only be performed if NAMs are not applicable to the test substance or deliver inconclusive results.

In 2021, defined approaches (DA) to assess skin sensitization of test substances using data generated by NAMs were adopted as OECD guideline (GL) 497 [6]. The NAMs used as information sources in the DA address three key events of the adverse outcome pathway (AOP) for skin sensitization: (i) molecular interaction with skin proteins; (ii) activation of keratinocytes; and (iii) activation of dendritic cells. The first key event can be addressed by an in chemico assay, the direct peptide reactivity assay (DPRA). The DPRA is based on quantification of the depletion of artificial model peptides caused by the reaction of test substances with these peptides. The method was initially developed for testing small organic molecules with electrophilic functions, which can react with nucleophilic thiol and/or amino groups of cysteine and lysine of skin proteins (and also of the DPRA model peptides) [7]. Most inorganic NM are larger than most small organic molecules and do not hold distinct electrophilic functions, which would covalently bind to cysteine and/or lysine. Hence, the application of DPRA for testing inorganic NM is mechanistically questionable. The second key event, inflammatory response and changes in gene expression of keratinocytes, can be addressed, e.g., by the KeratinoSens^{TM} and LuSens assays with immortalized human keratinocyte cell lines. Both cell lines were genetically modified by inserting a luciferase reporter gene for monitoring the activation of the Keap1–Nrf2–ARE pathway. Luciferase signal increases due to Nrf2-mediated activation of antioxidant response element (ARE), initiated by oxidative or electrophile stress detected by the Keap-1 protein [8,9]. The third key event, the activation of dendritic cells, can be addressed by the so-called human cell line activation test (h-CLAT). The h-CLAT assay is performed with the human monocytic leukemia cell line THP-1 exposed to the test substance. The activation of these cells is measured via the changes in the expression of two surface markers, CD54 and CD86, stained with fluorescent antibodies and quantified in a flow cytometer [10].

According to the first DA of the OECD GL No. 497, the "2 out of 3" (2o3) DA, any two of the three tests addressing these key events of the AOP determine the overall results. The second DA, the ITSv2 DA, uses information from the DPRA and h-CLAT combined with protein binding alerts generated using the OECD QSAR Toolbox. The overall result of the DA is discerning potential skin sensitizers and non-sensitizers (hazard assessment) in accordance with the UN GHS sensitizer categories. The DA GL based on AOP for skin sensitization was initially adopted to assess the skin sensitization potential of small organic molecules [4,11,12].

Thus far, only a few studies on NM with skin sensitization NAMs have been conducted. Likewise, only a few animal studies on NM's skin sensitization are available [13–20]. Physical and chemical properties of inorganic NM differ from those of small organic molecules. Most NMs are larger and often not monodisperse. There is no defined molecular weight of NM and a distribution of different particle sizes [1]. Larger NM penetrate only a few layers into the stratum corneum of intact skin but do not become bioavailable; thus, no interaction with living cells is expected [21,22]. Most inorganic NMs would interact with proteins via mechanisms other than covalent binding, e.g., non-covalent metal coordination bonds, or redox reactions. Moreover, inorganic NM can release metal ions and generate reactive oxygen species (ROS). NM forms protein coronas, which could alter the structures of the proteins. These could be molecular initiating events related to altered immune functions [3,11,23–28], but they may not be well captured by the current NAMs. In addition to these mechanistic stipulations, there are also technical hurdles: NMs are not directly applicable to these NAMs as they are not soluble (as long as they are NM), which requires a test item preparation different from the dissolution of small organic molecules and modifications to avoid interferences with optical read-outs of the NAMs.

To assess the applicability of the NAM-based DA for skin sensitization, we have tested nine widely used inorganic NM with three skin sensitization NAMs, which were adopted as OECD TGs (No. 442C, D, and E).

2. Materials and Methods

NM and quartz (SiO_2 DQ12) were tested in three NAMs: DPRA, LuSens, and h-CLAT (OECD TGs No. 442C [29], D [30], and E [31]) with modifications in test substance preparation (described below) to adapt these three assays for their applicability to NM. The results of these three assays were evaluated based on the criteria of DA OECD GL No. 497 [6], i.e., including borderline ranges [32].

2.1. Nanomaterials

In the present study, eight inorganic NM and one micron-sized material (SiO_2 DQ12) were assessed (Table 1). They consist of different metals or metalloids (Ba, Ce, Si, Ti, and Zn) as oxides or sulfates and represent different sizes and crystallinities (SiO_2 Aerosil, Levasil, and DQ12).

Table 1. Test substances and their properties.

Test Substance	CAS-Number	Physicochemical Properties	Skin Sensitization Potential
CeO_2 NM-212	1306-38-3	40 nm (SEM) very low water solubility (<0.001 wt.%) #	Cerium oxide is reported non-sensitizing in the GPMT [33]
$BaSO_4$ NM-220	7727-43-7	25 nm (SEM) # low water solubility (<0.05 wt.%) #	bulk material was reported non-sensitizing in the LLNA [1]
SiO_2 Levasil 200 (40%)	7631-86-9	5–50 nm (REM/TEM) * soluble in water	reported non-sensitizing in the GPMT * and the LLNA [1]
Quartz SiO_2 DQ12	n.a.	500–750 nm * crystalline quartz practically insoluble *	reported non-sensitizing in the LLNA [1]
SiO_2 Aerosil R972	68611-44-9	16 nm + hydrophobized colloidal silica water solubility > 1 mg/L *	Aerosil R812 and R 8200 were reported non-skin sensitizing in the GPMT [1]
SiO_2 Aerosil 200	7631-86-9	9 nm (TEM/SEM) [25] hydrophilic fumed silica water solubility > 1 mg/L *	bulk material was reported non-sensitizing in the LLNA [1]
TiO_2 NIST® SRM®	13463-67-7	19–37 nm + very low water solubility (0.001 g/L) *	reported non-sensitizing in the LLNA *
TiO_2 P25 Aeroxide	13463-67-7	21 nm * hydrophilic titanium dioxide	reported non-sensitizing in the patch test and the GMPT *
ZnO Z-Cote HP1	1314-13-2	190 nm [34]	ZnO nanomaterials were reported as non-sensitizing in patch test [1]

* Data provided in the safety sheet or certificate of analysis. # Value determined in the Department of Analytical and Material Science, BASF SE. + Value given in the manufacturer's website. [1] Data published in the registration dossier released by ECHA.

2.2. Test Substance Preparation

The preparation of the test substances was carried out based on the procedures described by the NANOGENOTOX standard operation procedure (SOP). This dispersion protocol was developed in the EU-project NanoREG and is one of the most widely used dispersion protocols. It was developed for the preparation of general batch dispersions for in vitro and in vivo toxicity testing to produce a highly dispersed state of a wide range of nanomaterials [35,36]. A fixed (gravimetric) test substance concentration of 2.56 mg/mL was prepared for every NM. Moreover, 25.6 mg of the respective test substance were weighed into a glass vessel and pre-wetted with 50 µL of ethanol. Thereafter, 9.95 mL of the respective vehicle (bovine serum albumin (0.05 *w*/*v*%) in deionized water for DPRA and cell culture medium for LuSens (with additional 4% DMSO) or h-CLAT assay) was added. The stock preparation was sonicated using a Branson Sonifier SFX 550 with 20% amplitude (400 W) for 16 min. Stock preparations were thoroughly shaken using a vortex prior to following dilution steps, and the homogeneity of the dispersion was assessed by visual inspection. All test substance stocks were prepared freshly for each experiment. The preparations were generally applied within two hours after weighing the test substances;

the test substance preparations were homogenous and stably dispersed by visual inspection unless otherwise noted for individual assays.

2.3. Test Concentrations

The stock dispersions of the test substance were prepared at 2.56 mg/mL as described above. The test guidelines prescribe either test substance concentrations with a defined molar concentration or a so-called "gravimetric approach" using mass per volume concentrations. Also, the constituents of NMs have a molar mass ($BaSO_4$ 233, CeO_2 172, SiO_2 60, TiO_2 80, and ZnO 81 g/mol), but at a given molar concentration, most of the test substance is not dissolved but only dispersed, i.e., packed inside the particles. Accordingly, the "gravimetric approach" could be applied by using an assumed average molecular weight of 200 g/mol for test substances without defined molar mass—but still assumed to be dissolved. Since this assumption does not hold, instead the relevant concentration metric of NM and other solid particles would be different; i.e., particle number per volume or particle surface area per volume have been proposed [37–39]. In addition, the nominal concentration of a NM in cell culture media (even for stable dispersions and using any concentration metric) may not be the effective concentration [40–42]. For other reasons, this is also true for dissolved molecules [43,44]. While the respective test guidelines define nominal maximum concentrations to be tested, the maximum concentration may be limited by cytotoxicity as described in the guidelines for test systems using living cells (LuSens and h-CLAT) and elsewhere [45–47]. Preparing dispersions of NM in cell culture media can greatly influence the results of in vitro tests [48]. Therefore, the dispersion protocols were standardized. The maximum concentration of a NM in a stable cell culture media dispersion is limited. We implemented one of the established protocols [35], which is using a maximum concentration of 2.56 mg/L. As a consequence, the maximum concentration of NM in the respective test system resulted from the dilution of this stock dispersion according to the respective protocols.

DPRA: Dilutions of test substances were prepared from the stock preparation (2.56 mg/mL) according to the standard pipetting procedure given in the DB-ALM protocol [49] (nominal, based on the constituents of the respective NM: 2.7 to 10.6 mM instead of stock solution concentration of 100 mM or 20 mg/mL for soluble organic test substances as required in OECD TG No. 442C) dilutions for C-peptide and K-peptide incubations, respectively, with 0.50 mM of the corresponding peptide.

LuSens: Dilutions of test substances were prepared from the stock preparation (2.56 mg/mL) with 640 μg/mL as the maximum concentration (while the TG requires a concentration of 2000 μM or 2000 μg/mL). Test substance concentrations were adjusted if cytotoxicity was observed (ZnO, CeO_2, both of the TiO_2 NM, SiO_2 Levasil 200, SiO_2 Aerosil 200, and SiO_2 DQ12). Tested concentrations are given in the Supplementary Materials. Consequently, only $BaSO_4$ and SiO_2 Aerosil R972 were tested at concentrations that were not cytotoxic and below 2000 μg/mL. Higher test concentrations were, however, not technically achievable with the concentration of 2.56 mg/mL of the stock dispersion, which was prepared according to the standardized protocol.

h-CLAT: The 1st experiment was always conducted with a maximum concentration of 1280 μg/mL, corresponding to the highest applicable dilution from the 2.56 mg/mL test substance stock preparation (according to the test guideline, the top final concentration is 1000 μg/mL). If this is non-cytotoxic, the maximum concentration should be re-determined. In any case, the final concentration in the plate should not exceed 5000 μg/mL for test chemicals dissolved or stably dispersed in saline or medium.

At the 1st experiment, ZnO, TiO_2 NIST® SRM®, TiO_2 Aeroxide P25, CeO_2 (all <50% viability), and $BaSO_4$ ($\leq$80% viability) were cytotoxic at the maximum concentration of 1280 μg/mL. The test concentrations were lowered accordingly in the following experiments. All SiO_2 test substances were not cytotoxic at the maximum concentration of 1280 μg/mL (viability $\geq$90%). The concentrations were, however, not increased in the following experiments as higher test concentrations were not achievable with the stock dispersion, which was prepared according to the standardized protocol. Cytotoxicity data and tested concentrations are given in the Supplementary Materials.

2.4. Direct Peptide Reactivity Assay (DPRA)

The DPRA experiments were performed following an adapted protocol based on OECD TG No. 442C. Following the stock preparations according to the above-described NANOGENOTOX protocol, the test substance dilutions were prepared (as triplicates) following the standard pipetting procedure (given in the DB-ALM protocol No. 154): (1) stock preparation was diluted 20-fold and incubated with 0.50 mM of the cysteine-containing peptide (C-peptide) (Ac-RFAACAA-COOH, 752 g/mol), and (2) stock preparation was diluted 1:4 and incubated with 0.50 mM of the lysine-containing peptide (K-peptide) (Ac-RFAAKAA-COOH, 776 g/mol) for 24 h at approx. 25 °C. The remaining non-depleted peptide concentration was determined by high-performance liquid chromatography (HPLC) with gradient elution and UV detection at 220 nm (corresponding to the absorption band of phenylalanine). In parallel, triplicates of the concurrent vehicle control and a positive control (PC) (50 mM ethylene glycol dimethacrylate (EGDMA)) were incubated with the peptides. Additionally, a co-elution control was carried out to detect possible interference of the test substance due to co-elution with the peptides via analysis of the ratio of peak areas at 220 nm and 258 nm. The mean depletion was evaluated based on TG No. 442C and GL No. 497 with criteria listed below (Table 2), defining negative, borderline, or positive reactivity.

Table 2. Evaluation criteria of DPRA, according to OECD TG No. 442C and GL No. 497, based on the C-peptide/test substance 1:10/K-peptide/test substance 1:50 prediction model (mean peptide depletion) and the C-peptide/test substance 1:10 prediction model (C-peptide-only depletion).

Mean Peptide Depletion [%]	C-Peptide Depletion [%] [1]	Reactivity	Evaluation
>42.47	>98.24	high reactivity	positive
>22.62; ≤42.47	>23.09; ≤98.24	moderate reactivity	positive
>6.38; ≤22.62	>13.89; ≤23.09	low reactivity	positive
>4.94; ≤8.32	>10.55; ≤18.47	borderline [2]	inconclusive
≤6.38	≤13.89	minimal or no reactivity	negative [3]

[1] If mean peptide depletion [%] could not be determined due to invalid K-peptide depletion (e.g., insolubility of the K-peptide samples or interference in the samples of the K-peptide), C-peptide depletion is considered for evaluation. [2] According to OECD GL No. 497, results in this range were considered borderline and evaluated as inconclusive due to their uncertainty. [3] For test substances that were not completely soluble by visual observation in the sample preparations containing the peptides immediately after preparation or after 24 h, the result may be under-predictive due to limited availability of the test substance. In this case, if the mean peptide depletion was below the borderline range, the reactivity was considered "inconclusive". Likewise, is a mean peptide depletion was above the borderline range, a test substance was considered "positive".

2.5. LuSens Assay

The LuSens assay was performed following an adapted protocol based on OECD No. TG No. 442D. The assay consisted of two parts: first, a cytotoxicity pre-experiment for determining the CV75 value (estimated concentration that affords 75% cell viability by linear regression) and a basis to determine the concentration range for the main experiment. If any interference between the test substance and MTT read-out was observed in the absorption spectrum at OD550nm, an alternative viability assay (CellTiter-Glo®, Promega GmbH, Walldorf, Germany) was performed to quantify ATP. Second, the main experiment consisted of measurement of the luciferase induction after 48 h exposure time (as triplicates) to the test substance with at least 8 concentrations for determining the keratinocyte activation. The maximum concentration tested in the main experiments was 640 µg/mL. For this, the cells were lyzed, and luciferase induction was evaluated by measuring luminescence signal after substrate addition (One Glo®, Promega). Cell viability was determined for each tested concentration with either an MTT or ATP assay. In parallel, triplicates of the concurrent vehicle control, a positive control (PC) (18 µg/mL ethylene glycol dimethacrylate (EGDMA)) and a negative control (450 µg/mL DL-lactic acid (LA)), were incubated with the cells. At least two independent experiments were performed for evaluation of the luciferase induction. Based on the OECD TG No. 442D and borderline ranges [32,50], mean fold induction in at least two independent experiments was evaluated with criteria listed below (Table 3) to conclude a negative, borderline, or positive keratinocyte activation potential.

Table 3. Evaluation criteria of LuSens assay according to OECD TG No. 442D and GL 497.

Mean Fold Induction	Reactivity	Evaluation
>1.76	activates keratinocytes	positive [1]
>1.28; ≤1.76	borderline [2]	inconclusive
≤1.28	does not activate keratinocytes	negative [3]

[1] If the luciferase induction was above 1.76-fold, statistically significant in at least 2 consecutive tested concentrations with a cell viability greater than 70%, and at least 3 test concentrations were non-cytotoxic, the outcome of the LuSens assay was considered positive. [2] Applying borderline range criteria for the LuSens assay [32,49], such as described in OECD GL 497, results in this range were considered borderline and evaluated as inconclusive due to their uncertainty. [3] A test substance was considered negative when the criteria mentioned above were not met up to the maximum concentration of 2000 µg/mL or up to the cytotoxicity limit (at least one concentration displaying viability below 70%).

2.6. *h-CLAT Assay*

The h-CLAT assay was performed following a modified protocol based on the OECD TG No. 442E with adaptations for NM implemented, as described above for the test substance preparation, and as follows. No pre-test for cytotoxicity assessment was performed. The 1st main experiment was always conducted with a maximum concentration of 1280 µg/mL, corresponding to the highest applicable concentration from the 2.56 mg/mL test substance stock preparation. For the main experiments, the human cell line THP-1 was treated with at least 8 testsubstance concentrations for 24 h. In parallel, the concurrent vehicle control, a positive control (PC) (4 µg/mL 1-chloro-2,4-dinitrobenzene (DNCB)), and a negative control (1000 µg/mL DL-LA) were incubated with the cells. To assess the matured cell population by flow cytometry density gradient centrifugation (on a Ficoll-Paque PLUS (GE Healthcare, 17-1440-02) with centrifugation at 1500 rpm for 15 min at room temperature, without break function in the deceleration phase) was performed to remove the test substance before applying to the flow cytometer to avoid interferences and damages to the instrument. Thereafter, cells were stained with PI along with FITC-labeled anti-human CD86 antibodies or FITC-labeled anti-human CD54 antibodies or the corresponding isotype control FITC labeled anti-mouse IgG1. The relative mean fluorescence (RFI) of CD86 and CD54 was evaluated by measuring the CD86 and CD54 signals by flow cytometric analysis. In parallel, the cytotoxicity of the test substance was measured using the fluorescence intensity of PI. At least two independent experiments were performed for evaluation of the dendritic cell activation, and test-substance concentrations were adjusted if evaluation criteria for viability (>50% in at least four tested concentrations) of an experiment were not met. The mean RFI induction in two independent experiments was evaluated based on OECD TG No. 442E and OECD GL 497, with criteria listed below (Table 4) defining a negative, borderline, or positive dendritic cell activation.

Table 4. Evaluation criteria of the h-CLAT assay according to OECD TG No. 442E and GL 497.

Mean RFI (CD86)	Mean RFI (CD54)	Reactivity	Evaluation
>184	>255	activates dendritic cells	positive [1]
>121; ≤184	>156; ≤255	borderline [2]	inconclusive
≤121	≤156	does not activate dendritic cells	negative [3]

[1] If the mean RFI value for CD86 was greater than 184 and/or the mean RFI value for CD54 is greater than 255 at any concentration in relation to vehicle control that does not reduce viability below 50% and reproduces in the same cell surface marker in at least an additional independent experiment, the h-CLAT assay was considered positive. [2] According to OECD GL 497, results in this range were considered borderline and evaluated as inconclusive due to their uncertainty. [3] A test substance was considered negative when the 'positive' criteria mentioned above were not met up to the maximum concentration of 5000 µg/mL for the vehicle culture medium or 1000 µg/mL for 0.2% DMSO in culture medium or up to the cytotoxicity limit (with viability less than 90% at the highest concentration tested). If these criteria were not met, no conclusion on dendritic could be derived under these test conditions.

2.7. Defined Approach

The results obtained in three NAMs were evaluated according to the AOP-based "2 out of 3" (2o3) and ITSv2 DAs for skin sensitization hazard identification described in OECD GL No. 497. In 2o3, any two of the three tests determine the overall results, i.e., if any two assays result positive, the overall test results yield the prediction of a sensitizer, while any two negative test results would yield the prediction of a test substance to be a non-sensitizer. To include experimental uncertainties for the predictivity of single assays, borderline ranges were considered for the evaluation of skin sensitization potential. In ITSv2, scores are assigned for the results of DPRA and h-CLAT experiments and for the results from the OECD QSAR Toolbox to yield an overall result that predicts the skin sensitization potency of chemicals: UN GHS category 1A (strong sensitizer), category 1B (other sensitizer), or not classified (non-sensitizer).

3. Results and Discussion

Nine particles, eight inorganic NM and DQ12, were tested in DPRA, LuSens, and h-CLAT assays. These are insoluble (as long as they are particles) and require modifications of the test protocols.

3.1. DPRA

The DPRA experiments were performed with a modified protocol based on OECD TG No. 442C, originally described by Gerberick et al. in 2004 [7]. The non-depleted peptide concentrations were determined by HPLC, and the depleted peptide concentrations evaluated (Table 5).

Table 5. Results and evaluation of the DPRA with test substances. Peptide depletion values for K-peptide and C-peptide, as well as the mean of both peptides, are given.

Test Substance	K-Peptide Depletion [%]	C-Peptide Depletion [%]	Mean Depletion [%] [1]	Result [2]
CeO_2 NM-212	0.60 ± 0.46	3.28 ± 0.94	1.94	negative [3]
$BaSO_4$ NM-220	0.00 ± 0.24	1.96 ± 0.82	0.98	negative [3]
SiO_2 Levasil 200 (40%)	0.25 ± 0.80	−0.36 ± 0.28	0.12	negative [3]
SiO_2 DQ12	−0.41 ± 0.97	−0.20 ± 0.64	0.00	negative [3]
SiO_2 Aerosil R972	−0.30 ± 0.63±	−2.00 ± 1.01	0.00	negative [3]
SiO_2 Aerosil 200	−0.55 ± 0.33	−0.88 ± 1.44	0.00	negative [3]
TiO_2 NIST® SRM®	−2.62 ± 0.84	−1.67 ± 2.67	0.00	negative [3]
TiO_2 P25 Aeroxide	−2.91 ± 0.39	1.16 ± 2.53	0.58	negative [3]
ZnO Z-Cote HP1	−3.09 ± 2.75	−6.76 ± 0.42	0.00	negative [3]

[1] Negative K- and C-peptide depletion values were set to 0.00 for the determination of the mean. [2] Mean peptide depletion values were evaluated according to the criteria mentioned in Table 2, regardless of chemical composition and solubility of the test substances and actual peptide—test substance ratios. [3] Considered "inconclusive" according to the test guideline as the test substance is inorganic and insoluble.

Mean peptide depletion was <2% for all nine test substances (at a stock concentration of 2.56 mg/mL). The tested stock concentration was below the required concentration (20 mg/mL in the gravimetric approach) and hence the ratio to the peptide according to TG442C [29]. The molar mass of the elemental composition of the selected NM was used to calculate the nominal concentrations, and the resulting test substance-to-peptide ratios were calculated based on the nominal concentrations.

Nevertheless, these "negative" or non-reactive results have to be considered "inconclusive" according to TG No. 442C [29], due to the following reasons: (1) Metals and inorganic substances should not be tested in DPRA, as they are "known to react with proteins via mechanisms other than covalent binding". (2) Insoluble substances cannot deliver firm conclusions on the lack of reactivity if no reactivity is observed, since it is "not sure if sufficient exposure can be achieved" and the effective concentration is unknown, in particular if sedimentation occurs despite prior sonification [51]. Studies on few inorganic materials have however demonstrated that indeed these materials can also yield positive results in the DPRA (compiled in [52]) or a modified version of the assay using LC-MS/MS detection (but adducts were not analyzed) testing multiple concentrations [53]. Further, Roberts et al. (2007) describe transition metals as being able to covalently bind proteins via covalent coordination bonds [54], and in a recent study with inorganic NM, positive results for C-peptide reactivity (and no reactivity for K-peptide) were reported for TiO_2 and CeO_2 nanoparticles [13,15]. The C-peptide, however, is prone to dimerization by oxidation, which would lead to peptide depletion without binding of the test substance, leading to an overestimation of reactivity, and metal ions had been described to cause dimerization of the thiol-group-containing nucleophile in the ADRA [55]. Therefore, positive results with thiol nucleophiles should be evaluated with caution, and the formation of reaction products of the peptides should be quantified to the depletion of the peptide. An approach to differentiate between dimers and reaction products has been proposed for the ADRA to identify false positives due to thiol nucleophile dimerization [56]. While no reactivity was observed for any of the NM assessed in the present study, adduct analysis and kinetics may clarify the applicability domain limitation stated in OECD TG No. 442C.

With adaptations, the DPRA was technically applicable to insoluble NM. According to the method's data interpretation procedure (DIP), it can only deliver conclusive results for non-dissolved test substances in the case of a peptide depletion above the defined threshold, whereas less peptide depletion is not assessed as being negative, but as inconclusive. To enable the investigation of inorganic test substances, which can react with proteins via mechanisms other than covalent binding, different analytical methods are required that can address these reactions and identify potential adducts (e.g., metal complex formation) or modified proteins in the corona of the NM.

3.2. LuSens

The LuSens experiments were performed following a modified protocol based on the OECD TG 442D, originally described by Ramirez et al. in 2014 [30]. For cytotoxicity measurements, MTT tests with a range of dilutions were performed. For CeO_2 NM-212, prior centrifugation was necessary, as large deviations were observed in MTT measurements. For SiO_2 DQ12 and TiO_2 NIST® SRM® ATP tests were conducted instead of MTT tests as an alternative cytotoxicity determination. After assessment of cytotoxicity, appropriate test substance concentrations were selected for incubations to assess luciferase induction. The resulting CV75 and EC 1.50 values were calculated, and keratinocyte activation was evaluated (Table 6).

Four NM yielded negative results, whereas two NM and DQ12 were tested positive. Three replicate experiments with $BaSO_4$ delivered inconsistent results: a negative and borderline result and an invalid experiment due to high cytotoxicity. Experiments with SiO_2 Aerosil R972 delivered borderline results; however, these results were considered inconclusive due to visible inhomogeneity and the test guideline requiring a homogenous test substance preparation. Overall, the LuSens assay was technically applicable with adaptations to the TG protocol.

Table 6. Results and evaluation of the LuSens assay with test substances. Values of the calculated test substance concentration with 75% cell viability (CV75), maximum fold-induction values (Imax), and the calculated concentration with 1.50-fold luciferase induction (EC 1.50) are given, if applicable or determined.

Test Substance	CV75 [μg/mL]	Imax	EC 1.50 [μg/mL]	Result
CeO_2 NM-212	532	0.71	n.a. [1]	negative
$BaSO_4$ NM-220	n.a. [1]	1.41	n.a. [1]	inconclusive [2]
Levasil 200 (40%)	61.5	2.77	33	positive
SiO_2 DQ12	498	3.32	<179 [3]	positive
SiO_2 Aerosil R972	n.a. [1]	1.81	n.a. [1]	inconclusive [2]
SiO_2 Aerosil 200	15	1.85	n.a. [1]	negative
TiO_2 NIST® SRM®	351	1.01	n.a. [1]	negative
TiO_2 P25 Aeroxide	222	0.89	n.a. [1]	negative
ZnO Z-Cote HP1	13.7	5.17	<7 [3]	positive

[1] n.a. = not applicable, no cytotoxic effects were observed in any of the concentrations, or the mean fold induction at the highest tested concentration (with a viability >70%) was lower than 1.50. [2] Considered "inconclusive" according to OECD TG442D as the test substance preparation was visually inhomogeneous (SiO_2 Aerosil R972) or tested below the maximum concentration without cytotoxicity ($BaSO_4$). [3] Not determined, the mean fold induction at the lowest tested concentration was higher than 1.50.

Despite the conclusive results according to TG442D, these results should be interpreted with caution, as the predictivity of the LuSens assay is based on the reaction of the test substance (or products formed from or by the test substance) with thiols of Keap-1, which is then activating the Keap1–Nrf2–ARE pathway [24]. The reactivity of solid, inorganic materials with these thiols is not well investigated. Interestingly, none of the NM exhibited any significant reactivity in the DPRA with the C-peptide containing a thiol group, while three test substances (SiO_2 Levasil 200 (40%), SiO_2 DQ12, and ZnO Z-Cote HP1) delivered clearly positive results in the LuSens assay. Two other SiO_2-based NM were tested negative and borderline; hence, no direct correlation of elemental composition could be driven, and the published data on SiO_2-based NM are also not consistent [17,57]. Recently, studies with NM in the KeratinoSens assay (addressing the same key event as the LuSens assay) were conducted. In one study, positive results for gold and silver nanoparticles were reported, whereas NiO and TiO_2 were tested negative [20]. In another study, CuO was tested positive, whereas CeO_2, TiO_2, ZnO, and other metal oxides (NM) were tested negative [16]. The negative results for the TiO_2-based NM and CeO_2 in our experiments agree with these studies' outcomes. For ZnO, DQ12, and SiO2 Levasil, our experiments indicated an activation of the Keap1–Nrf2–ARE pathway. This pathway is contributing to skin sensitization in the context of other KEs of the AOP. Independent of this, activation of this pathway indicates oxidative stress, which can indeed be caused by inorganic NM, either by catalyzing the formation of ROS or by releasing metal ions. Previous studies investigated the surface reactivity and ROS production of NM [23,25] by different methods (electron spin resonance (ESR) spectroscopy, ferric reducing ability of serum (FRAS), and protein carbonylation). The relative activity of one NM compared to another varied with different methods, and some nanoforms of a NM showed results different from another nanoform [58]. In several assays, DQ12, ZnO, and some silica showed higher activities (based on the same mass concentration) than $BaSO_4$, CeO_2, and TiO_2 NM. The LuSens assay may serve as another assay to assess the surface reactivity and ROS production of NM, which is one criterion to group NM [59,60] and define sets of similar nanoforms [58,61]. This approach has already been explored [23].

3.3. h-CLAT

The h-CLAT experiments were conducted following a modified protocol based on the OECD TG No. 442E, without cytotoxicity measurements prior to the main experiment. Hence, the first main experiment was always conducted with the highest achievable concentration (diluted from the stock preparation), and the concentration was adjusted after assessment of cytotoxicity. After 24-h exposure time, density gradient centrifugation was performed to separate NM from cells prior to flow cytometry to avoid interferences. To investigate the influence of this step, experiments with and without centrifugation were conducted with SiO_2 Levasil 200 (40%). No apparent influence of centrifugation on cell viability was observed while particles were removed. Cells were stained with the corresponding fluorescently labeled antibodies. The relative mean fluorescence (RFI) of CD86 and CD54 were measured, which were then used to calculate EC150 and EC200 values for CD86 and CD54, respectively, and evaluated to assess the dendritic cell activation potential of NM (Table 7).

Table 7. Results and evaluation of the h-CLAT assay with test substances. Values of maximum fold-induction (Imax) and the calculated test substance concentrations resulting in a RFI of 150% for CD86 or 200% for CD54 (EC150 and EC200) are given, if applicable or determined.

Test Substance	EC150 (CD86)	Imax (CD86)	EC200 (CD54)	Imax (CD54)	Evaluation
CeO_2 NM-212	n.a.	140	n.a.	100	negative/borderline for CD86 in one run [1]
$BaSO_4$ NM-220	n.a.	121	n.a.	97	negative
SiO_2 Levasil 200 (40%)	n.a.	87	n.a.	201	inconclusive [2]
SiO_2 DQ12	n.a.	176	<144 µg/mL	970	positive
SiO_2 Aerosil R972	n.a.	138	n.a.	120	inconclusive [2]
SiO_2 Aerosil 200	n.a.	171	n.a.	235	inconclusive [3]
TiO_2 NIST® SRM®	n.a.	128	n.a.	132	negative/borderline for CD86 in one run [1]
TiO_2 P25 Aeroxide	<357 µg/mL	213	n.a.	152	positive
ZnO Z-Cote HP1	n.a.	163	<144 µg/mL	2585	positive

[1] Results of valid experiments yielded negative (without borderline range) and negative or borderline if the borderline ranges provided in GL 497 are considered. Hence, at least one more independent experiment would be needed to derive a conclusive prediction in the h-CLAT. [2] Considered "inconclusive" according to the test guideline as the maximum concentration was below the maximum stated in the guideline and the actually tested maximum concentration was not cytotoxic. [3] Without considering the borderline range, the first run was positive for CD86, the second negative for both, and the third run was positive for CD54. Hence, at least one more independent experiment would be needed to derive a conclusive prediction in the h-CLAT. n.a. = not applicable.

Two NM and DQ12 yielded clearly positive results for one of the CD markers, whereas only one NM, $BaSO_4$, was tested negative for the upregulation of both surface markers. Most of the NM yielded negative or borderline results for one of the surface markers. According to DIP of the TG, the tests should be performed up to a defined concentration (2 mg/mL) or up to the cytotoxicity limit. However, application at this concentration was not possible following the NANOGENETOX protocol. Hence, all NM should have been tested at a concentration limited by their cytotoxicity. Three NM, SiO_2 Aerosil 200, SiO_2 Aerosil R972, and SiO_2 Levasil 200, yielded negative or borderline results for CD54; however, in these experiments, the cytotoxicity limit was not reached; thus, the results are considered inconclusive. A previous study with silica NM investigated with LuSens and h-CLAT assays found no activity in either assay and not cytotoxicity even at the highest concentrations [17]. In another assay adopted as part of OECD TG 442E, namely the GARDskin assay, 13 metal compounds were tested, and the others concluded a similar predictive capacity as compared to defined organic molecules [62].

In conclusion, the h-CLAT assay was technically applicable with the addition of the centrifugation step to the TG protocol. The significance of negative or borderline results obtained with this assay is, however, limited by the achievable maximum concentration with a given dispersion protocol. Moreover, the biological relevance of this assay towards the potential skin sensitization of NM is questionable.

3.4. Assessment of the Skin Sensitization Potential with DAs

The results obtained in three NAMs (according to OECD TG No. 442C, D, and E), addressing three KE in skin sensitization AOP, were evaluated according to the 2o3 and DA for skin sensitization described in OECD GL No. 497. First, all results obtained in the three assays were considered in the evaluation (Table 8), even though some are inconclusive according to the TG since the test substances are inorganic materials, are not completely soluble, and/or sufficient test concentrations of dispersed particles could not be achieved by standard protocols (please see footnotes for details).

Table 8. Evaluation results and ITSv2 score of NAMs, as well as evaluation of the 2o3 DA with DPRA, LuSens, and h-CLAT assays for assessing the skin sensitization potential of test substances. Information from the OECD Toolbox is not available [1].

Test Substance	DPRA [1]	LuSens [2]	h-CLAT	Evaluation 2o3 [5]
CeO_2 NM-212	n [1] inconclusive	n	n/br inconclusive	non sensitiser [5] inconclusive
$BaSO_4$ NM-220	n [1] inconclusive	inconclusive [2]	n	non sensitiser [5] inconclusive
SiO_2 Levasil 200 (40%)	n [1] inconclusive	p	inconclusive [3]	inconclusive
SiO_2 DQ12	n [1] inconclusive	p	p	sensitiser
SiO_2 Aerosil R972	n [1] inconclusive	inconclusive [2]	inconclusive [3]	inconclusive
SiO_2 Aerosil 200	n [1] inconclusive	n	inconclusive [4]	non sensitiser [5] inconclusive
TiO_2 NIST® SRM®	n [1] inconclusive	n	n/br inconclusive	non sensitiser [5] inconclusive
TiO_2 P25 Aeroxide	n [1] inconclusive	n	p	non sensitiser [5] inconclusive
ZnO Z-Cote HP1	n [1] inconclusive	p	p	sensitiser

[1] According to TG 442C, inorganic materials are out of the applicability domain of DPRA, and the insolubility of the NM makes the negative results inconclusive. [2] Considered "inconclusive" according to OECD TG 442D as the test substance preparation was visually inhomogeneous (SiO_2 Aerosil R972) or tested below the maximum concentration without cytotoxicity ($BaSO_4$). [3] Considered "inconclusive" according to the test guideline as the maximum concentration was below the maximum stated in the guideline and the actually tested maximum concentration was not cytotoxic. [4] Without considering the borderline range, the first run was positive for CD86, the second negative for both, and the third run was positive for CD54. Hence, at least one more independent experiment would be needed to derive a conclusive prediction in the h-CLAT. Considering the borderline range, the first run was borderline in both markers, and the following two runs were negative in CD86 and borderline in CD54. [5] Where two predictions are provided, the first evaluation disregards inconclusive results due to borderline evaluation in individual assays and limitations in the DPRA (see footnote 1 to this table). Abbreviations: n = negative; p = positive; br = borderline.

In the 2o3 DA, of the nine materials tested, two NM were evaluated inconclusive due to solubility/inhomogeneity issues (SiO_2 Levasil 200 (40%) and SiO_2 Aerosil R972), and two (SiO_2 DQ12 and ZnO Z-Cote HP1) were identified as skin sensitizers. The remaining five NM possibly have no skin sensitization potential; however, the final evaluation has to be considered inconclusive because (i) negative results of the DPRA with inorganic and insoluble test substances are inconclusive, and (ii) LuSens and h-CLAT assays partly delivered inconclusive results due to borderline results and inconsistencies between single experiments.

OECD GL497 also provides DAs comprising the DPRA, h-CLAT, and the OECD toolbox as information sources in ITSv2. Using the present dataset, evaluation according to ITSv2 DA was challenging due to negative/inconclusive results in the experimental information sources and the unavailability of protein binding alerts from the OECD toolbox. Two NM could be evaluated with ITSv2. They are assessed as being skin sensitizers by this DA. Both DAs consistently indicate skin sensitization potential for DQ12 and ZnO NM. Both test substances tested negative in animal studies, and ZnO was also found negative in humans [63,64]. However, both materials have substantial ROS-production capabilities, and the results obtained with NAMs may rather reflect this activity than specific skin sensitization potential [23,25]. According to in vivo skin sensitization studies (TG No. 442B: Skin Sensitization Local Lymph Node Assay, LLNA: BrdU-ELISA), TiO_2, CeO_2, SiO_2, and ZnO NM do not induce skin sensitization in the mice [17,19,65]. LLNA data of small organic molecules are, however, not well correlated to skin sensitization in humans, and in vitro methods should be assessed towards human relevance [66]. Moreover, with the LLNA, test substances are applied to the intact skin, and the negative results could merely reflect the absence of sufficient skin penetration. Bypassing the skin barrier for a more conservative hazard assessment would require a Guinea Pig Maximization Test using an adjuvant to increase skin penetration [67]. This test has been performed for CeO_2 NM-212, SiO_2 Levasil 200, materials related to SiO_2 R972, and TiO_2 P25; all four NM were found negative (see Table 1 for references).

A principal mechanism by which solid NM could cause skin sensitization is via the release of ions (or other soluble substances) with a skin sensitization potential. None of the ions constituting the NM tested in this study are classified as skin sensitizers. However, dissolution of skin sensitizing constituents from respective NM (e.g., nickel nanoparticles) could give relevant info towards a potential for skin sensitization. The solubility can give first-tier information on the expected concentration of released ions; the solubility values of the nanoparticles in this study are listed in Table 1. More refined methods to measure dissolution kinetics of NM are available [68].

4. Conclusions

NM could technically be applied to skin sensitization NAMs (DPRA, LuSens, and h-CLAT assays) with some adaptations. The dispersion of the NM is crucial. In this study, NM dispersions were prepared using a standardized protocol to obtain highly dispersed NM, which were applied to the test systems directly after preparation. Some of the TG are using molar concentrations that are not defined for NM. Gravimetric approaches in the TG are based on mixtures and formulations consisting of small molecules with the assumption of an average molecular weight of 200 g/mol, thus not directly applicable for larger structures such as solid particles. An adequate concentration metric for NM testing in skin sensitization NAMs needs to be defined.

None of the NM-depleted peptides in DPRA. Nevertheless, all results had to be considered inconclusive due to the insolubility of the NM and because inorganic materials are out of the applicability domain of the DPRA.

In cell-based assays, further adaptations were required due to interference with the read-out. Centrifugation steps were performed to separate NM from cells prior to optical read-out. Some NM interfered with the MTT assay, and ATP assays were used as an alternative for viability measurement; PrestoBlue assays could be another alternative [69]. Three test substances were positive in the LuSens assay. These test substances showed a higher ROS production in other assays, and the LuSens assay may well be suitable to detect NM-induced ROS formation in mammalian cells. Three test substances were positive in the h-CLAT assay for dendritic cell activation. Two of them, ZnO and DQ12, were also positive in the LuSens assay. Consequently, these two substances were assessed as overall positive according to the DA. They were, however, not showing skin sensitization potential in animal studies nor—for ZnO—in humans. For seven NM, no consistent overall assessment could be derived from the three NAM tests. All are inorganic and insoluble and hence outside the applicability domain of the DPRA. Four of them did not form stable dispersions or did not achieve the maximum concentrations defined in the TG of the LuSens or h-CLAT. Three NM gave borderline results in the h-CLAT or non-consistent results in the LuSens and h-CLAT. For all NM, the current NAM TG does not define adequate concentration metrics. Moreover, the relevance of the tested KE towards skin sensitization by inorganic, insoluble matter is unknown. Relevant mechanisms of NM skin sensitization should be identified, and respective NAMs (with adequate concentration metrics) should be developed and validated for reliability and human relevance (ideally using human reference data). The current NAMs, which were developed for small organic molecules, seem not to be a good fit for NM. The currently available data from humans, animal studies, and in vitro do not indicate a skin sensitization potential for inorganic NM without skin sensitizing constituents.

Supplementary Materials: The following supporting information can be downloaded at: https://www.mdpi.com/article/10.3390/toxics12080616/s1.

Author Contributions: Conceptualization: R.L. and A.A.H.; methodology: B.B. and S.N.K.; formal analysis: B.W.; investigation, B.W., B.B. and S.N.K.; data curation: B.W., S.N.K. and A.A.H.; writing—original draft preparation: A.A.H.; writing—review and editing: B.W., S.N.K. and R.L.; visualization: A.A.H.; supervision: R.L. and D.F.-W.; funding acquisition: R.L. All authors have read and agreed to the published version of the manuscript.

Funding: This research was funded by BMBF project Aerosafe (031L0128C).

Institutional Review Board Statement: Not applicable.

Informed Consent Statement: Not applicable.

Data Availability Statement: Additional details are provided in the Supplementary Materials.

Acknowledgments: The authors thank Carina Oberfrank, Natascha Partosa, and Jutta Steinbrenner for expertly performing the DPRA, LuSens, and h-CLAT assays and Wendel Wohlleben for the nanomaterial characterization and comments to the manuscript.

Conflicts of Interest: All authors are employees of BASF, a chemical company that produces and markets nanomaterials. B.W., S.N.K. and R.L. developed the LuSens assay and the "2 out of 3" approach.

References

1. Khan, I.; Saeed, K.; Khan, I. Nanoparticles: Properties, applications and toxicities. *Arab. J. Chem.* **2019**, *12*, 908–931.
2. Jagiello, K.; Sosnowska, A.; Stępnik, M.; Gromelski, M.; Płonka, K. *Nano-Specific Alternative Methods in Human Hazard/Safety Assessment under Different EU Regulations, Considering the Animal Testing Bans Already in Place for Cosmetics and Their Ingredients*; QSAR Lab Ltd.: Gdańsk, Poland, 2023.
3. EFSA Scientific Committee; More, S.; Bampidis, V.; Benford, D.; Bragard, C.; Halldorsson, T.; Hernández-Jerez, A.; Hougaard Bennekou, S.; Koutsoumanis, K.; Lambré, C.; et al. Guidance on risk assessment of nanomaterials to be applied in the food and feed chain: Human and animal health. *EFSA J.* **2021**, *19*, e06768.
4. Gomes, S.I.; Scott-Fordsmand, J.J.; Amorim, M.J. Alternative test methods for (nano)materials hazards assessment: Challenges and recommendations for regulatory preparedness. *Nano Today* **2021**, *40*, 101242. [CrossRef]
5. Di Cristo, L.; Janer, G.; Dekkers, S.; Boyles, M.; Giusti, A.; Keller, J.G.; Wohlleben, W.; Braakhuis, H.; Ma-Hock, L.; Oomen, A.G.; et al. Integrated approaches to testing and assessment for grouping nanomaterials following dermal exposure. *Nanotoxicology* **2022**, *16*, 310–332.
6. OECD. *Guideline No. 497: Defined Approaches on Skin Sensitisation*; OECD: Paris, France, 2023.
7. Gerberick, G.F.; Vassallo, J.D.; Bailey, R.E.; Morrall, S.W.; Lepoittevin, J.-P. Development of a peptide reactivity assay for screening contact allergens. *Toxicol. Sci.* **2004**, *81*, 332–343. [CrossRef] [PubMed]
8. Dinkova-Kostova, A.T.; Holtzclaw, W.D.; Wakabayashi, N. Keap1, the sensor for electrophiles and oxidants that regulates the phase 2 response, is a zinc metalloprotein. *Biochemistry* **2005**, *44*, 6889–6899. [CrossRef] [PubMed]
9. Gautam, A.; Singh, D.; Vijayaraghavan, R. Dermal Exposure of Nanoparticles: An Understanding. *J. Cell Tissue Res.* **2011**, *11*, 2703–2708.
10. Ashikaga, T.; Yoshida, Y.; Hirota, M.; Yoneyama, K.; Itagaki, H.; Sakaguchi, H.; Miyazawa, M.; Ito, Y.; Suzuki, H.; Toyoda, H. Development of an in vitro skin sensitization test using human cell lines: The human Cell Line Activation Test (h-CLAT): I. Optimization of the h-CLAT protocol. *Toxicol. In Vitro* **2006**, *20*, 767–773. [CrossRef] [PubMed]
11. Hofer, S.; Hofstätter, N.; Punz, B.; Hasenkopf, I.; Johnson, L.; Himly, M. Immunotoxicity of nanomaterials in health and disease: Current challenges and emerging approaches for identifying immune modifiers in susceptible populations. *WIREs Nanomed. Nanobiotechnology* **2022**, *14*, e1804. [CrossRef]
12. Roach, K.A.; Stefaniak, A.B.; Roberts, J.R. Metal nanomaterials: Immune effects and implications of physicochemical properties on sensitization, elicitation, and exacerbation of allergic disease. *J. Immunotoxicol.* **2019**, *16*, 87–124. [CrossRef]

13. Bezerra, S.F.; Rodrigues, B.d.S.; da Silva, A.C.G.; de Ávila, R.I.; Brito, H.R.G.; Cintra, E.R.; Veloso, D.F.M.C.; Lima, E.M.; Valadares, M.C. Application of the adverse outcome pathway framework for investigating skin sensitization potential of nanomaterials using new approach methods. *Contact Dermat.* **2021**, *84*, 67–74. [CrossRef]
14. Gautam, R.; Yang, S.; Maharjan, A.; Jo, J.; Acharya, M.; Heo, Y.; Kim, C. Prediction of Skin Sensitization Potential of Silver and Zinc Oxide Nanoparticles Through the Human Cell Line Activation Test. *Front. Toxicol.* **2021**, *3*, 649666. [CrossRef]
15. Kim, E.-N.; Seo, J.-A.; Kim, B.-H.; Jeong, G.-S. Defining the reactivity of nanoparticles to peptides through direct peptide reactivity assay (DPRA) using a high pressure liquid chromatography system with a diode array detector. *Toxicol. Res.* **2023**, *39*, 485–495. [CrossRef] [PubMed]
16. Kim, S.-H.; Lee, D.; Lee, J.; Yang, J.-Y.; Seok, J.; Jung, K.; Lee, J. Evaluation of the skin sensitization potential of metal oxide nanoparticles using the ARE-Nrf2 Luciferase KeratinoSensTM assay. *Toxicol. Res.* **2021**, *37*, 277–284. [CrossRef] [PubMed]
17. Kim, S.-H.; Lee, D.H.; Choi, S.; Yang, J.-Y.; Jung, K.; Jeong, J.; Oh, J.H.; Lee, J.H. Skin Sensitization Potential and Cellular ROS-Induced Cytotoxicity of Silica Nanoparticles. *Nanomaterials* **2021**, *11*, 2140. [CrossRef]
18. Lee, D.H.; Kim, S.-H.; Lee, J.H.; Yang, J.-Y.; Seok, J.-H.; Jung, K.; Lee, J.K. Flow cytometric evaluation of the potential of metal oxide nanoparticles for skin sensitization using 5-Bromo-2-deoxyuridine. *Toxicol. Res.* **2021**, *37*, 369–377. [CrossRef]
19. Maharjan, A.; Gautam, R.; Lee, G.; Kim, D.; Lee, D.; Acharya, M.; Kim, H.; Heo, Y.; Kim, C. Assessment of skin sensitization potential of zinc oxide, aluminum oxide, manganese oxide, and copper oxide nanoparticles through the local lymph node assay: 5-bromo-deoxyuridine flow cytometry method. *J. Toxicol. Environ. Health Part A* **2024**, 1–11. [CrossRef]
20. OECD. *Guidance Document No. 382: Study Report on Applicability of the Key Event-Based TG 442D for In Vitro Skin Sensitisation Testing of Nano-Materials, in Series on Testing and Assessment No. 382*; OECD: Paris, France, 2023.
21. Monteiro-Riviere, N.A.; Wiench, K.; Landsiedel, R.; Schulte, S.; Inman, A.O.; Riviere, J.E. Safety Evaluation of Sunscreen Formulations Containing Titanium Dioxide and Zinc Oxide Nanoparticles in UVB Sunburned Skin: An In Vitro and In Vivo Study. *Toxicol. Sci.* **2011**, *123*, 264–280. [CrossRef]
22. Monteiro-Riviere, N.A. Safety of Nanoparticle Skin Penetration. In *Percutaneous Penetration Enhancers Chemical Methods in Penetration Enhancement: Nanocarriers*; Dragicevic, N., Maibach, H.I., Eds.; Springer: Berlin/Heidelberg, Germany, 2016; pp. 363–376.
23. Seleci, D.A.; Tsiliki, G.; Werle, K.; Elam, D.A.; Okpowe, O.; Seidel, K.; Bi, X.; Westerhoff, P.; Innes, E.; Boyles, M.; et al. Determining nanoform similarity via assessment of surface reactivity by abiotic and in vitro assays. *NanoImpact* **2022**, *26*, 100390. [CrossRef]
24. Aleksic, M.; Meng, X. Protein Haptenation and Its Role in Allergy. *Chem. Res. Toxicol.* **2024**, *37*, 850–872. [CrossRef]
25. Bahl, A.; Hellack, B.; Wiemann, M.; Giusti, A.; Werle, K.; Haase, A.; Wohlleben, W. Nanomaterial categorization by surface reactivity: A case study comparing 35 materials with four different test methods. *NanoImpact* **2020**, *19*, 100234. [CrossRef]
26. Boraschi, D.; Alijagic, A.; Auguste, M.; Barbero, F.; Ferrari, E.; Hernadi, S.; Mayall, C.; Michelini, S.; Pacheco, N.I.N.; Prinelli, A.; et al. Addressing Nanomaterial Immunosafety by Evaluating Innate Immunity across Living Species. *Small* **2020**, *16*, e2000598. [CrossRef] [PubMed]
27. Cao, M.; Chen, C. Bioavailability of nanomaterials: Bridging the gap between nanostructures and their bioactivity. *Natl. Sci. Rev.* **2022**, *9*, nwac119. [CrossRef] [PubMed]
28. Tirumala, M.G.; Anchi, P.; Raja, S.; Rachamalla, M.; Godugu, C. Novel Methods and Approaches for Safety Evaluation of Nanoparticle Formulations: A Focus Towards In Vitro Models and Adverse Outcome Pathways. *Front. Pharmacol.* **2021**, *12*, 612659. [CrossRef]
29. OECD. *Test No. 442C: In Chemico Skin Sensitisation*; OECD: Paris, France, 2024.
30. OECD. *Test No. 442D: In Vitro Skin Sensitisation*; OECD: Paris, France, 2024.
31. OECD. *Test No. 442E: In Vitro Skin Sensitisation*; OECD: Paris, France, 2024.
32. Kolle, S.N.; Mathea, M.; Natsch, A.; Landsiedel, R. Assessing Experimental Uncertainty in Defined Approaches: Borderline Ranges for *In Chemico* and *In Vitro* Skin Sensitization Methods Determined from Ring Trial Data. *Appl. In Vitro Toxicol.* **2021**, *7*, 102–111. [CrossRef]
33. Singh, C.; Friedrichs, S.; Ceccone, G.; Gibson, P.; Jensen, K.; Levin, M.; Goenaga Infante, H.; Carlander, D.; Rasmussen, K. NM-211, NM-212, NM-213. Characterisation and test item preparation. *European Union Joint Research Centre, JRC89825*. 2014. Available online: https://policycommons.net/artifacts/2163300/cerium-dioxide-nm-211-nm-212-nm-213/2918811/ (accessed on 17 July 2024).
34. Yin, H.; Coleman, V.A.; Casey, P.S.; Angel, B.; Catchpoole, H.J.; Waddington, L.; McCall, M.J. A comparative study of the physical and chemical properties of nano-sized ZnO particles from multiple batches of three commercial products. *J. Nanoparticle Res.* **2015**, *17*, 96. [CrossRef]
35. Alstrup Jensen, K.; Kembouche, Y.; Christiansen, E.; Jacobsen, N.; Wallin, H.; Guiot, C.; Spalla, O.; Witschger, O. Final Protocol for Producing Suitable Manufactured Nanomaterial Exposure Media. 2011. Available online: https://www.anses.fr/en/system/files/nanogenotox_deliverable_6.pdf (accessed on 17 July 2024).
36. Alstrup Jensen, K.; Booth, A.; Kembouche, Y.; Loeschner, K.; Boraschi, D. NANoREG D 2.06 SOP for Probe-Sonicator Calibration of Delivered Acoustic Power and De-Agglomeration Efficiency for In Vitro and In Vivo Toxicological Testing; 2018. Available online: https://www.rivm.nl/ (accessed on 20 August 2024).
37. Oberdörster, G.; Oberdörster, E.; Oberdörster, J. Concepts of Nanoparticle Dose Metric and Response Metric. *Environ. Health Perspect.* **2007**, *115*, A290.

38. Rushton, E.K.; Jiang, J.; Leonard, S.S.; Eberly, S.; Castranova, V.; Biswas, P.; Elder, A.; Han, X.; Gelein, R.; Finkelstein, J.; et al. Concept of Assessing Nanoparticle Hazards Considering Nanoparticle Dosemetric and Chemical/Biological Response Metrics. *J. Toxicol. Environ. Health Part A* **2010**, *73*, 445–461. [CrossRef]
39. Ma-Hock, L.; Sauer, U.G.; Ruggiero, E.; Keller, J.; Wohlleben, W.; Landsiedel, R. The Use of Nanomaterial In Vivo Organ Burden Data for In Vitro Dose Setting. *Small* **2021**, *17*, 2005725. [CrossRef]
40. DeLoid, G.; Cohen, J.M.; Darrah, T.; Derk, R.; Rojanasakul, L.; Pyrgiotakis, G.; Wohlleben, W.; Demokritou, P. Estimating the effective density of engineered nanomaterials for in vitro dosimetry. *Nat. Commun.* **2014**, *5*, 3514. [CrossRef]
41. Keller, J.G.; Quevedo, D.F.; Faccani, L.; Costa, A.L.; Landsiedel, R.; Werle, K.; Wohlleben, W. Dosimetry in vitro—Exploring the sensitivity of deposited dose predictions vs. affinity, polydispersity, freeze-thawing, and analytical methods. *Nanotoxicology* **2021**, *15*, 21–34.
42. Cohen, J.M.; Teeguarden, J.G.; Demokritou, P. An integrated approach for the in vitro dosimetry of engineered nanomaterials. *Part. Fibre Toxicol.* **2014**, *11*, 20. [CrossRef]
43. Groothuis, F.A.; Heringa, M.B.; Nicol, B.; Hermens, J.L.; Blaauboer, B.J.; Kramer, N.I. Dose metric considerations in in vitro assays to improve quantitative in vitro–in vivo dose extrapolations. *Toxicology* **2013**, *332*, 30–40. [CrossRef] [PubMed]
44. Dimitrijevic, D.; Fabian, E.; Nicol, B.; Funk-Weyer, D.; Landsiedel, R. Toward Realistic Dosimetry In Vitro: Determining Effective Concentrations of Test Substances in Cell Culture and Their Prediction by an In Silico Mass Balance Model. *Chem. Res. Toxicol.* **2022**, *35*, 1962–1973. [CrossRef]
45. Soenen, S.J.; Rivera-Gil, P.; Montenegro, J.-M.; Parak, W.J.; De Smedt, S.C.; Braeckmans, K. Cellular toxicity of inorganic nanoparticles: Common aspects and guidelines for improved nanotoxicity evaluation. *Nano Today* **2011**, *6*, 446–465. [CrossRef]
46. Wohlleben, W.; Kolle, S.N.; Hasenkamp, L.-C.; Böser, A.; Vogel, S.; von Vacano, B.; van Ravenzwaay, B.; Landsiedel, R. Artifacts by marker enzyme adsorption on nanomaterials in cytotoxicity assays with tissue cultures. *J. Phys. Conf. Ser.* **2011**, *304*, 012061. [CrossRef]
47. Landsiedel, R.; Birk, B.; Demuth, P.; Fabian, E.; Hewitt, N.J.; Hollnagel, H.M.; Scheel, J. The Use of Toxicokinetic Information for Setting Concentrations of In Vitro Toxicity Tests and for Interpreting Their Results: A Proposed Workflow. *Appl. In Vitro Toxicol.* **2024**, *10*, 15–26. [CrossRef]
48. Schulze, C.; Kroll, A.; Lehr, C.-M.; Schäfer, U.F.; Becker, K.; Schnekenburger, J.; Schulze-Isfort, C.; Landsiedel, R.; Wohlleben, W. Not ready to use—Overcoming pitfalls when dispersing nanoparticles in physiological media. *Nanotoxicology* **2008**, *2*, 51–61. [CrossRef]
49. European Commission, Joint Research Centre EURL ECVAM dataset on alternative methods to animal experimentation (DB-ALM). *European Commission Joint Research Centre (JRC)*; 2019. [Dataset] PID. Available online: http://data.europa.eu/89h/b7597ada-148d-4560-9079-ab0a5539cad3 (accessed on 12 July 2024).
50. Gabbert, S.; Mathea, M.; Kolle, S.N.; Landsiedel, R. Accounting for Precision Uncertainty of Toxicity Testing: Methods to Define Borderline Ranges and Implications for Hazard Assessment of Chemicals. *Risk Anal.* **2022**, *42*, 224–238. [CrossRef]
51. Pradhan, S.; Hedberg, J.; Blomberg, E.; Wold, S.; Wallinder, I.O. Effect of sonication on particle dispersion, administered dose and metal release of non-functionalized, non-inert metal nanoparticles. *J. Nanoparticle Res.* **2016**, *18*, 285. [CrossRef]
52. Urbisch, D.; Mehling, A.; Guth, K.; Ramirez, T.; Honarvar, N.; Kolle, S.; Landsiedel, R.; Jaworska, J.; Kern, P.S.; Gerberick, F.; et al. Assessing skin sensitization hazard in mice and men using non-animal test methods. *Regul. Toxicol. Pharmacol.* **2015**, *71*, 337–351. [CrossRef]
53. Hemming, J.D.C.; Hosford, M.; Shafer, M.M. Application of the direct peptide reactivity assay (DPRA) to inorganic compounds: A case study of platinum species. *Toxicol. Res.* **2019**, *8*, 802–814. [CrossRef]
54. Roberts, D.W.; Patlewicz, G.; Kern, P.S.; Gerberick, F.; Kimber, I.; Dearman, R.J.; Ryan, C.A.; Basketter, D.A.; Aptula, A.O. Mechanistic applicability domain classification of a local lymph node assay dataset for skin sensitization. *Chem. Res. Toxicol.* **2007**, *20*, 1019–1030. [CrossRef]
55. Fujita, M.; Yamamoto, Y.; Watanabe, S.; Sugawara, T.; Wakabayashi, K.; Tahara, Y.; Horie, N.; Fujimoto, K.; Kusakari, K.; Kurokawa, Y.; et al. Cause of and countermeasures for oxidation of the cysteine-derived reagent used in the amino acid derivative reactivity assay. *J. Appl. Toxicol.* **2019**, *39*, 191–208. [CrossRef] [PubMed]
56. Choudhury, R.P.; Singh, A.; Mathai, E.; Sudhakar, D.; Tourneix, F.; Alépée, N.; Gautier, F. The dimer effect: A refinement approach towards skin sensitization assessment *in-chemico* using Amino acid Derivative Reactivity Assay. *J. Appl. Toxicol.* 2024; Epub ahead of print. [CrossRef]
57. McLean, P.; Marshall, J.; García-Bilbao, A.; Beal, D.; Katsumiti, A.; Carrière, M.; Boyles, M.S. A comparison of dermal toxicity models; assessing suitability for safe(r)-by-design decision-making and for screening nanomaterial hazards. *Toxicol. In Vitro* **2024**, *97*, 105792. [CrossRef]
58. Janer, G.; Ag-Seleci, D.; Sergent, J.A.; Landsiedel, R.; Wohlleben, W. Creating sets of similar nanoforms with the ECETOC NanoApp: Real-life case studies. *Nanotoxicology* **2021**, *15*, 1016–1034.
59. Arts, J.H.E.; Hadi, M.; Irfan, M.-A.; Keene, A.M.; Kreiling, R.; Lyon, D.; Maier, M.; Michel, K.; Petry, T.; Sauer, U.G.; et al. A decision-making framework for the grouping and testing of nanomaterials (DF4nanoGrouping). *Regul. Toxicol. Pharmacol.* **2015**, *71*, S1–S27. [CrossRef] [PubMed]

60. Wohlleben, W.; Hellack, B.; Nickel, C.; Herrchen, M.; Hund-Rinke, K.; Kettler, K.; Riebeling, C.; Haase, A.; Funk, B.; Kühnel, D.; et al. The nanoGRAVUR framework to group (nano)materials for their occupational, consumer, environmental risks based on a harmonized set of material properties, applied to 34 case studies. *Nanoscale* **2019**, *11*, 17637–17654. [CrossRef] [PubMed]
61. Forreryd, A.; Gradin, R.; Larne, O.; Rajapakse, N.; Deag, E.; Johansson, H. The GARD™ skin assay: Investigation of the applicability domain for metals. *ALTEX* **2023**, *40*, 425–438.
62. Loosli, F.; Rasmussen, K.; Rauscher, H.; Cross, R.K.; Bossa, N.; Peijnenburg, W.; Arts, J.; Matzke, M.; Svendsen, C.; Spurgeon, D.; et al. Refinement of the selection of physicochemical properties for grouping and read-across of nanoforms. *NanoImpact* **2022**, *25*, 100375. [CrossRef] [PubMed]
63. Kim, K.-B.; Kim, Y.W.; Lim, S.K.; Roh, T.H.; Bang, D.Y.; Choi, S.M.; Lim, D.S.; Kim, Y.J.; Baek, S.-H.; Kim, M.-K.; et al. Risk assessment of zinc oxide, a cosmetic ingredient used as a UV filter of sunscreens. *J. Toxicol. Environ. Health Part B* **2017**, *20*, 155–182. [CrossRef]
64. ECHA, REACH Registration Dossier. Reg. in 2010. Available online: https://echa.europa.eu/information-on-chemicals/registered-substances (accessed on 15 July 2024).
65. Park, Y.-H.; Jeong, S.H.; Yi, S.M.; Choi, B.H.; Kim, Y.-R.; Kim, I.-K.; Kim, M.-K.; Son, S.W. Analysis for the potential of polystyrene and TiO_2 nanoparticles to induce skin irritation, phototoxicity, and sensitization. *Toxicol. In Vitro* **2011**, *25*, 1863–1869. [CrossRef]
66. Natsch, A.; Landsiedel, R.; Kolle, S.N. A triangular approach for the validation of new approach methods for skin sensitization. *ALTEX Altern. Anim. Exp.* **2021**, *38*, 669–677. [CrossRef]
67. Keller, J.G.; Peijnenburg, W.; Werle, K.; Landsiedel, R.; Wohlleben, W. Understanding Dissolution Rates via Continuous Flow Systems with Physiologically Relevant Metal Ion Saturation in Lysosome. *Nanomaterials* **2020**, *10*, 311. [CrossRef]
68. OECD. *Test No. 406: Skin Sensitisation*; OECD: Paris, France, 2022.
69. Emter, R.; Natsch, A. A fast Resazurin-based live viability assay is equivalent to the MTT-test in the KeratinoSens assay. *Toxicol. In Vitro* **2015**, *29*, 688–693. [CrossRef] [PubMed]

toxics

MDPI

Article

Deriving a Continuous Point of Departure for Skin Sensitization Risk Assessment Using a Bayesian Network Model

Fleur Tourneix [1], Leopold Carron [1], Lionel Jouffe [2], Sebastian Hoffmann [3] and Nathalie Alépée [1,*]

[1] L'Oréal, Research & Innovation, 1Eugène Schueller, 93600 Aulnay-sous-Bois, France
[2] Bayesia S.A.S., Parc Cérès, Bâtiment N 21, rue Ferdinand Buisson, 53810 Changé, France
[3] seh consulting + services, Stembergring 15, 33106 Paderborn, Germany
* Correspondence: nathalie.alepee@loreal.com

Abstract: Regulations of cosmetic ingredients and products have been the most advanced in embracing new approach methodologies (NAMs). Consequently, the cosmetic industry has assumed a forerunner role in the development and implementation of animal-free next-generation risk assessment (NGRA) that incorporates defined approaches (DAs) to assess the skin sensitization potency of ingredients. A Bayesian network DA predicting four potency categories (SkinSens-BN) was constructed against reference Local Lymph Node Assay data for a total of 297 substances, achieving a predictive performance similar to that of other DAs. With the aim of optimally informing risk assessment with a continuous point of departure (PoD), a weighted sum of the SkinSens-BN probabilities for four potency classes (non-, weak, moderate, and strong/extreme sensitizer) was calculated, using fixed weights based on associated LLNA EC3-values. The approach was promising, e.g., the derived PoDs for substances classified as non-sensitizers did not overlap with any others and 77% of PoDs were similar or more conservative than LLNA EC3. In addition, the predictions were assigned a level of confidence based on the probabilities to inform the evaluation of uncertainty in an NGRA context. In conclusion, the PoD derivation approach can substantially contribute to reliable skin sensitization NGRAs.

Keywords: skin sensitization; risk assessment; point of departure; Bayesian network; defined approach

1. Introduction

Initiated by societal pressure and ethical concerns, political chemical safety programs around the globe are aiming at moving away from animal-based solutions and are calling for hazard and risk assessment approaches based on new approach methodologies (NAMs). Regulations of cosmetic ingredients and products have been the most advanced in abandoning animal experiments and embracing NAM solutions, especially in Europe [1,2]. Consequently, the cosmetic industry has assumed a forerunner role in the development and implementation of animal-free approaches to assess the primarily hazard of chemical ingredients.

Substantial efforts have been made to pave the way to advance NAMs for skin sensitization, which have led to substantial progress [3]. Based on the qualitative mechanistic understanding operationalized in the OECD skin sensitization adverse outcome pathway (AOP) [4], in vitro and in chemico test methods have been developed to address the first three key events (KEs) of the AOP. Based on thorough validation and independent assessment, several NAMs have been included in OECD Test Guidelines (TGs). TG 442C contains three in chemico reactivity test methods modeling the molecular initiating event (MIE) or KE1 "covalent binding of a chemical to skin protein" [5,6]: the Direct Peptide Reactivity Assay (DPRA), the Amino acid Derivative Reactivity Assay, and the kinetic DPRA (kDPRA), which were reviewed by Alépée, et al. [7]. TG 442D includes the KeratinoSens™ and the very similar LuSens, two in vitro test methods modeling KE2 "Keratinocyte activation" [8], while TG 442E features the four cell-based test methods U-SENS™, h-CLAT, IL8-Luc, and the GARD™skin that model KE3 "Dendritic cell activation" [9].

Although addressing the MIE could in theory be sufficient to cover a sequential chain of KEs, modeling errors inherent to the NAMs have led to the common understanding that a combination of NAMs covering at least two KEs of the skin sensitization AOP are required to provide high predictivity [10]. Consequently, a plethora of combinations of skin sensitization NAMs and also including other information sources, the so-called defined approaches (DAs), have been developed, including the sequential stacking tier strategy DAs developed somewhat later [11,12]. The majority of DAs have been reviewed by Kleinstreuer, et al. [13]. Recently, two DAs with a relatively simple decision tree approach to either predict skin sensitization hazard or United Nations (UN) Globally Harmonized System (GHS) classification categories have been adopted by the OECD [14]. The first three DAs included in this guideline use combinations of OECD-validated in chemico and in vitro test data, in some cases along with in silico information. The DAs currently described in this guideline are: (i) the "2 out of 3" (2o3) defined approach for hazard identification and (ii) two versions of the integrated testing strategy (ITSv1) for hazard and UN GHS potency categorization, both of which use the same test methods to address KE1 and KE3, but differ in the in silico predictions they incorporate (ITSv1: Derek Nexus; ITSv2: (OECD QSAR Toolbox 4.5)). The other Das have been developed for risk assessment purposes, i.e., they provide predictions of four categories or of a continuous value (see, for example, [13,15–17]). DAs can be applied as or transformed into a NAM-based point of departure (PoD) to be used in the next-generation risk assessment (NGRA) framework for skin sensitization, replacing the traditionally used animal-based or human-based PoDs [18,19].

Among them, a DA based on a Bayesian network, usually referred to as the Bayesian integrated testing strategy (ITS) for skin sensitization potency assessment, has been developed to derive a potency prediction of being a non-, weak, moderate, or extreme/strong sensitizer based on the Bayes factor [20]. Bayesian networks are probabilistic by definition, as they describe relationships between variables of the system of interest by conditional probabilities, which together form the joint probability distribution of the system. Bayesian networks can be graphically described by directed acyclic graphs (DAGs) comprising input variables, latent variables, and connections between dependent variables [21]. The advantages of Bayesian approaches have been acknowledged for risk assessment in general and for skin sensitization in particular, comprising the representation of the underlying mechanistic/dependent processes through DAGs, the ability to compensate for missing input data, and the inherent ability to address uncertainty [20,22–25]. In a risk assessment context, the selection of a category, either based on the maximum posterior probability or the Bayes factor, and subsequently of the lowest value in a category as the PoD, results in a limited number of potential PoDs. In addition, such a PoD comes with an unknown level of associated uncertainty, as the distance between the lowest category value and the true value can fall anywhere in the range that the category spans, as highlighted in a case study [18].

Building on the advantages of the Bayesian integrated testing strategy (ITS) for skin sensitization, a Bayesian network model (SkinSens-BN) was built to assess potency classification. This new DA offers enhanced flexibility by expanding the scope of the input data used and the covered chemistry (i.e., mainly cosmetic ingredients). While predicting four potency categories, which can be reduced to obtain predictions of UN GHS categories and skin sensitization hazard, the SkinSens-BN posterior probabilities were used to predict, in addition, a continuous PoD and to derive a categorical indicator of confidence associated with the predicted results.

2. Materials and Methods

2.1. Data Inputs

2.1.1. NAMs

The 13 inputs included in the Bayesian network were selected to inform several biological events relevant for the skin sensitization mode of action. Building upon previous work (for example, summarized by [13,26]), bioavailability was addressed by three physicochemical properties (molecular weight (MW) calculated from the structure, octanol–water partition coefficient (clogP) from Biobyte v.5.2, and volatility from EPISuite v. 4.1 categorized according to [27]), metabolism by Tissue Metabolism Simulator (TIMES-SS, v2.29.1.28 model v23.28), and the reactivity mode of action by the ToxTree module "Skin sensitization reactivity domains" (version 2.6.13) [28]. The three AOP key events were covered with the in vitro/chemico test methods DPRA, KeratinoSens™ [29] and the U-SENS™ [30,31], the latter two also informing cytotoxicity. Local Lymph Node Assay (LLNA) data extracted from the OECD database were used as a reference [32]. Further details are provided in Supplemental Table S1 "Inputs".

In total, 297 chemicals were selected using pre-defined criteria, such as the availability of LLNA test results, which were used to determine the reference result for determining a substance's potency. Data were retrieved from two sources: 219 substances, including 70 proprietary substances (identity not revealed) from Tourneix, Alépée, Detroyer, Eilstein, Martinozzi Teissier, Nardelli, Noçairi, Pauloin, Piroird and Del Bufalo [12], which included 184 cosmetic ingredients, inter alia, 72 dyes, 22 preservatives, and 40 fragrances; and 78 from the OECD database [32]. A total of 3861 input data were retrieved. NAM data were also collected from additional sources [12,33]. The data are summarized in Supplemental Table S1 "Data". Training set input data were complete, except for 40 substances with inconclusive (12) or partly missing (28) DPRA data, one substance with partly missing KeratinoSens™ data, 14 substances with inconclusive volatility data, and 39 substances with inconclusive TIMES-SS predictions. Input data for the test set were complete, except for six substances not tested in the U-SENS™, four substances with inconclusive or missing DPRA data, four substances for which no or partly missing KeratinoSens™ data were available, and six substances with inconclusive TIMES-SS predictions.

2.1.2. In Vivo Reference Data

As the Bayesian network was to be constructed against reference data, Local Lymph Node Assay (LLNA) data were extracted from the OECD database [32]. Supplemental Table S1 "Data" includes the EC3 value for each substance, i.e., the interpolated dose that stimulates a three-fold increase in lymph node cell proliferative activity compared to the vehicle control. For non-sensitizers, the EC3 value was set at 100%, as conducted previously (for example, [18,34]). In addition, the skin sensitization categories of the UN GHS, i.e., Cat. 1A for EC3 $\leq$ 2% (strong/extreme sensitizers), Cat. 1B for EC3 between 2 and 100%, and no category (NS), and a 4-class categorization that divides the GHS Cat. 1B into weak (EC3 $\geq$ 10%) and moderate skin sensitizers (2% < EC3 < 10%), are presented (Figure 1) [35,36].

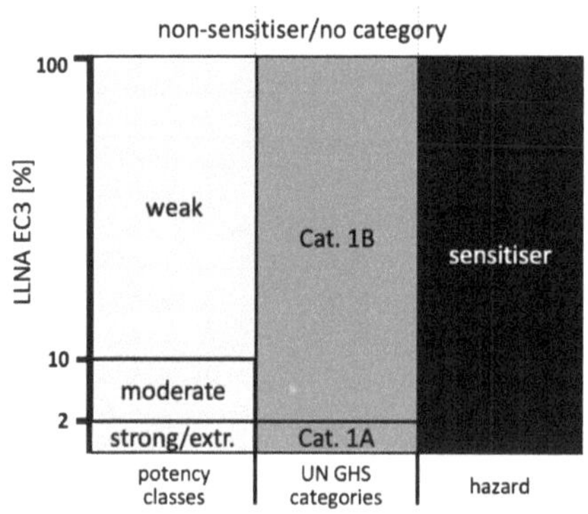

Figure 1. Relationship of skin sensitization categories for potency (4 classes), UN GHS categories, and hazard derived from LLNA EC3 values.

2.2. Bayesian Network

2.2.1. Construction and Training

The Bayesian network was generated with BayesiaLab v10.2, a commercial software (available at https://www.bayesia.com (accessed on 17 July 2024)). The input parameters used are listed in Supplemental Table S1 "Inputs", structured by the biological event they inform. Each quantitative input variable was discretized into two or three categories using by using a Minimum Description Length approach for finding the value of the thresholds and their number [37]. For the prediction of the four LLNA potency classes as described above, each quantitative input variable was discretized into two or three categories using data-driven thresholds. For example, the U-SENS™ EC150 thresholds were 3.86 and 51.79, resulting in three categories. All discretization thresholds for the quantitative input variables are provided in Supplemental Table S1 "Inputs".

The SkinSens-BN network, based on data-driven and expert knowledge, is structurally similar to the one by Jaworska, Natsch, Ryan, Strickland, Ashikaga and Miyazawa [20], who described the rationale for the network structure, the direct dependencies, and the direction between nodes/inputs. The arrows connect the conditionally dependent variables. This representation allows to reduce the complexity of the network, i.e., the number of probabilities to estimate. Expert knowledge was used to create meaningful latent variables that allow to create probabilistic summaries of the associated manifest variables. The rationale of the selection of input variables that cover the three first key events of the AOP for skin sensitization and of the structure is provided in Supplemental Table S1 "Input".

The 297 substances were divided into training and test sets based on the two major sources used for substance selection and data retrieval. The network was trained with the input data of 219 chemicals [11], which are provided in Supplemental Table S1 "Data", to predict the posterior distribution of the probabilities of a chemical belonging to each potency class given the observations. The resulting network was tested with the 78 substances from the OECD database [32]. These were primarily selected to obtain the coverage of the LLNA potency spectrum and of the physico-chemical properties (MW, clogP, and volatility similar to that of the training set (Supplemental Figure S1)). However, the two sets differed in the proportion of UN GHS Cat. 1A substances (lower in the test set) and the clogP (higher in the test).

Four latent variables (bioavailability, metabolism, cysteine, and U-SENS™) were used to structure the network by connecting related inputs, e.g., the three physico-chemical inputs' molecular weight, clogP and volatility informing bioavailability. The DAG representing the network was further structured manually.

2.2.2. Performance Analysis

The Skin-Sens-BN obtained with the training set was internally validated by 5-fold cross-validation, stratified to obtain identical distribution in the 4 LLNA potency classes. Cross-validation resulted in an average accuracy of 61%, which was considered to sufficiently demonstrate the robustness and generalizability of the predictive performance of the BN.

The predictive capacity for four potency classes (non-sensitizer, weak, moderate, and strong/extreme), determined by the most likely predicted class, the three UN GHS (1A and 1B vs. No Category), and for skin sensitization hazard (NS vs. S) was assessed by comparison with the LLNA reference results using contingency tables. For a four-class prediction (NS/weak/moderate/strong–extreme), randomly assigning each item to a class would result in a 25% accuracy rate. In the current study, the achieved percentage surpasses this probability and even more the 50% chance of landing on either side for the binary S/NS prediction. As highlighted schematically in Figure 1, classes/categories are simply combined moving from potency to UN GHS to hazard. Pertinent predictive parameters, i.e., accuracy, and specificity and sensitivity for hazard classification were calculated.

2.3. *Confidence Categorization*

In order to derive an indicator of confidence of a chemical belonging to a potency class, the predicted probabilities, called probability profile, were converted to Generalized Bayes Factors (GBFs) applying the same formula used by Jaworska, Natsch, Ryan, Strickland, Ashikaga and Miyazawa [20], i.e., calculating the ratio of the posterior odds and the prior odds in the training set per class. Subsequently, we transformed the GBFs to what was termed "weight of evidence" (W = 10 × log10 (GBF)), which is measured in deciban, with one deciban being "about the smallest change in W that is directly perceptible to human intuition" [38]. The confidence in the prediction was categorized based on the maximum W-values across the four potency classes. Based on Jeffreys' decision rule [39], W-values between −5 and 5 were associated with "weak" confidence, W-values between −10 and −5 and between 5 and 10 with "moderate" confidence, and W-values smaller −10 or larger than 10 with "high" confidence.

In other words, the SkinSens-BN model provides for each substance an indication of the confidence in the predicted potency class result based on the data observation.

2.4. *PoD Derivation*

The final node of the Bayesian network returns the discrete probability p for a substance belonging to each of the four classes (non-sensitizer (NS), "weak", "moderate" and "strong or extreme" sensitizer). Using fixed weights for each of the classes that were based on associated LLNA EC3-values, i.e., 100 for NS, 10 for weak, 2 for moderate, and 0.2 for strong or extreme) the point of departure (PoD) was defined as the following sum of weighted probabilities:

$$PoD_{BN}\ [\%] = p(NS) \times 100 + p(weak) \times 10 + p(moderate) \times 2 + p(strong\ or\ extreme) \times 0.2 \quad (1)$$

The weights relate to LLNA EC3-value, with 100 representing a non-sensitizing result in the LLNA, 10 representing to lowest EC3 considered to be weak, 2 representing the EC3 used to discriminate UN GHS categories 1A and 1B, and 0.2 considered as a representative value for strong and extreme sensitizer [36]. The PoD sum assumes its maximum when p(NS) = 100 and the other probabilities are 0. In this case, the PoD is 100, corresponding to a negative LLNA. The minimum is obtained for p(strong or extreme) = 100 (and the other probabilities being 0), resulting in a PoD of 0.2%. As this minimum is larger than the LLNA EC-value of 0.2%, the approach will, in comparison to the LLNA, systematically underpredict the PoD for highly potent sensitizers. This limitation is acknowledged, but considered acceptable, as substances with a low PoD are rarely used as cosmetic ingredients. On the other end of the spectrum, PoD_{BN} will practically always be <100%.

3. Results

3.1. SkinSens-BN and Its Predictive Performance

3.1.1. The Network Structure

The Bayesian network, referred to as SkinSens-BN, was constructed using 13 inputs informing various biologically relevant events, including mechanistic key events as operationalized in the skin sensitization AOP, and 219 defined substances, for which most inputs were available, and four latent variables. A graphical representation of the SkinSens-BN is shown in Figure 2. The inputs are displayed as circles. The four latent variables "Bioavailability", "Metabolism", "Cysteine", and "U-SENS" are indicated as rounded rectangles, while the final node "LLNA potency prediction" is represented as a target. Arrows connect the dependent variables, with the arrow direction indicating the relationship.

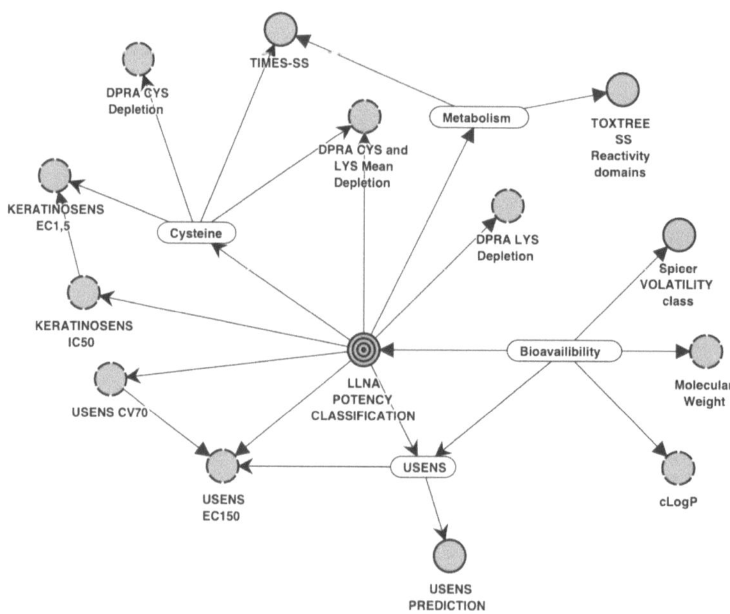

Figure 2. The directed acyclic graph (DAG) of the SkinSens-BN. The gray-shaded circles indicate the inputs (continuous border line: qualitative input; dotted border line: discretized quantitative input), rounded rectangles indicate latent variables, arrows connect dependent inputs, and the final node "LLNA potency prediction" is represented as a "target".

3.1.2. Predictivity: Training and Test Sets, Individual and Combined

The SkinSens-BN predictivity for the training set obtained by comparison to the LLNA reference data was calculated for discriminating four potency classes, the UN GHS categories, and a binary hazard classification. The results are summarized at the top of Table 1. The accuracy of prediction was 64% for the four potency categories, increasing to 68% for UN GHS categories and to 84% for hazard classes. For hazard, the specificity was 80% (55/69), the sensitivity 87% (130/150), and the balanced accuracy, i.e., the average of specificity and sensitivity, 84%. The number of substances over-predicted and under-predicted was very similar in each of the sub-tables, indicating the SkinSens-BN equally weighted mispredictions, i.e., not reducing mispredictions in one direction at the cost of the other directions.

In comparison to the training set, the test set predictions were lower for the potency classes and the UN GHS category, as indicated by reduced accuracies by 11% and 5%, respectively (Table 1, middle). The lower test set predictivity of the potency classes is primarily caused by the overprediction of weak sensitizers and mispredictions of moderate sensitizers. A potential reason explaining mispredictions is an imbalanced distribution of chemistry or other important factors between the training and test sets. For example, of the eight acrylates, three were in the training set and five in the test set, including four weak sensitizers that were overpredicted as strong/extreme. Note that three out of these four overpredicted weak sensitizers (GHS Cat. 1B) were also overpredicted as GHS Cat. 1A by both ITSv1 and ITSv2 [32]. Interestingly, the accuracy for hazard was higher (by 2%), as was the sensitivity (130/150 = 87% in the training set vs. 47/50 = 94% in the test set), while the specificity (55/69 = 80% in the training set vs. 20/28 = 71% in the test set) and balanced accuracy (83.2% in the training set vs. 82.7% in the test set) were lower. Table 1 also includes the predictivity when combining the training and test sets (at the bottom).

Table 1. Contingency tables for training set predictions at the top and test set predictions in the middle and the combined sets at the bottom (gray shades: indicator of misprediction severity; bold: correct predictions; NS: non-sensitizer; mod.: moderate; extr.: extreme; S: sensitizer (weak/moderate/strong/extreme)).

LLNA Reference Data Classes/Categories

Training Set — SkinSens-BN

A: 4 potency classes	NS	weak	mod.	strong/extr.	Σ
NS	**55**	12	7	1	75
weak (EC3 ≥ 10%)	7	**20**	8	6	41
mod. (2% ≤ EC3 < 10%)	4	2	**14**	6	26
strong/extr. (EC3 < 2%)	3	5	18	**51**	77
Σ	69	39	47	64	219
accuracy	64% (140/219)				

B: UN GHS categories	no Cat.	Cat. 1B	Cat. 1A	Σ
NS	**55**	19	1	75
weak/mod.	11	**44**	12	67
strong/extr.	3	23	**51**	77
Σ	69	86	64	219
accuracy	68% (150/219)			

C: hazard	NS	S	Σ
NS	**55**	20	75
S	14	**130**	144
Σ	69	150	219
accuracy	84% (185/219)		

Test set — SkinSens-BN

A: 4 potency classes	NS	weak	mod.	strong/extreme	Σ
NS	**20**	1	2	0	23
weak (EC3 ≥ 10%)	6	**7**	6	1	20
mod. (2% ≤ EC3 < 10%)	0	2	**5**	3	10
strong/extr. (EC3 < 2%)	2	7	7	**9**	25
Σ	28	17	20	13	78
accuracy	53% (41/78)				

B: UN GHS categories	no Cat.	Cat. 1B	Cat. 1A	Σ
NS	**20**	3	0	23
weak/mod.	6	**20**	4	30
strong/extr.	2	14	**9**	25
Σ	28	37	13	78
accuracy	63% (49/78)			

C: hazard	NS	S	Σ
NS	**20**	3	23
S	8	**47**	55
Σ	28	50	78
accuracy	86% (67/78)		

Training and test set — SkinSens-BN

A: 4 potency classes	NS	weak	mod.	strong/extreme	Σ
NS	**75**	13	9	1	98
weak (EC3 ≥ 10%)	13	**27**	14	7	61
mod. (2% ≤ EC3 < 10%)	4	4	**19**	9	36
strong/extr. (EC3 < 2%)	5	12	25	**60**	102
Σ	97	56	67	77	297
accuracy	61% (181/279)				

B: UN GHS categories	no Cat.	Cat. 1B	Cat. 1A	Σ
NS	**75**	22	1	98
weak/mod.	17	**64**	16	97
strong/extr.	5	37	**60**	102
Σ	97	123	77	297
accuracy	67% (199/297)			

C: hazard	NS	S	Σ
NS	**75**	23	98
S	22	**177**	199
Σ	97	200	297
accuracy	85% (252/297)		

3.1.3. Confidence Assessment

The confidence in the SkinSens-BN prediction was determined in relation to the maximum GBFs across the four potency classes, which were derived from the probability profile, i.e., the posterior distribution, and the prior. The maximum GBF was transformed into a W-value, which was interpreted using a simplification of the Jeffrey's decision. The level of confidence was grouped into the three categories of high, moderate, and low. This level of confidence is intended to provide risk assessors with an indication of confidence when using SkinSens-BN results, informing the next risk assessment step. To summarize the results, these were grouped by potency category with the highest GBF.

Of the 98 substances predicted as NS, the majority (71/98 = 72.4%) were associated with a "high" level of confidence. "Weak" substances were predominant (51/61 = 83.6%), with an assigned "moderate" level of confidence, as were substances with "moderate" potency (29/36 = 80.5%). The level of confidence of substances predicted as extreme/strong sensitizers was most evenly distributed, with a "high" level of confidence assigned to 48.0% (49/102) of the substances.

Table 2 exemplifies the approach of transforming the probability profile into a W-value, from which the level of confidence is derived by using three substances that were weak sensitizers in the LLNA. "Weak" was also the most likely predicted class in the SkinSens-BN probability profile. However, the individual probability profiles differ in shape, in particular the probability of the weak class (p(weak)). While p(weak) was very high for geraniol, it was just slightly higher than p(moderate) for hydroxycitronellal. This difference was also reflected in the W-values, leading to a high confidence for geraniol, a moderate confidence for N,N-dibutylaniline, and a low confidence for hydroxycitronellal.

Table 2. SkinSens-BN probability profile of three example substances, with the corresponding W-values, predicted classes, and confidence levels.

Substance Name	Geraniol		N,N-dibutylaniline		Hydroxycitronellal	
LLNA EC3 (UN GHS cat.)	**26% (1B)**		**19.6% (1B)**		**33% (1B)**	
SkinSens-BN	**Prob. Profile**	**W**	**Prob. Profile**	**W**	**Prob. Profile**	**W**
p(NS)	0.1539	−4.02	0.0247	−1.26	0.0017	−2.43
p(weak)	0.7240	10.96	0.5687	7.97	0.3912	4.84
p(moderate)	0.1144	−3.38	0.2930	1.69	0.3810	3.41
p(strong/ext.)	0.0078	−1.72	0.1135	−5.08	0.2261	−1.45
predicted class	weak		weak		weak	
confidence	high		moderate		low	

3.2. Derivation of a Continuous PoD with SkinSens-BN and Its Comparison to EC3

The discrete SkinSens-BN a posteriori probability distribution for the four potency classes, i.e., the probability profile, was used to construct an approach to derive a continuous PoD (PoD_{BN}). For each of the 297 substances, a sum of these probabilities associated with fixed weights, which were based on LLNA EC3 values associated with each class, was calculated. The probability profiles for all substances and the PoD_{BN} are provided in Supplemental Table S1. The PoD_{BN} ranged from the absolute PoD minimum of 0.20%, which was obtained for four substances (lauryl gallate, tetrachlorosalicylanilide, dinitrochlorobenzene, and 4-nitrobenzyl bromide), to the maximum PoD_{BN} of 99.76% (Figure 3), with a median of 14.23%, a lower quartile of 1.52%, and an upper quartile of 78.84%.

To exemplify the approach, we selected three substances from the test set. Their identities, LLNA EC3 values, UN GHS category, probability profiles, predicted class, i.e., derived from the max. of the probability profile, and PoD estimates are summarized in Table 3.

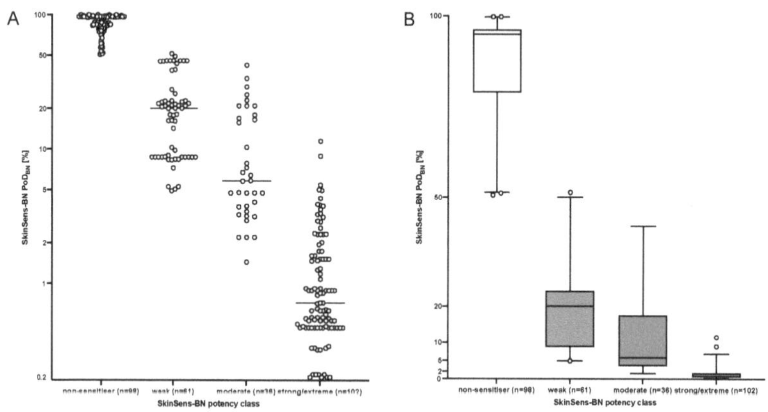

Figure 3. Comparison of PoD_{BN} (y-axis) and the predicted most likely potency class (x-axis), i.e., the class with the highest probability, with the number of substance (n) in the class: (**A**) as a dotplot (with median line) and (**B**) as a boxplot.

Table 3. PoD_{BN} derivation for three example substances (bold: class with the highest probability).

Substance Name		Lilial	4-Methoxy-α-methyl benzenpropanal	3,4-Dihydro-coumarin
LLNA EC3		8.6%	23.6%	5.6%
UN GHS category		1B	1B	1B
SkinSens-BN probability profile	p(NS)	0.0244	0.1605	0.1614
	p(weak)	**0.5685**	**0.5185**	0.3946
	p(moderate)	0.2934	0.2736	**0.4284**
	p(strong/extreme)	0.1138	0.0474	0.0156
predicted class (PoD as lower bound of predicted class)		Weak (10%)	Weak (10%)	Moderate (2%)
confidence		moderate	moderate	low
PoD_{BN}		8.73%	21.79%	20.94%

For Lilial, all PoD (LLNA EC3, based on predicted class and PoD_{BN}) were very similar. With an EC3 of 8.6%, the PoD_{BN} was almost identical (8.7%), while the PoD derived from the max. of the probability profile was 10%, i.e., the lower bound of the weak category. For 4-Methoxy-α-methyl benzenpropanal, the PoD_{BN} was similar to the EC3, both approx. a factor two higher compared to the approach of assigning it to the most likely potency class, i.e., a category PoD of 10%. For 3,4-Dihydro-coumarin, the class-based PoD and the EC3 were similar, while the PoD_{BN} was higher.

This comparison of the PoD_{BN} to the most likely potency class was conducted for all 297 substances. The results are summarized in Figure 3. Of the 297 substances, for 98, the most likely potency class was non-sensitizers. The PoD_{BN} in this class ranged from 50.62% to 99.76%, with a median of 94.94%, and was higher than all other PoD_{BN}, except for one proprietary substance predicted in the weak potency class. The PoD_{BN} of the 102 substances in the strong/extreme potency class ranged from 0.20% to 11.47%, with a median of 0.7% and an upper quartile of 1.64%. The values clearly overlapped with the PoD_{BN} of the substances in the moderate potency class (33 substances), while the PoD_{BN} of 5 of those 33 overlapped with the weak potency class. Weak and moderate predicted potency classes, with medians 20.07% and 5.81%, respectively, showed wider distributions and overlapped considerably.

The large overlap indicated that potency grouping by the max. value in the probability profile may suggest a certainty in the result that is not reflected when considering the entire profile.

Figure 3 also demonstrates that the PoD_{BN} will generally be higher than a PoD derived from the lowest threshold value of the predicted potency class. This is, for example, indicated by 22 substances in the strong/extreme potency class with a PoD_{BN} > 2%, including 1 substance with a PoD_{BN} of 11.5% and by only 1 substance in the moderate potency class with a PoD < 2%. In contrast, 21 of the 61 substances in the weak class had a PoD < 10% (min. of 4.89%).

Next, the continuous PoDs were compared with the corresponding LLNA reference EC3-values, assigning an EC3 of 100 to non-sensitizers. A dotplot of all PoD_{BN}-EC3 pairs is shown in Figure 4. The data were clearly positively correlated, with highly statistically significant (p-values < 0.001) Pearson and Spearman correlation coefficients of 0.75 and 0.73, respectively, compared to the LLNA EC3. The data points below the line of identity indicate substances for which the PoD_{BN} is lower, and those above the line of identity indicate substances for which the PoD_{BN} is higher. For example, the cluster of data points (triangles) in the upper left corner of Figure 4A had a clearly higher PoB_{BN} and the substances with EC3-values between 2 and 10 were frequently overpredicted (black dots in Figure 4B low the line of identity).

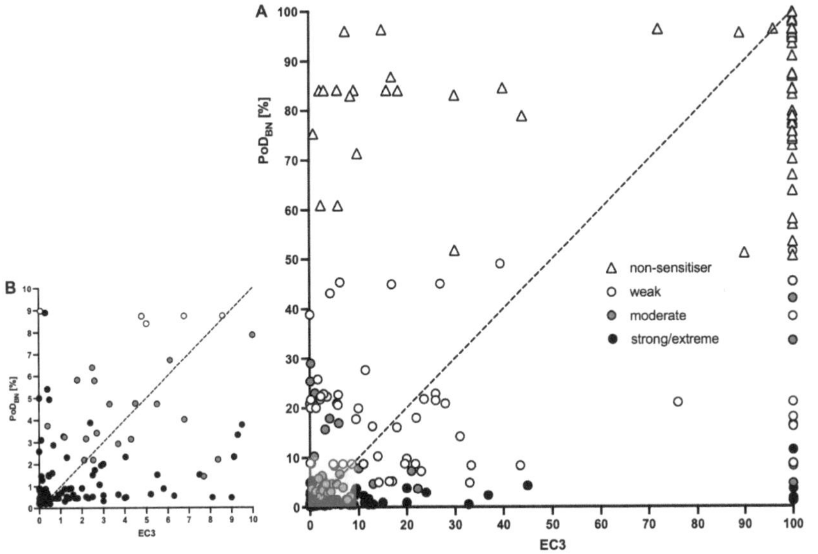

Figure 4. (**A**): PoD_{BN} (%) of the SkinSens-BN (y-axis) compared to the EC3 (%) from LLNA (x-axis) for the 297 chemicals. Shape and colors represent the potency category based on the max. value of the probability profile (diagonal line: line of identity). (**B**): Magnification of the gray-shaded area of (**A**).

To further quantify the similarity, the PoD_{BN}/EC3 ratios were investigated. Values below 1 indicated substances with PoD_{BN} < EC3, i.e., more conservative-derived PoD, and values above 1 indicated substances with PoD_{BN} > EC3, i.e., less conservative-derived PoD. The ratios are represented as a histogram in Figure 5A. Ratios between 0.316 and 3.16, i.e., maximum $10^{0.5}$-fold difference in PoD_{BN} and EC3 in either direction, were considered as "similar". This approach is based on the median LLNA EC3 standard deviation when using the same vehicle, i.e., 0.25, and the calculations presented by Hoffmann (2015). This group comprised 58.9% (175/297) of all substances. For 17.9% (53/297) of the substances, the PoD_{BN} was at least 3.16-times lower than EC3, i.e., more conservative, while the PoD_{BN} was less conservative for 23.2% (69/297). The occurrence of more severe less conservative ratios can partly be explained by the difference in the scaling of the two parameters at the lower end (minimum PoD_{BN}: 0.2% vs. minimum EC3: 0.0003%). The least conservative PoD derived with the SkinSens-BN was derived for oxazolone, with a ratio of 6960 (20.88/0.003), which was also the most severely underpredicted substance in the linear regression-based PoD models by Natsch and Gerberick [15]. As a summary measure, the geometric mean fold error was calculated as 3.55 for all substances and as 3.97 for the test set only, indicating a slight decrease in performance.

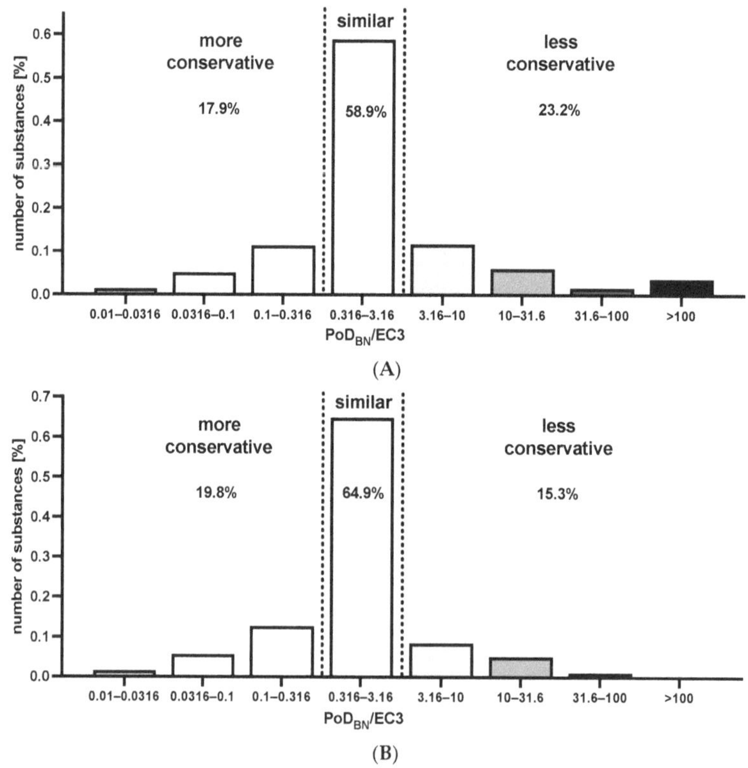

Figure 5. (**A**): Histogram of the ratio of the PoD_{BN} and the LLNA EC3 for the substances (n = 297). (**B**): Histogram of the ratio of the PoD_{BN} and the LLNA EC3 for substances with an EC3 $\geq$ 0.2 (n = 268). Substances with a ratio < 0.316 had a more conservative PoD_{BN} and substances with a ratio > 3.16 had a less conservative PoD_{BN} (gray shades).

To explore this effect of different scaling, a histogram without the substances with LLNA EC-values < 0.2, i.e., 29 most extreme skin sensitizers in the LLNA, is presented in Figure 5B. The absolute number of substances, for which the PoD_{BN} was more conservative, remained the same, i.e., 53, resulting in 19.8% due to the smaller total amount of substances. Except for a proprietary substance (OA39:EC3 = 0.1 and PoD_{BN} = 0.48), also the "similar" substances were not affected. However, as expected, the number of substances, for which the PoD_{BN} was less conservative, was substantially reduced by 28. It remains to be explored how to best address these scaling differences, e.g., by not deriving a PoD_{BN} for substances with a probability profile maximum value for the extreme/strong potency category that is associated with at the least moderate confidence.

4. Discussion

Skin sensitization is the human health effect for which a generally applicable, systematic, and exclusively NAM-based risk assessment approach is most advanced. An NGRA framework has been proposed that provides guidance for a tiered and transparent integration of relevant information, while allowing for flexibility [18,19]. This NGRA framework has been applied in various case studies to demonstrate its applicability and to initiate a constructive dialogue with stakeholders [22,40–43]. A centerpiece of the NGRA is the defined approaches (DAs) that integrate data from experimental NAMs that address at least two key events of the skin sensitization AOP [4]. Several DAs of different levels of complexity have been developed. These comprise two decision-tree-based DAs for hazard identification and classification according to the UN GHS, which were recently adopted by the OECD [14]. In addition, DAs using more sophisticated statistical approaches providing results that can, either directly or transformed, be used to derive a PoD for risk assessment have been proposed [20,44,45]. Among these, the Bayesian network by Jaworska, Natsch, Ryan, Strickland, Ashikaga and Miyazawa [20] is of particular interest due to its strengths, such as the provision of a probability profile across the potency classes, which can be used to quantify uncertainty associated with predictions, and the ability to cope with missing data. In contrast, the property that such network predictions are categorical limits the ability to derive a more precise PoD than the lower LLNA EC3-values associated with each class.

Building on the strengths, a Bayesian network similar to the one of Jaworska, Natsch, Ryan, Strickland, Ashikaga and Miyazawa [20] was developed, called SkinSens-BN. The main differences were the adaptation to some new input parameters and an increased total number of substances used that covered a broad spectrum of physico-chemical properties and LLNA EC3-values, including non-sensitizers. Overall, the predictive performance was considered in the range of the OECD-adopted DAs. The skin sensitization hazard was predicted as good, as reported for the '2of3' DA [14,46]. Small differences in sensitivity and specificity were evened out, as indicated by the almost identical accuracies and balanced accuracies that ranged from 83% to 85%. In terms of UN GHS category predictions, the SkinSens-BN was compared to the "Integrated Testing Strategy (ITS)" DA, also included in the OECD TG 497. For both ITS versions, an accuracy of 71% was reported, with no misprediction over two categories and some inconclusive predictions. The respective accuracy of the SkinSens-BN was 67%. Six substances were mispredicted by two classes, five LLNA non-sensitizers as GHS Cat. 1A and one LLNA GHS Cat. 1A as a non-sensitizer (highlighted in Supplemental Table S1 "Data"). However, only one of those substances was part of the data used for calculating the ITS predictive performance in OECD TG 497. For these comparisons, it needs to be kept in mind that the number of substances was more than twice as high and that no inconclusive predictions were present for the SkinSens-BN.

Regarding the prediction of four potency classes, which was 53% for the test set and 61% overall, a comparison with the Bayesian integrated testing strategy (ITS) for skin sensitization potency assessment would be most informative [20]. However, an independent evaluation is, to our knowledge, not available. An evaluation of the Bayesian integrated testing strategy (ITS) for skin sensitization potency assessment regarding three LLNA potency classes, which are similar to the GHS categories, obtained an accuracy of 68% with 115 substances, a predictivity very similar to the GHS predictivity of the SkinSens-BN [13]. Indications such as the seemingly lower predictivity for acrylates could be conducted as follow-up to further improve the SkinSens-BN, and may allow the further improvement in the predictive performance for four classes. However, it needs to keep in mind that the construction of the SkinSens-BN was primarily a means to an end. Nevertheless, the successful construction of a Bayesian network can contribute to building trust in the general approach and demonstrates the flexibility of the approach in terms of the inputs and chemistry to be covered.

Once it was confirmed that the SkinSens-BN performed promisingly, the primary goal of constructing an approach to derive a continuous PoD was addressed. A weighted sum combining fixed weights-associated LLNA EC3-values for each of the classes with the respective probabilities of their probability profile as weights was constructed. This approach, which is generally applicable, resulted in continuous PoDs in the theoretical range from 0.2 to 100%. The comparison of the SkinSens-BN PoD and the respective LLNA EC3-values showed that, for 77% of the 297 substances, a similar or more conservative PoD was derived.

Options to further characterize and improve the SkinSens-BN include conducting a sensitivity analysis to characterize the impact that individual inputs have on the prediction, assessing the impact of missing inputs, and tuning it more toward human relevance, as conducted by Natsch [47]. The PoD_{BN} issue of underpredictions of strong and extreme sensitizers could be solved by the combination with an approach that can reliably identify extreme sensitizers in a first step, for example, by the use of NAMs targeting the identification of strong and extreme sensitizers, as the kDPRA [48], or by exploring targeted prediction models, as described for the ADRA by Alépée, Tourneix, Singh, Ade and Gregoire [7]. In addition, the adaptation of the PoD_{BN} algorithm and the adjustment of the PoD_{BN} based on analog information or a tiered decision process that specifically addresses extreme potency prediction, e.g., by not deriving a PoD_{BN} for substances with a probability profile maximum value for the extreme/strong potency category that is associated with at the least moderate confidence, could be explored. Although it needs to be considered that strong and extreme skin sensitizers are only very rarely, if at all, used as cosmetic ingredients, the improvement in an adapted strategy for these potency classes could be applicable to other industries, such as agrochemicals, botanicals, and medical and wearable devices, which are faced with the same concerns [49–52].

In addition, similarly to the approach of rating the confidence in predictions using the Bayes factor applied by Jaworska, Natsch, Ryan, Strickland, Ashikaga and Miyazawa [20], the probability profile was the basis to assign the prediction of confidence levels. The GBFs were transformed into W-values, which were categorized to provide the three confidence levels of "low", "moderate", and "high". In essence, the approach resulted in a higher confidence the more probability was assigned to one of the four potency classes, or, in other words, the higher the distribution peak. On average, non-sensitizer potency predictions were associated with the highest confidence and moderate potency predictions with the lowest confidence.

The confidence rating is expected to substantially inform the next risk assessment steps in a weight of evidence approach. The overall risk assessment outcome is evaluated as a weight of evidence considering the calculated PoD [and, in the case of SkinSens-BN, the probability profile], the confidence in the use of NAM input data within the DAs (applicability domain), and the relative conservatism in the transformation of the DA outcome to a PoD (the most likely predicted class). For cases with insufficient confidence to reach a decision, it may, in combination with the detailed evaluation of individual inputs, point to the next steps in the process, which are needed to increase the confidence. Alternatively, the margin of exposure could be increased. This is intended to explore the usefulness of the SkinSens-BN-derived PoDs with associated confidence levels in several case studies, which may also identify potential avenues to improve the approach.

In conclusion, the development of the SkinSens-BN model and its related PoD derivation approach, in the context of the NGRA, clearly indicates that quantitative risk assessments of skin sensitization can be achieved without a reliance on data from studies conducted on animals.

Supplementary Materials: The following supporting information can be downloaded at: https://www.mdpi.com/article/10.3390/toxics12080536/s1, Figure S1: Comparison of EC3 values, GHS categories, and three physico-chemical properties of the training and the test sets using the Kruskal–Wallis test, with *p*-values; Table S1: Summary of all inputs used to construct the Bayesian network [5,27–31].

Author Contributions: Conceptualization, N.A. and F.T.; methodology, L.C. and F.T.; software, L.J.; validation, L.J.; formal analysis, F.T., N.A. and S.H.; investigation, F.T., L.J. and L.C.; resources, L.J. and S.H.; data curation, F.T.; writing—original draft preparation, S.H., F.T. and N.A.; writing—review and editing, S.H., F.T. and N.A.; visualization, L.C. and S.H.; supervision, N.A.; project administration, N.A. and F.T.; funding acquisition, N.A. All authors have read and agreed to the published version of the manuscript.

Funding: This research received no external funding. Lionel Jouffe and Sebastian Hoffmann were compensated for their contribution by L'Oréal.

Institutional Review Board Statement: Not applicable.

Informed Consent Statement: Not applicable.

Data Availability Statement: The authors confirm that the data supporting the findings of this study are available within the article and its Supplementary Materials.

Conflicts of Interest: Authors Sebastian Hoffmann and Lionel Jouffe were employed by the compagnies seh consulting + services and Bayesia S.A.S respectively. The remaining authors declare the research was conducted in the absence of any commercial or financial relationships that could be construed as a potential conflict of interest.

References

1. European Commission Regulation (EC) No 1223/2009 of the European parliament and the council of 30 November 2009 on cosmetic products. *Off. J. Eur. Union* **2009**, *L342*, 59–209.
2. SCCS. *The SCCS Notes of Guidance for the Testing of Cosmetic Ingredients and Their Safety Evaluation*; 12th Revision; SCCS/1647/22; SCCS: Brussels, Belgium, 2023.
3. Gądarowska, D.; Kalka, J.; Daniel-Wójcik, A.; Mrzyk, I. Alternative Methods for Skin-Sensitization Assessment. *Toxics* **2022**, *10*, 740. [CrossRef] [PubMed]
4. OECD. *The Adverse Outcome Pathway for Skin Sensitisation Initiated by Covalent Binding to Proteins*; OECD: Paris, France, 2014.
5. Natsch, A.; Emter, R. Reaction Chemistry to Characterize the Molecular Initiating Event in Skin Sensitization: A Journey to Be Continued. *Chem. Res. Toxicol.* **2017**, *30*, 315–331. [CrossRef] [PubMed]
6. OECD. *Test No. 442C: In Chemico Skin Sensitisation*; OECD: Paris, France, 2023.
7. Alépée, N.; Tourneix, F.; Singh, A.; Ade, N.; Grégoire, S. Off to a good start? Review of the predictivity of reactivity methods modelling the molecular initiating event of skin sensitization. *ALTEX—Altern. Anim. Exp.* **2023**, *40*, 606–618. [CrossRef] [PubMed]
8. OECD. *Test No. 442D: In Vitro Skin Sensitisation*; OECD: Paris, France, 2022.
9. OECD. *Test No. 442E: In Vitro Skin Sensitisation*; OECD: Paris, France, 2023.
10. Jowsey, I.R.; Basketter, D.A.; Westmoreland, C.; Kimber, I. A future approach to measuring relative skin sensitising potency: A proposal. *J. Appl. Toxicol.* **2006**, *26*, 341–350. [CrossRef] [PubMed]
11. Tourneix, F.; Alépée, N.; Detroyer, A.; Eilstein, J.; Ez-Zoubir, M.; Teissier, S.M.; Noçairi, H.; Piroird, C.; Basketter, D.; Del Bufalo, A. Skin sensitisation testing in practice: Applying a stacking meta model to cosmetic ingredients. *Toxicol. In Vitro* **2020**, *66*, 104831. [CrossRef] [PubMed]
12. Tourneix, F.; Alépée, N.; Detroyer, A.; Eilstein, J.; Martinozzi Teissier, S.; Nardelli, L.; Noçairi, H.; Pauloin, T.; Piroird, C.; Del Bufalo, A. Assessment of a defined approach based on a stacking prediction model to identify skin sensitization hazard. *Toxicol. In Vitro* **2019**, *60*, 134–143. [CrossRef] [PubMed]
13. Kleinstreuer, N.C.; Hoffmann, S.; Alepee, N.; Allen, D.; Ashikaga, T.; Casey, W.; Clouet, E.; Cluzel, M.; Desprez, B.; Gellatly, N.; et al. Non-animal sensitization methods to predict skin (II): An assessment of defined approaches. *Crit. Rev. Toxicol.* **2018**, *48*, 359–374. [CrossRef]
14. OECD. *Guideline No. 497: Defined Approaches on Skin Sensitisation*; OECD: Paris, France, 2023. [CrossRef]
15. Natsch, A.; Gerberick, G.F. Integrated skin sensitization assessment based on OECD methods (I): Deriving a point of departure for risk assessment. *ALTEX—Altern. Anim. Exp.* **2022**, *39*, 636–646. [CrossRef] [PubMed]
16. Natsch, A.; Gerberick, G.F. Integrated skin sensitization assessment based on OECD methods (II): Hazard and potency by combining kinetic peptide reactivity and the "2 out of 3" Defined Approach. *ALTEX—Altern. Anim. Exp.* **2022**, *39*, 647–655. [CrossRef]
17. Reynolds, J.; Gilmour, N.; Baltazar, M.T.; Reynolds, G.; Windebank, S.; Maxwell, G. Decision making in next generation risk assessment for skin allergy: Using historical clinical experience to benchmark risk. *Regul. Toxicol. Pharmacol.* **2022**, *134*, 105219. [CrossRef]
18. Gilmour, N.; Alépée, N.; Hoffmann, S.; Kern, P.S.; Van Vliet, E.; Bury, D.; Miyazawa, M.; Nishida, H.; Cosmetics, E. Applying a next generation risk assessment framework for skin sensitisation to inconsistent new approach methodology information. *ALTEX—Altern. Anim. Exp.* **2023**, *40*, 439–451. [CrossRef] [PubMed]

19. Gilmour, N.; Kern, P.S.; Alépée, N.; Boislève, F.; Bury, D.; Clouet, E.; Hirota, M.; Hoffmann, S.; Kühnl, J.; Lalko, J.F.; et al Development of a next generation risk assessment framework for the evaluation of skin sensitisation of cosmetic ingredients *Regul. Toxicol. Pharmacol.* **2020**, *116*, 104721. [CrossRef] [PubMed]
20. Jaworska, J.S.; Natsch, A.; Ryan, C.; Strickland, J.; Ashikaga, T.; Miyazawa, M. Bayesian integrated testing strategy (ITS) for skin sensitization potency assessment: A decision support system for quantitative weight of evidence and adaptive testing strategy *Arch. Toxicol.* **2015**, *89*, 2355–2383. [CrossRef] [PubMed]
21. Conrady, S.; Jouffe, L. *Bayesian Networks & BayesiaLab—A Practical Introduction for Researchers*; Bayesia USA: Nashville, TN, USA, 2015; p. 382.
22. Gilmour, N.; Reynolds, J.; Przybylak, K.; Aleksic, M.; Aptula, N.; Baltazar, M.T.; Cubberley, R.; Rajagopal, R.; Reynolds, G.; Spriggs, S.; et al. Next generation risk assessment for skin allergy: Decision making using new approach methodologies. *Regul. Toxicol. Pharmacol.* **2022**, *131*, 105159. [CrossRef] [PubMed]
23. Jaworska, J.; Hoffmann, S. Integrated Testing Strategy (ITS)—Opportunities to better use existing data and guide future testing in toxicology. *ALTEX—Altern. Anim. Exp.* **2010**, *27*, 231–242. [CrossRef] [PubMed]
24. Maertens, A.; Golden, E.; Luechtefeld, T.H.; Hoffmann, S.; Tsaioun, K.; Hartung, T. Probabilistic risk assessment—The keystone for the future of toxicology. *ALTEX—Altern. Anim. Exp.* **2022**, *39*, 3–29. [CrossRef] [PubMed]
25. Mentzel, S.; Grung, M.; Tollefsen, K.E.; Stenrød, M.; Petersen, K.; Moe, S.J. Development of a Bayesian network for probabilistic risk assessment of pesticides. *Integr. Environ. Assess. Manag.* **2022**, *18*, 1072–1087. [CrossRef]
26. Strickland, J.; Zang, Q.; Kleinstreuer, N.; Paris, M.; Lehmann, D.M.; Choksi, N.; Matheson, J.; Jacobs, A.; Lowit, A.; Allen, D.; et al. Integrated decision strategies for skin sensitization hazard. *J. Appl. Toxicol.* **2016**, *36*, 1150–1162. [CrossRef] [PubMed]
27. Spicer, C.W.; Gordon, S.M.; Kelly, T.J.; Holdren, M.W.; Mukund, R. *Hazardous Air Pollutant Handbook*; CRC Press: Boca Raton, FL, USA, 2002.
28. Patlewicz, G.; Jeliazkova, N.; Safford, R.J.; Worth, A.P.; Aleksiev, B. An evaluation of the implementation of the Cramer classification scheme in the Toxtree software. *SAR QSAR Environ. Res.* **2008**, *19*, 495–524. [CrossRef] [PubMed]
29. Natsch, A.; Emter, R. Nrf2 activation as a key event triggered by skin sensitisers: The development of the stable KeratinoSens reporter gene assay. *Altern. Lab. Anim.* **2016**, *44*, 443–451. [CrossRef] [PubMed]
30. Alépée, N.; Piroird, C.; Nardelli, L. U-SENS™: A U937 Cell Line Activation Test for Skin Sensitization. In *Alternatives for Dermal Toxicity Testing*; Eskes, C., van Vliet, E., Maibach, H.I., Eds.; Springer International Publishing: Cham, Switzerland, 2017; pp. 311–330.
31. Piroird, C.; Ovigne, J.M.; Rousset, F.; Martinozzi-Teissier, S.; Gomes, C.; Cotovio, J.; Alepee, N. The Myeloid U937 Skin Sensitization Test (U-SENS) addresses the activation of dendritic cell event in the adverse outcome pathway for skin sensitization. *Toxicol. In Vitro* **2015**, *29*, 901–916. [CrossRef] [PubMed]
32. OECD. *Supporting Document to the OECD Guideline 497 on Defined Approaches for Skin Sensitisation*; Series on Testing and Assessment; No. 336; OECD: Paris, France, 2021.
33. Hoffmann, S.; Kleinstreuer, N.; Alepee, N.; Allen, D.; Api, A.M.; Ashikaga, T.; Clouet, E.; Cluzel, M.; Desprez, B.; Gellatly, N.; et al. Non-animal methods to predict skin sensitization (I): The Cosmetics Europe database. *Crit. Rev. Toxicol.* **2018**, *48*, 344–358. [CrossRef] [PubMed]
34. Hoffmann, S. LLNA variability: An essential ingredient for a comprehensive assessment of non-animal skin sensitization test methods and strategies. *ALTEX—Altern. Anim. Exp.* **2015**, *32*, 379–383. [CrossRef]
35. ECETOC. *Contact Sensitisation: Classification According to Potency*; European Centre for Ecotoxicology and Toxicology of Chemicals: Brussels, Belgium, 2003.
36. United Nations. *Globally Harmonized System of Classification and Labelling of Chemicals (GHS)*; United Nations: San Francisco, CA, USA, 2015.
37. Fayyad, U.M.; Irani, K.B. Multi-Interval Discretization of Continuous-Valued Attributes for Classification Learning. In Proceedings of the International Joint Conference on Artificial Intelligence, Chambéry, France, 28 August-3 September 1993.
38. Good, I.J. Studies in the History of Probability and Statistics. XXXVII A. M. Turing's statistical work in World War II. *Biometrika* **1979**, *66*, 393–396. [CrossRef]
39. Jeffreys, H. Some Tests of Significance, Treated by the Theory of Probability. *Math. Proc. Camb. Philos. Soc.* **1935**, *31*, 203–222. [CrossRef]
40. Assaf Vandecasteele, H.; Gautier, F.; Tourneix, F.; Vliet, E.V.; Bury, D.; Alépée, N. Next generation risk assessment for skin sensitisation: A case study with propyl paraben. *Regul. Toxicol. Pharmacol.* **2021**, *123*, 104936. [CrossRef] [PubMed]
41. Gautier, F.; Tourneix, F.; Assaf Vandecasteele, H.; van Vliet, E.; Bury, D.; Alépée, N. Read-across can increase confidence in the Next Generation Risk Assessment for skin sensitisation: A case study with resorcinol. *Regul. Toxicol. Pharmacol.* **2020**, *117*, 104755. [CrossRef] [PubMed]
42. OECD. *Case Study on the Use of Integrated Approaches for Testing and Assessment for Skin Sensitisation: Demonstrating the Next Generation Risk Assessment Framework Using Geraniol*; Series on Testing and Assessent No. 368; OECD: Paris, France, 2022.
43. OECD. *Case Study on the Use of Integrated Approaches for Testing and Assessment for Skin Sensitisation of Diethanolamine: Application of a Next Generation Risk Assessment Framework*; Series on Testing and Assessment No. 374; OECD: Paris, France, 2023.

44. Hirota, M.; Ashikaga, T.; Kouzuki, H. Development of an artificial neural network model for risk assessment of skin sensitization using human cell line activation test, direct peptide reactivity assay, KeratinoSens™ and in silico structure alert parameter. *J. Appl. Toxicol.* **2018**, *38*, 514–526. [CrossRef] [PubMed]
45. Reynolds, G.; Reynolds, J.; Gilmour, N.; Cubberley, R.; Spriggs, S.; Aptula, A.; Przybylak, K.; Windebank, S.; Maxwell, G.; Baltazar, M.T. A hypothetical skin sensitisation next generation risk assessment for coumarin in cosmetic products. *Regul. Toxicol. Pharmacol.* **2021**, *127*, 105075. [CrossRef] [PubMed]
46. Urbisch, D.; Mehling, A.; Guth, K.; Ramirez, T.; Honarvar, N.; Kolle, S.; Landsiedel, R.; Jaworska, J.; Kern, P.S.; Gerberick, F.; et al. Assessing skin sensitization hazard in mice and men using non-animal test methods. *Regul. Toxicol. Pharmacol.* **2015**, *71*, 337–351. [CrossRef] [PubMed]
47. Natsch, A. Integrated skin sensitization assessment based on OECD methods (III): Adding human data to the assessment. *ALTEX—Altern. Anim. Exp.* **2023**, *40*, 571–583. [CrossRef] [PubMed]
48. Natsch, A.; Haupt, T.; Wareing, B.; Landsiedel, R.; Kolle, S.N. Predictivity of the kinetic direct peptide reactivity assay (kDPRA) for sensitizer potency assessment and GHS subclassification. *ALTEX—Altern. Anim. Exp.* **2020**, *37*, 652–664. [CrossRef]
49. Kolle, S.N.; Flach, M.; Kleber, M.; Basketter, D.A.; Wareing, B.; Mehling, A.; Hareng, L.; Watzek, N.; Bade, S.; Funk-Weyer, D.; et al. Plant extracts, polymers and new approach methods: Practical experience with skin sensitization assessment. *Regul. Toxicol. Pharmacol.* **2023**, *138*, 105330. [CrossRef] [PubMed]
50. Ruparel, N.; Islas-Robles, A.; Hilberer, A.; Cantrell, K.; Madrid, M.; Ryan, C.; Gerberick, G.F.; Persaud, R. Deriving a point of departure for assessing the skin sensitization risk of wearable device constituents with in vitro methods. *Food Chem. Toxicol.* **2024**, *189*, 114725. [CrossRef] [PubMed]
51. Strickland, J.; Truax, J.; Corvaro, M.; Settivari, R.; Henriquez, J.; McFadden, J.; Gulledge, T.; Johnson, V.; Gehen, S.; Germolec, D.; et al. Application of Defined Approaches for Skin Sensitization to Agrochemical Products. *Front. Toxicol.* **2022**, *4*, 852856. [CrossRef] [PubMed]
52. Svobodová, L.; Rucki, M.; Vlkova, A.; Kejlova, K.; Jírová, D.; Dvorakova, M.; Kolarova, H.; Kandárová, H.; Pôbiš, P.; Heinonen, T.; et al. Sensitization potential of medical devices detected by in vitro and in vivo methods. *ALTEX—Altern. Anim. Exp.* **2021**, *38*, 419–430. [CrossRef] [PubMed]

toxics

MDPI

Article

Evaluation of the Safety of Cosmetic Ingredients and Their Skin Compatibility through In Silico and In Vivo Assessments of a Newly Developed Eye Serum

Anca Maria Juncan [1,2,3,*], Luca-Liviu Rus [3], Claudiu Morgovan [3] and Felicia Loghin [1]

1 Department of Toxicology, Faculty of Pharmacy, "Iuliu Hațieganu" University of Medicine and Pharmacy, 6 Pasteur Str., 400349 Cluj-Napoca, Romania; floghin@umfcluj.ro
2 SC Aviva Cosmetics SRL, 71A Kövari Str., 400217 Cluj-Napoca, Romania
3 Preclinic Department, Faculty of Medicine, "Lucian Blaga" University of Sibiu, 2A Lucian Blaga Str., 550169 Sibiu, Romania; liviu.rus@ulbsibiu.ro (L.-L.R.); claudiu.morgovan@ulbsibiu.ro (C.M.)
* Correspondence: juncan.anca@umfcluj.ro or ancamaria.juncan@ulbsibiu.ro

Abstract: The term "risk assessment" is often substituted with "safety assessment", to demonstrate the safe properties of cosmetic ingredients and formulations. With respect to the actual legislative framework, the proper use of in silico evaluation could offer a representative non-animal substitute for the toxicity evaluation of cosmetic ingredients. The in silico assessment needs to be integrated with other lines of proof (in vitro and/or in vivo data) in the form of a complex methodology in order to demonstrate the safety evaluation of cosmetic ingredients/products. The present study aimed to develop and characterize a new cosmetic formulation, designed for the skin care of the periorbital area. Quality control comprising stability, physicochemical, and microbiological evaluation was performed. Another objective of this study was to present a screening model for the safety evaluation of the cosmetic formulation by identifying individual ingredients, and to confirm the skin compatibility based on in vivo evaluation. The results demonstrated the in silico and in vivo safety profile of the cosmetic ingredients used in the present formulation. In silico evaluation, using a novel, specific software applicable for the risk evaluation of ingredients and formulations, showed that the incorporated ingredients were non-mutagenic and non-sensitizing, and considering the margin of safety (MoS), the cosmetic raw materials could be considered safe. Skin compatibility was confirmed by the patch test performed under dermatological control, evidencing the "non-irritating" potential of the developed cosmetic formulation.

Keywords: cosmetic ingredients; anti-ageing eye serum; safety assessment; skin compatibility; alternative methods; in silico evaluation

1. Introduction

The continuous evolution of the cosmetics industry, together with the development and improvement of the legislation designed to protect the consumer, has contributed to the increasing credibility of cosmetology and cosmetics. Whereas in the past cosmetic products were recommended and mainly used for beauty or hygienic purposes, nowadays they are applied for more complex and demanding objectives, such as maintaining an optimum skin condition, performing a eutrophic function, and finally fulfilling an aesthetic criterion, while being safe and effective. Three main aspects are considered of major importance regarding the formulation and development of cosmetic products—quality, safety, and efficacy [1,2].

According to Regulation (EC) No. 1223/2009 [2], which legally governs cosmetics in the European Union, it must be assumed that cosmetics are safe for human use, and that an assessment of this safety was performed, this assumption being an essential characteristic of the Cosmetics Regulation. Based on this regulation, the consumers are protected against a potential risk (Article 3 regarding safety), by establishing specific rules for the safety assessment of cosmetics (Article 10 referring to safety assessment, to ensure compliance with Article 3) [2,3]. An essential legal requirement is to ensure that the cosmetic formulation has been assessed according to a safety report (Cosmetic Product Safety Report (CPSR), part B) [2].

However, cosmetic formulations contain a mixture of different ingredients with various properties, like emollients, preservatives, humectants, surfactants, fragrances, colorants, and sometimes vegetal extracts [4,5]. Accordingly, even in cosmetics are generally safe for consumers, some allergic effects can be reported [5].

The cosmetics industry is highly motivated to apply non-animal approaches, also known as new approach methodologies (NAMs) or next-generation risk assessment (NGRA), including in vitro, in chemico, and in silico evaluations. These refer to toxicological endpoints and can be applied according to the Cosmetics Regulation No. 1223/2009, which prohibits animal testing, as well marketing of cosmetic ingredients and finished products [6,7].

NAMs used in the safety evaluation of cosmetic raw materials are nowadays at different stages of implementation, with some already in routine use, while others need more evidence to support their application (Figure 1) [8]. In silico data are useful in regulatory risk assessment when used as supporting evidence in an overall safety evaluation of cosmetic formulations. Figure 1 presents different NAMs, emphasizing in silico methods and clinical evaluation, considered for the safety assessment of a newly developed formulation, as described in this study.

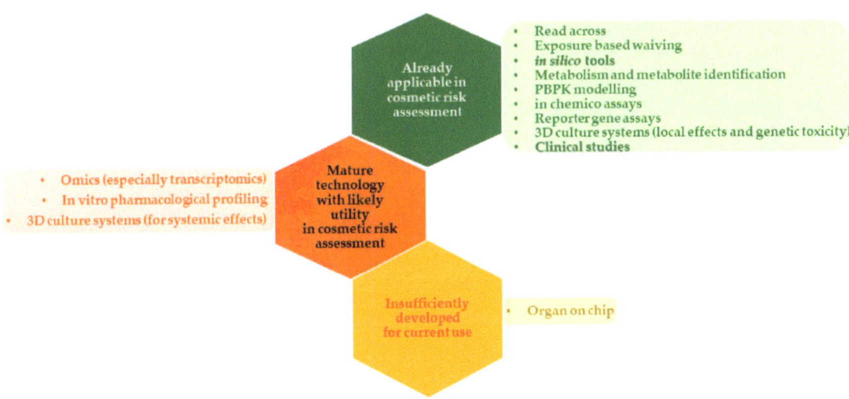

Figure 1. Status of new approach methodologies (NAMs) for the risk assessment of cosmetic ingredients (in silico tools and clinical studies are evidenced as these methods were considered for a comprehensive safety assessment of a novel cosmetic formulation).

Even if NAMs are currently widely used for the risk assessment of cosmetic ingredients and formulations, there is still a high demand for a combined risk evaluation methodology, as alternative tests are not always fully applicable for multicomponent ingredients and cosmetic formulations, this representing a main disadvantage compared, e.g., to in vivo evaluation [9].

The first objective of the present study was the formulation of an anti-ageing eye serum, specially designed for the periorbital skin area, which incorporates innovative active ingredients such as low-molecular-weight (LMW) HA, claiming an anti-ageing effect, and medium-molecular-weight (MMW) HA, possessing regenerating properties, together with an anti-ageing botanical complex based on horse chestnut flowers. Quality control of the formulation, evaluating the physicochemical characteristics and microbiology, and including a challenge (preservative efficacy evaluation), was performed in this study.

Another objective of this study was to carry out an in silico safety assessment, in order to predict the hazards and to evaluate the safety, by characterizing individual cosmetic ingredients of the developed cosmetic formulation. Finally, we sought to confirm the skin compatibility and tolerance through an in vivo assessment.

2. Materials and Methods

2.1. Selection and Safety-Level Data of Ingredients incorporated into the Cosmetic Formulation

A classification of the raw materials was performed, in order to select the ingredients for the cosmetic formulation (anti-ageing eye serum). Based on material safety data sheets (MSDSs), the following ingredients according to their INCI (International Nomenclature of Cosmetic Ingredients) and commercial denomination were selected for the present formulation:

(i) Butylene glycol cocoate (Cocoate BG, Gattefosse, France) is a multifunctional cosmetic ingredient, which functions as an emollient and solubilizer. It also provides excellent skin compatibility and possesses excellent sensorial properties [10].

(ii) Glycerin (Elton Corporation S.A., Ilfov, Romania) is a safe ingredient, widely used in cosmetics, possessing various functions like serving as a skin conditioning agent, humectant, skin protectant, hair conditioning agent, viscosity-decreasing agent, fragrance ingredient, and denaturant [11,12]. Also, glycerin was demonstrated to have optimal skin tolerability even on atopic dry skin [13].

(iii) As broad-spectrum cosmetic preservatives against bacteria, yeasts, and molds, phenoxyethanol (and) ethylhexylglycerin (Euxyl PE 9010, Schülke&Mayr GmbH, Norderstedt, Germany) were selected [14].

(iv) Low-molecular-weight HA (LMW-HA) (20–50 kDa) (PrimalHyal 50, Givaudan, France) improves skin biomechanical properties, like skin roughness and firmness [15,16], while medium-molecular-weight HA (MMW-HA) (100–300 kDa) (PrimalHyal 300, Givaudan, France) has the capacity to reinforce the skin's natural defense, along with possessing a regenerating effect [16,17]. Depending on its molecular weight, hyaluronic acid (HA) has different effects in skin care formulations, and in association with other active ingredients, supplementary benefits can be claimed [16].

(v) Fructose (and) glycerin (and) water (and) *Aesculus hippocastanum* (*Horse chestnut*) extract (Gatuline Link n Lift, Gattefosse, France) is a natural complex based on flowers of *Aesculus hippocastanum* (*horse chestnut*), with anti-ageing effects, improving skin texture and especially reducing crow's feet and under-eye wrinkles (length, surface, and volume) [18–20].

Available information on the safe use level and ingredient concentration was retrieved from the Cosmetic Ingredient Database (CosIng) [21] and Cosmetic Ingredient Review (CIR) [22], independent entities responsible for the safety evaluation of individual cosmetic ingredients [23]. Also, COSMILE Europe, a cosmetic ingredient database launched in February 2023 by Cosmetics Europe, was considered in order to find available information on the cosmetic ingredients' properties, functions, whether they are of synthetic or natural origin, and in which types of formulation they are appropriate [24]. Table 1 presents the general information and safety-level data of cosmetic raw materials in the formulated eye serum according to the CosIng, CIR, and COSMILE databases.

Table 1. General information and safety-level data of incorporated cosmetic ingredients according to the CIR, CosIng, and/or COSMILE databases.

INCI Name	CAS Nr.	Description	Cosmetic Restriction	Maximum Conc. in Ready for Use Preparation	Function(s)	SCCS Opinions	Ingredient Status Database (CosIng */CIR **/ COSMILE ***)	Reference
Aqua	7732-18-5	Water	N	N	Solvent	NA	CosIng/COSMILE	NA
Butylene Glycol Cocoate	73138-39-3	Coconut oil fatty acids, 2-hydroxybutyl ester	N	N	Surfactant—emulsifying Emulsion stabilizing	NA	CosIng/CIR/COSMILE	[25]
Glycerin	56-81-5	Glycerol	N	N	Denaturant Hair conditioning	NA	CosIng/CIR/COSMILE	[26]
Phenoxyethanol	122-99-6	2-Phenoxyethanol	Annex V of the EU Cosmetics Regulation (1223/2009/EU)	1%	Antimicrobial Preservative	Opinion concerning restrictions on materials listed in annex VI of Directive 76/768/EEC on Cosmetic Products **** Opinion on phenoxyethanol	CosIng/CIR/COSMILE	[27–29]
Ethylhexylglycerin	70445-33-9	3-[2-(Ethylhexyl)oxyl]-1,2-propandiol	N	N	Deodorant Skin conditioning	NA	CosIng/CIR/COSMILE	[30]
Hydrolysed Hyaluronic Acid		Hydrolyzed hyaluronic acid is the hydrolysate of hyaluronic acid derived by an acid, enzyme or another method of hydrolysis	N	N	Hair conditioning Humectant	NA	CosIng/CIR/COSMILE	[31]
Fructose	57-48-7	/	N	N	Humectant	NA	CosIng/CIR/COSMILE	[32]
Aesculus hippocastanum(*Horse chestnut*) Extract	8053-39-2	*Aesculus hippocastanum* flower extract is the extract of the flowers of *Horse chestnut*, *Aesculus hippocastanum L.*, Hippocastanaceae	N	N	Skin conditioning	NA	CosIng/COSMILE	NA

* CosIng—Cosmetic Ingredient Database, ** CIR—Cosmetic Ingredient Review, *** COSMILE Europe—Cosmetics Europe database; SCCS—Scientific Committee on Consumer Safety; **** the previous Directive 76/768/EC (adopted on 27 July 1976) was replaced by the EU Cosmetics Regulation (1223/2009/EU), adopted in 2009 and fully in force since July 2013; N—not applicable; NA—not available.

2.2. Development and Manufacturing Procedure of the Anti-Ageing Eye Serum

Phase A: Butylene glycol cocoate was heated at 75–80 °C.

Phase B: The aqueous phase, which incorporates ultrapure water (PURELAB® Option Q7 (Type I), ELGA LabWater, High Wycombe, UK), glycerin, and the preservative, was heated to 75 °C. Homogenization was performed to completely disperse the components of the aqueous phase.

Phase A was added to Phase B under continuous stirring, using a T 50 digital ULTRA-TURRAX equipped with a dispersing element S 50 N-G 45 G (IKA, Staufen, Germany) (1600 rpm for 15 min).

Phase C: The active complex fructose, glycerin, water, *Aesculus hippocastanum* (*horse chestnut*) extract, together with LMW and HMW hydrolyzed HA (previously dissolved in 10 mL of water for complete dissolution), were added under constant stirring (600 rpm for 5 min) to the emulsion obtained from phases A and B, previously completely cooled to 40 °C.

2.3. Quality Control of the Anti-Ageing Eye Serum

Complying with the requirements of Regulation 1223/2009, various evaluations were conducted for the developed formulation: (a) stability evaluation; (b) physicochemical control: organoleptic testing (appearance, color, odor), pH, density and viscosity evaluation; (c) microbiological control and preservative efficacy test (challenge test) [33–39].

2.3.1. Stability of the Cosmetic Formulation

For accelerated stability testing, the formulation was alternatively stored for 30 days at 4 °C (16 h) (LKUv 1610 MediLine, Liebherr, Germany), 20 °C (8 h), and 40 °C (16 h) (natural convection drying oven SLN-32 (STD), Pol-Eko, Wodzisław Śląski, Poland) [33,36,37].

2.3.2. Physicochemical Testing of the Developed Formulation

Another criterion for the quality evaluation was the physicochemical control. Thus, organoleptic testing (appearance, color, odor) (ISO 6658:2005 p. 5.4.2 [40]) was carried out and the pH (PB-234 ed. I of 03.10.2013r.), density (20 °C) (PB-155 ed. I of 2 May 2012), and viscosity (Brookfield DV-III Ultra, spindle Sc4-18/RPM: 250 (o/min/shear rate 330 (1/s)) were determined.

2.3.3. Microbiological Quality and Challenge Testing of the Cosmetic Formulation

The microbiological control of the formulation was performed applying standard methods: enumeration and detection of aerobic mesophilic bacteria [41], yeast and mold counts [42], and *Staphylococcus aureus* [43], *Candida albicans* [44], *Escherichia coli* [45], and *Pseudomonas aeruginosa* [46] detection.

The efficacy of the preservation system, respectively, phenoxyethanol and ethylhexylglycerin incorporated in the developed formulation, was also evaluated according to the international cosmetics challenge test standard (PN EN ISO 11930:2012 [47]) [33].

2.4. Safety Assessment of the Anti-Ageing Eye Serum

2.4.1. In Silico Approaches for Safety Evaluation of Cosmetic Ingredients and Risk Assessment of the Developed Anti-Ageing Serum

For this purpose, we used a special software, a system dedicated to the specific field of cosmetics, SpheraCosmolife (SpheraCosmolife v. 0.24), which is a module of LIFE VERMEER delivered by Kode Chemoinformatics together with the Istituto di Ricerche Farmacologiche Mario Negri IRCCS, Milan, Italy [48]. The software, applicable for integrated hazard and exposure assessment of cosmetic ingredients and formulations as part of risk evaluation, was implemented for the in silico assessment of the novel developed anti-ageing serum.

The evaluation process is defined within the Scientific Committee on Consumer Safety (SCCS) Notes of Guidance (NoG) for the testing of cosmetic ingredients and their safety evaluation [49]. The margin of safety (MoS), considering the systemic exposure dose, including several models for exposure and risk prediction, and also the threshold of toxicological concern (TTC), are determined with the aid of this software. For a substance to be considered safe, the MoS must be higher than 100, as specified in the SCCS NoG and initially proposed by the World Health Organization (WHO), when defining an interspecies and intraspecies factor each of 10. The software also indicates other toxicological features (e.g., mutagenicity and skin sensitization), to assure an overall assessment of the potential risk of the incorporated cosmetic ingredients [50].

2.4.2. Clinical Safety Evaluation of the Developed Cosmetic Formulation—Dermatological Semi-Open Test

The present study aimed to evaluate the sensitizing/irritant potential and the skin tolerance. After the application of the formulation, under a patch test, the probability of erythema or skin edema appearance was evaluated. This study included 25 healthy Caucasian people, with phototype I-IV according to the Fitzpatrick scale. Subjects with a known history of a dermatological, medical, and/or physical condition that could influence the outcome of this study were not included. Subjects' skin conditions were considered as general inclusion criteria in this study (skin without irritations and changes requiring pharmacological treatment, subjects using any treatment on the test site, having any active skin disease that could interfere with the purpose of this study). None of the subjects reported previous hypersensitivity or adverse reactions to the individual ingredients of the cosmetic formulation.

The patch (12 mm diameter Finn Chamber, SmartPractice, Phoenix, AZ, USA) was applied (for 48 h) on the arm or interscapular area. A "blank" control sample and a control sample with water were used to avoid inaccurate interpretations related to skin irritations. The skin reaction was examined by a dermatologist 30 min after patch removal. Other evaluations were performed 72 h and 96 h after application.

For clinical safety evaluation, evaluation parameters of skin reactions were assessed by (I) clinical erythema assessment on a six-point severity scale (0 = no erythema; 0.5 = light erythema; 1 = erythema and/or papules; 2 = erythema and/or papules and/or vesicles; 3 = erythema and/or papules and/or vesicles and/or blisters; 4 = erythema, bullous and/or ulcerative reaction and/or papules and/or vesicles and/or blisters), and by (II) clinical edema assessment based on a five-point severity scale (0 = no edema; 1 = very light edema, hardly visible; 2 = light edema; 3 = moderate edema; and 4 = strong edema (extended swelling even beyond the application area)) at all evaluation time points.

The results are expressed based on the average irritation index (X_{av}). The formulation could be considered "not-irritating", "slightly irritating", "moderately irritating", or "highly irritating" [33,51,52].

An informed consent form (ICF), including information about the purpose of this study, methodology, and possible side effects was filled in by the volunteer subjects. This study was conducted by an external laboratory (J.S. Hamilton Poland Sp. z o.o., Gdynia, Poland) in accordance with the recommendations of the Cosmetics Regulation and current guidelines [48–50].

3. Results

3.1. Anti-Ageing Eye Serum Formulation

MSDSs of cosmetic ingredient categories incorporated in the formulation were accessed for their INCI denomination, physicochemical characteristics, toxicological evaluation, compatibility with other cosmetic ingredients, and concentration level in cosmetic formulations. Available safe-level information for ingredients and safety information considering use restriction according to the current legal framework were obtained from the CosIng, CIR, and COSMILE databases. Considering all this information, a selection of cosmetic ingredients was performed. Table 2 presents the developed anti-ageing eye serum considering the commercial and INCI denomination of the ingredients, their function in the cosmetic formulation, the supplier, and the concentration limit.

Table 2. The anti-ageing eye serum formulation.

Commercial Name	INCI	Function	Supplier	INCI-KEY * (%)
Aqua	Water	Solvent		A
Cocoate BG	Butylene Glycol Cocoate	Emollient/solubilizer	Gattefossé	E
Glycerol	Glycerin	Denaturant/humectant/ solvent	ELTON	E
Euxyl PE 9010	Phenoxyethanol and Ethylhexylglycerin	Preservative	Schülke & Mayr GmbH	F
PrimalHyal™ 50	Hydrolyzed Hyaluronic Acid	Antistatic/humectant/ skin conditioning /moisturizing	Givaudan Active Beauty	F
PrimalHyal™ 300	Hydrolyzed Hyaluronic Acid	Antistatic/humectant/ skin conditioning/moisturizing	Givaudan Active Beauty	F
Gatuline Link n Lift	Fructose (and) Glycerin (and) Water (and) *Aesculus hippocastanum* (*Horse chestnut*) Extract	Active ingredient/ anti-ageing	Gattefossé	E

* INCI Key A > 50%; 1% < E ≤ 5%; 0.1% < F < 1%.

3.2. Quality Control of the Anti-Ageing Eye Serum—Physicochemical Characterization and Microbiological Evaluation

Considering the quality characteristics, several evaluations were conducted, in order to demonstrate the physicochemical and pharmacotechnical properties of the developed anti-ageing eye serum. The results demonstrated the stability of the developed formulation under the performed study and conditions, and they showed that it possesses adequate physicochemical properties. Table 3 presents the results of the physicochemical tests of the anti-ageing eye serum, initially and after 30 days, while the formulation was maintained alternatively at different temperatures (4, 20, and 40 °C).

Table 3. Anti-ageing eye serum physicochemical properties.

Parameter	Unit	Results	
		Initial	After 30 Days
Viscosity at 20 °C (Brookfield DV-III Ultra)	mPa·s	6.43 ± 0.07	7.47 ± 0.08
Density at 20 °C (PB-155 ed. I of 02.05.2012)	g/cm^3	1.016 ± 0.003	1.015 ± 0.003
Organoleptic testing (ISO 6658:2005 p. 5.4.2)			
Appearance		W/O mixture *	Liquid **
Color		Beige	Light yellow
Odor		Specific of cosmetic ingredients	Characteristic of ingredients
Consistency		Liquid	Liquid
pH (PB-234 ed. I of 03.10.2013r.)		5.5 ± 0.2	5.6 ± 0.2

* W/O—water in oil; **—without mechanical impurities.

The microbiological quality of the eye serum was confirmed, and the determinations are presented in Table 4. Also, the challenge test performed evidenced the effectiveness of the antimicrobial protection of the developed cosmetic formulation (Figure 2).

Table 4. Microbiological evaluation of the developed anti-ageing serum.

Parameter	Standard	Result (CFU/g)	Admissibility Limit (CFU/g)	Concordance
Enumeration and detection of aerobic mesophilic bacteria	ISO 21149:2017 [41]	<10	<100	√
Yeast and mold count	ISO 16212:2017 [42]	<10	<10	√
Staphylococcus aureus detection	ISO 22718:2016 [43]	-	-	√
Candida albicans detection	ISO 18416:2016 [44]	-	-	√
Escherichia coli detection	ISO 21150:2016 [45]	-	-	√
Pseudomonas aeruginosa detection	ISO 22717:2016 [46]	-	-	√

"-"—absent; CFU—colony-forming units.

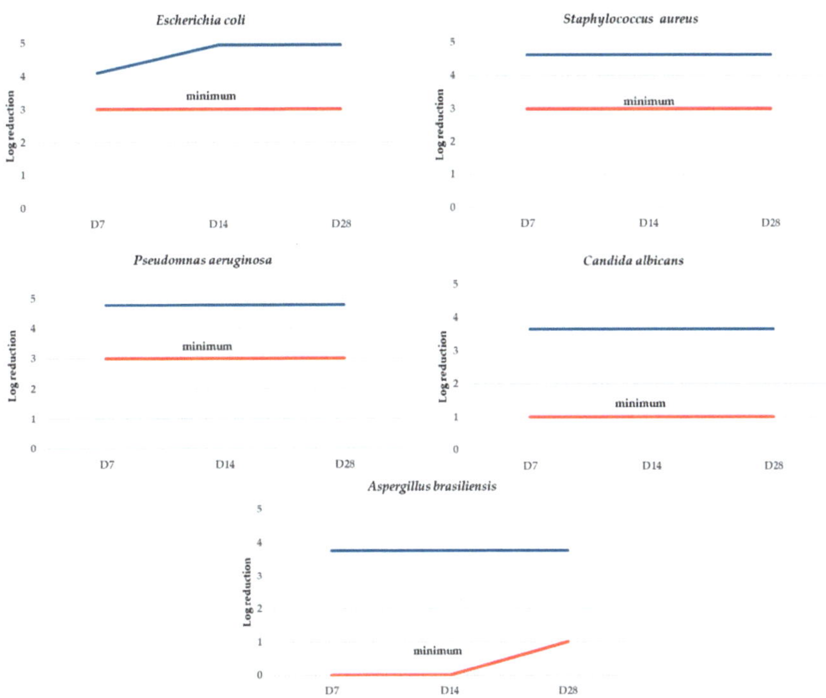

Figure 2. Anti-ageing serum challenge test results. D7—after 7 days; D14—after 14 days; D28—after 28 days. – – – antimicrobial protection, – – – standard minimum effectiveness.

3.3. Safety Assessment of the Developed Anti-Ageing Eye Serum

3.3.1. In Silico Assessment for the Safety Evaluation of Cosmetic Ingredients and Risk Assessment of the Formulation

The SpheraCosmolife software provided a summary table (Table 5) of the results for the ingredients incorporated in the eye serum. The results depend on the product type and on the presumed concentrations of the ingredients. The software shows ingredients present in any of the Annexes of the Cosmetics Regulation. Also, it presents the mutagenicity (Ames test), skin sensitization, the dermal absorption according to the Kroes approach, the MoS, and the TTC.

Table 5. Hazard and exposure specifications of the cosmetic ingredients incorporated into the anti-ageing eye serum.

Ingredient ID	CAS	INCI	Conc. % (*w*/*w*)	Annex	Mutagenicity	Skin Sensitization	Dermal Abs.	MoS	TTC *
Deionized Water	7732-18-5	Aqua	92.55	-	-	-	-	-	-
Cocoate BG	73138-39-3	Butylene Glycol Cocoate	1.00	-	-	-	-	-	-
Glycerin	56-81-5	Glycerin	4.00	-	NON-Mutagen	NON-Sensitizer	80%	334.07	**0.046 mg/kg bw/day**
Phenoxyethanol	122-99-6	Phenoxyethanol	0.90	V	NON-Mutagen	NON-Sensitizer	80%	460.28	**0.046 mg/kg bw/day**
Ethylhexylglycerin	70445-33-9	Ethylhexylglycerin	0.10	-	NON-Mutagen	NON-Sensitizer	80%	9737.47	**0.0023 mg/kg bw/day**
Hydrolyzed Hyaluronic Acid	9004-61-9	Hydrolyzed Hyaluronic Acid	0.70	-	NON-Mutagen	Sensitizer	10%	9,713,408.1	**0.0023 mg/kg bw/day**
Fructose	57-48-7	Fructose	2.50	-	NON-Mutagen	NON-Sensitizer	80%	2120.28	**0.046 mg/kg bw/day**
Aesculus hippocastanum (*Horse Chestnut*) Extract	8053-39-2	*Aesculus hippocastanum* (*Horse Chestnut*) Extract	0.50	-	-	-	-	-	-
	experimental value		good reliability		moderate reliability		low reliability NON-Senzitizer		low reliability Senzitizer

Product type: **anti-ageing eye serum**. *—TTC values according to Cramer class classification and adapted for cosmetic ingredients, as calculated by SpheraCosmolife: Class I—low toxicity (0.046 mg/kgbw/d) and Class II—medium toxicity (0.023 mg/kgbw/d).

For instance, in the developed cosmetic formulation, phenoxyethanol is the only restricted ingredient listed in Annex V of the Cosmetic Regulation (the maximum admissible concentration is 1%). A lower concentration than 1% (in our case 0.9%) must be introduced in the software and used, so that the formulation is safe and complies with the legislation. Moreover, in the "details" section regarding the provided regulatory aspects, the software checked if the ingredient is classified according to the Classification, Labelling and Packaging (CLP) regulation (EC N°1272/2008) (Figure 3).

Molecules found in the annex lists: 1

Found molecule no. 1	
CAS	122-99-6
INCI	PHENOXYETHANOL
Found in annex	ANNEX V: LIST OF PRESERVATIVES ALLOWED IN COSMETIC PRODUCTS
Annex details	ANNEX V: Reference Number: 29 ANNEX V: MAX Concentration in ready for use preparation: 1.0% ANNEX V: NOTES: 0125/99 - Opinion concerning Restrictions on Materials listed in annex VI of Directive 76/768/EEC on Cosmetic Products
Found Safer Chemicals classification	Yellow [Triangle]
Safer Chemicals details	Safer functional use: Preservatives-antioxidants
Found in CLP list	YES
CLP details	CLP HARMONIZED CLASSIFICATION (Link): https://echa.europa.eu/it/information-on-chemicals/cl-inventory-database/-/discli/details/57792 HAZARD CLASS AND CATEGORY CODE (CLASSIFICATION): Acute Tox. 4 * Eye Irrit. 2 HAZARD STATEMENT CODE (CLASSIFICATION): H302 H319 PICTOGRAM, SIGNAL WORD CODE (LABELLING): GHS07 Wng HAZARD STATEMENT CODE (LABELLING): H302 H319

Figure 3. Regulatory information provided by the SpheraCosmolife software for phenoxyethanol (*—Phenoxyethanol could cause serious eye irritation, considering the hazard statement according to ECHA).

The safety of phenoxyethanol, which was considered as a preservative complex together with ethylhexylglycerin in the developed formulation, was also supported by the MoS value calculated by the applied software, which in this case was much higher than 100 (MoS = 460.28), meaning it can be concluded that phenoxyethanol is safe at the proposed concentration.

The SpheraCosmolife software provided valuable data for mutagenicity and skin sensitization hazard identification and characterization, especially for hydrolyzed HA (LMW- and MMW-HA) at a concentration of 0.7% in a cosmetic formulation specifically recommended for the periorbital area. The software reported values for HA regarding mutagenicity (Ames integrated model) and skin sensitization (Caesar, DT model, and integrated model). Figure 4 presents information on the hazard assessment for the above-mentioned example. The provided values are highlighted with green or light green when they predict the ingredient as safe, while the other values are marked with pale red and predict a low reliability for skin sensitization.

Hazard identification

Mutagenicity - Ames test (Janus workflow prediction)	**NON-Mutagen (moderate reliability)**
Skin sensitization (Consensus of Caesar and JRC models)	**Sensitizer (low reliability)**
Skin sensitization (Caesar model prediction)	**NON-Sensitizer (LOW reliability)**
Skin sensitization (IRFMN/JRC model prediction)	**Sensitizer (LOW reliability)**

Figure 4. The output of the SpheraCosmolife software for the hazard identification for hydrolyzed HA (0.7% in the anti-ageing eye serum).

3.3.2. Anti-Ageing Serum Skin Tolerance—Dermatological Semi-Open Test

A very good skin tolerance was demonstrated for the developed cosmetic formulation. According to the patch test performed under dermatological control, none of the subjects included in this study reported any allergic reactions and/or irritation at T_1 (48 h after product application) or at T_2 (72 h after product application). Considering nonpositive skin reactions of the subjects at T_1/T_2, no further evaluation was performed at T_3 (96 h after application). The sum of negative reactions (erythema and edema) represents the average irritation index (X_{av}), calculated as the average of readings obtained on the subject panel (n = 25). The irritant potential of the tested formulation according to X_{av} was evaluated considering a four-scale classification ($X_{av} < 0.50$—"not irritating", $0.50 \leq X_{av} < 2.00$—"slightly irritating", $2.00 \leq X_{av} < 5.00$—"moderately irritating", and $5.00 \leq X_{av}$—"highly irritating"). Based on this evaluation, it can be concluded that the developed anti-ageing serum meets the requirements of the skin compatibility test and can be described as "not irritating" ($X_{av} < 0.50$) (Table 6).

Table 6. Skin responses of the 25 subjects included in the semi-open test for the assessment of the irritating and sensitizing effects of the developed anti-ageing eye serum, expressed as the average irritation index (X_{av}).

	T_1 (48 h after Application)	T_2 (72 h after Application)	T_3 (96 h after Application)
Erythema	0	0	Examination ignored
Edema	0	0	Examination ignored
X_{av}	0	0	

0 values in columns T_1 and T_2 refer to the sum of negative reactions (n = 25). X_{av}—represents the average irritation index (sum of negative reactions (erythema and edema)).

4. Discussion

The purpose of the Cosmetic Regulation is to protect consumers from potential health hazards and help them make informed decisions when purchasing cosmetic products. According to the present regulation, three main aspects are essential regarding a cosmetic formulation: quality, safety, and efficacy [2,3,38].

Hyaluronic acid represents a popular active ingredient, which is used in skin care products due to its topical benefits recognized as moisturizing, regenerating, or anti-ageing, depending on the molecular weight. Used in combination with other active ingredients, HA or HA derivates incorporated into cosmetic formulations can claim supplementary effects [16]. In the developed formulation presented in this study, LMW-HA and MMW-HA, together with a botanical active complex, were incorporated to sustain the claimed effect of the eye serum. LMW-HA stimulates tight junction protein synthesis, increasing moisturization, and sustains collagen type I synthesis, providing skin firmness. MMW-HA increases fibroblasts and keratinocytes' proliferation, sustaining the skin cellular regeneration process [15,17,53]. The extract based on Aesculus hippocastanum rich in flavonoids provides periorbital-wrinkle-reducing properties, claiming an anti-ageing effect by sustaining the dermo-epidermal junction integrity, increasing the synthesis of type IV and VII collagen, and holding a smoothing capacity [18,20,54].

Significant data about the ingredients' safety levels and concentration limits were first accessed through the CosIng, CIR, and COSMILE databases, according to the selection and incorporation of cosmetic ingredients in the developed eye serum, as described in this study.

The quality of a cosmetic preparation is related to the provision of specifications with the expected compliance, such as aspect, color and odor, physicochemical properties, stability, control of contaminants (for, e.g., pathogenic micro-organisms), etc. [55].

In the study performed, a stability test was conducted by exposing samples of the developed cosmetic formulation under accelerated temperature conditions, followed by evaluations of various parameters to observe any physical, chemical, or microbiological changes.

Also, interactions between the cosmetic formulation and the primary packaging should be considered for a complete safety assessment and according to current legislation [38,55].

Safety can be considered the provision of safe products for the general population under normal use conditions, which can be based on three main aspects, namely, ingredient safety, finished product safety, and cosmetovigilance (post-/in-market safety) [3].

The assurance of the safety of a formulation starts from the safe use of ingredients and can comprise two major points: (I) a risk-based assessment of the ingredients, and (II) compliance of the cosmetic product formulation with specific ingredient restrictions according to the current legislative framework [23].

In the European Union, the safety of a cosmetic formulation depends on the safety of the ingredients [56]. Each ingredient incorporated into the cosmetic formulation must prove a toxicological profile [39] broadly described in the MSDS. Generally, the SCCS cosmetic product safety evaluation relies on the principles and practices of the risk assessment process commonly applicable to individual cosmetic ingredients. This risk evaluation procedure is subdivided into four parts: (i) hazard identification, (ii) dose–response assessment, (iii) exposure assessment, and (iv) risk characterization [57,58].

The systemic effects of each ingredient, when evaluating all significant routes of absorption (dermal, oral, or inhalation) and the calculated MoS according to the No Observed Adverse Effect Level (NOAEL), must be considered for the safety profile of a cosmetic ingredient [39]. The SpheraCosmolife software provided data considering the exposure to the individual ingredients, which depended on the inputs in the system, that is, the formulation type and ingredient concentrations. The exposure was calculated for diverse scenarios (based on the parameters defined by the SCCS NoG): (i) absorption of 100% (oral or inhalation), (ii) absorption of 50% (a default value for dermal exposure), or (iii) the more realistic case for dermal absorption based on skin permeation models (the software provides the output of two models for skin permeation, choosing the worst scenario, considering the most conservative of those two values). With the aid of the software, we reported values (experimental or predicted) related to mutagenicity and skin sensitization, and the calculated MoS and the TTC were obtained, assessing the risk associated with the formulation's ingredients and demonstrating their safe use in the developed anti-ageing eye serum.

A special case of a restricted cosmetic ingredient in the developed formulation that should be taken into consideration is phenoxyethanol, a widely used preservative up to 1% in leave-on and rinse-off cosmetics. It presents high skin penetration, with an existing SCCS opinion confirming the safe use of this substance [27]. However, phenoxyethanol is mentioned in Annex V of the Cosmetics Regulation. Phenoxyethanol is not classified as a skin sensitizer by the European Chemicals Agency, although it showed rare sensitizer properties [59]. When used as a preservative complex with ethylhexylglycerin, phenoxyethanol at a final concentration of 0.9% was demonstrated to be safe according to its toxicity estimation in the developed formulation and in compliance with the legislation.

Skin sensitization, particularly followed by the development of allergic contact dermatitis, is probably the most important adverse reaction to cosmetic ingredients and/or products, considering their dermal route [60,61]. A substantial demand for risk assessment of cosmetic ingredients is to demonstrate that human exposure does not cause skin sensitization [60,62]. Evaluation of skin sensitization was previously performed based on a guinea pig assay (Magnusson and Kligman, Buehler), which was afterwords replaced by a more accurate method, namely the mouse local lymph node assay (LLNA). Nowadays, NAMs are widely applied for the risk evaluation of skin sensitizers [7,60]. In this sense, a very interesting research study was performed and presented by Kalicinska et al., describing an in silico model for the prediction of the sensitizing potential of cosmetic ingredients, especially haptens [63]. According to the values provided by the SpheraCosmolife software, considering the product type and the concentrations used of LMW- and MMW-HA, low skin sensitization was evidenced.

Finished product safety evaluations ensure that the product is safe, based on the safety profiles of the ingredients, as well as considering the product type and its normal conditions of use and exposure, complemented by a confirmatory safety assessment of the cosmetic formulation. The image in Figure 5 provides an overview of all aspects of the safety assessment for the developed formulation denominated an anti-ageing eye serum, including the special criteria of "ingredients' safety" and "compatibility of the finished product", including the "experience from the market" (market surveillance, cosmetovigilance), which should be considered according to the legal framework.

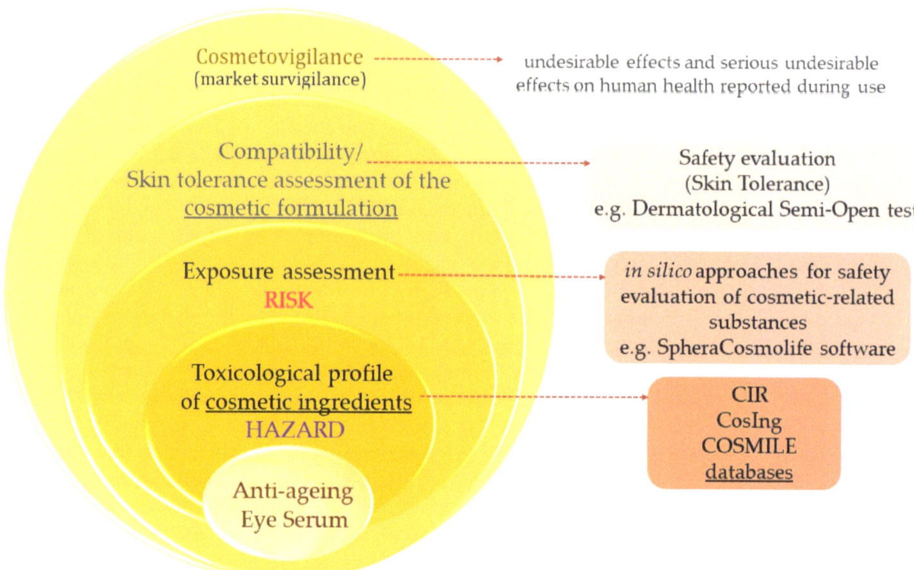

Figure 5. Fundamental aspects of safety assessment of the developed cosmetic formulation.

The above-described safety details provide useful information supporting the formulation's consumer safety. The results of the screening evaluation methodology applied for the assessment of the novel cosmetic formulation confirmed the safety of the individual ingredients and the formulation and the skin compatibility.

Further in vivo efficacy assessment, for, e.g., skin microrelief evaluation and/or periorbital wrinkles' length and depth reduction, confirmed the cosmetic claim of the anti-ageing formulation, completing our comprehensive characterization and evaluation according to the current legal framework.

5. Conclusions

The safety of cosmetic formulations represents an essential and mandatory legislative requirement and also a major industry priority. Cosmetic safety evaluation, implying testing of the ingredients and continuous improvement of the assessment capabilities, is a constant focus of innovation and research activities. The databases that we used review relevant data regarding the safety profiles of the cosmetic ingredients selected for the newly developed anti-ageing eye serum, and it can be stated that they are safe in their current usage. Compliance with the microbiological specifications was confirmed and we found that the physicochemical characteristics were adequate for the developed formulation. With the aid of the SpheraCosmolife software, risk characterization of the cosmetic raw materials used in the developed anti-ageing serum was performed, according to the process defined within the SCCS NoG. The MoS and TTC were calculated, together with the evaluation of other toxicological properties (mutagenicity and skin sensitization), confirming the safety of the cosmetic ingredients used. Phenoxyethanol, incorporated as a preservative in the developed cosmetic formulation, is the only restricted ingredient, listed in Annex V of the Cosmetic Regulation, but when used at a correct concentration level below 1%, it is confirmed to be safe. The SpheraCosmolife software provided valuable data regarding skin sensitization, especially for hydrolyzed HA, which in the proposed concentration of 0.7% and incorporated in a cosmetic product formulated for the periorbital area, showed low sensitizing properties. In vivo evaluation proved the skin compatibility through a lack of a sensitizing/irritant effect under a patch test for the developed cosmetic formulation.

Based on the complexity of these data, a screening safety evaluation method was performed by identifying individual ingredients, together with skin compatibility confirmation through in vivo evaluation. A combined risk assessment approach is still needed, considering that alternative testing methods are not always fully appropriate for multicomponent ingredients/formulations, with this representing the main disadvantage compared to in vivo evaluation.

Author Contributions: Conceptualization, A.M.J., C.M., L.-L.R., and F.L.; methodology, A.M.J., C.M., and L.-L.R.; software, A.M.J., C.M., and L.-L.R.; resources, A.M.J., C.M., and L.-L.R.; writing—original draft preparation, A.M.J., C.M., and L.-L.R.; writing—review and editing, A.M.J., C.M., L.-L.R., and F.L.; visualization, A.M.J., C.M., L.-L.R., and F.L.; supervision, F.L.; project administration, A.M.J. All authors have read and agreed to the published version of the manuscript.

Funding: This research was funded by the Ministry of Regional Development, Public Administration and European Funds (MDRAPFE), grant no. 16/AXA1/1.2.1C/04.10.2017 (DERMAGEFOR—"Development of some new dermatocosmetic formulations based on innovative active ingredients for the skin anti-ageing treatment").

Institutional Review Board Statement: The study protocol and the informed consent described in this study were approved by the Institutional Review Board of J.S. Hamilton Poland Sp. z o.o. (protocol no. 300879/19/JSHR).

Informed Consent Statement: Informed consent was obtained from all subjects involved in this study.

Data Availability Statement: Data are contained within the article.

Conflicts of Interest: Author Anca Maria Juncan is owner of the company SC Aviva Cosmetics SRL. The remaining authors declare that the research was conducted in the absence of any commercial or financial relationships that could be construed as potential conflicts of interest.

References

1. Ferreira, M.; Matos, A.; Couras, A.; Marto, J.; Ribeiro, H. Overview of Cosmetic Regulatory Frameworks around the World. *Cosmetics* **2022**, *9*, 72. [CrossRef]
2. Regulation (EC) No 1223/2009 of the European Parliament and of the Council of 30 November 2009 on Cosmetic Products. Available online: https://eur-lex.europa.eu/legal-content/EN/TXT/PDF/?uri=CELEX:32013R0655 (accessed on 17 February 2024).
3. Renner, G.; Audebert, F.; Burfeindt, J.; Calvet, B.; Caratas-Perifan, M.; Leal, M.E.; Gorni, R.; Long, A.; Meredith, E.; O'Sullivan, Ú.; et al. Cosmetics Europe Guidelines on the Management of Undesirable Effects and Reporting of Serious Undesirable Effects from Cosmetics in the European Union. *Cosmetics* **2017**, *4*, 1. [CrossRef]
4. Juncan, A.M.; Fetea, F.; Socaciu, C. Application of Fourier Transform Infrared Spectroscopy for the Characterization of 525 Sustainable Cosmetics and Ingredients with Antioxidant Potential. *Environ. Eng. Manag. J.* **2014**, *13*, 105–113. [CrossRef]
5. Goebel, C.; Kosemund-Meynen, K.; Gargano, E.M.; Politano, V.; von Bölcshazy, G.; Zupko, K.; Jaiswal, N.; Zhang, J.; Martin, S.; Neumann, D.; et al. Non-Animal Skin Sensitization Safety Assessments for Cosmetic Ingredients—What Is Possible Today? *Curr. Opin. Toxicol.* **2017**, *5*, 46–54. [CrossRef]
6. Gellatly, N.; Sewell, F. Regulatory Acceptance of in Silico Approaches for the Safety Assessment of Cosmetic-Related Substances. *Comput. Toxicol.* **2019**, *11*, 82–89. [CrossRef]
7. Vinardell, M.P.; Mitjans, M. Alternative Methods to Animal Testing for the Safety Evaluation of Cosmetic Ingredients: An Overview. *Cosmetics* **2017**, *4*, 30. [CrossRef]

8. Cosmetics Europe—The Personal Care Association Non-animal Approaches to Safety Assessment of Cosmetic Products-Edge Science and Constant Innovation: The Keys to Success. Available online: https://cosmeticseurope.eu/search/?q=in+silico535 (accessed on 21 March 2024).
9. Bialas, I.; Zelent-Kraciuk, S.; Jurowski, K. The Skin Sensitisation of Cosmetic Ingredients: Review of Actual Regulatory Status. *Toxics* **2023**, *11*, 392. [CrossRef] [PubMed]
10. Butylene Glycol Cocoate. Available online: https://explore.azelis.com/en_GB/ro_pc/cocoate-bg (accessed on 23 March 2024).
11. Lodén, M.; Wessman, W. The influence of a cream containing 20% glycerin and its vehicle on skin barrier properties. *Int J Cosmet Sci.* **2001**, *23*, 115–119. [CrossRef] [PubMed]
12. van Zuuren, E.J.; Fedorowicz, Z.; Arents, B.W.M. Emollients and Moisturizers for Eczema: Abridged Cochrane Systematic Review Including GRADE Assessments. *Br. J. Dermatol.* **2017**, *177*, 1256–1271. [CrossRef]
13. Loden, M.; Anderson, A.-C.; Anderson, C.; Bergbrant, I.-M.; Frödi, T.; Öhman, H.; Sandström, M.-H.; Särnhult, T.; Voog, E.; Stenberg, B.; et al. A Double-Blind Study Comparing the Effect of Glycerin and Urea on Dry, Eczematous Skin in Atopic Patients Patients. *Acta Derm. Venerol.* **2002**, *82*, 45–47. [CrossRef]
14. Phenoxyethanol (and) Ethylhexylglycerin Euxyl PE 9010 Preservative. Available online: https://www.ulprospector.com/en/eu/PersonalCare/Detail/33934/1318586/euxyl-PE-9010?st=1&sl=359373241&crit=a2V5d29yZDpbRXV4eWwgUEUg (accessed on 23 March 2024).
15. Hydrolyzed Hyaluronic Acid PrimalHyalTM 50. Available online: https://www.ulprospector.com/en/eu/PersonalCare/Detail/830/724175/PrimalHyal-50?st=1&sl=359377519&crit=a2V5d29yZDpbUHJpbWFsSHlhbCA1 (accessed on 23 March 2024).
16. Juncan, A.M.; Moisă, D.G.; Santini, A.; Morgovan, C.; Rus, L.L.; Vonica-Ţincu, A.L.; Loghin, F. Advantages of Hyaluronic Acid and Its Combination with Other Bioactive Ingredients in Cosmeceuticals. *Molecules* **2021**, *26*, 4429. [CrossRef] [PubMed]
17. Hydrolyzed Hyaluronic Acid PrimalHyalTM 300. Available online: https://www.ulprospector.com/en/eu/PersonalCare/Detail/830/724176/PrimalHyal-300?st=1&sl=359384606&crit=a2V5d29yZDpbUHJpbWFsSHlhbOK (accessed on 23 March 2024).
18. Fructose (and) Glycerin (and) Water (and) Aesculus Hippocastanum (Horse Chestnut) Extract GATULINE® LINK N LIFT. Available online: https://www.ulprospector.com/en/eu/PersonalCare/Detail/3983/718995/GATULINE-LINK-N-LIFT?st=1&sl=359431155&crit=a2V5d29yZDpbR2F0dWxp (accessed on 23 March 2024).
19. Clairet, A.; Bardin, V.; Trevisan, S.; Jomier, M. Horse Chestnut Flower Extract Redesigns Eye Contour. *Personal Care*, 9 May 2018; pp. 37–40.
20. Wilkinson, J.A.; Brown, A.M. Horse Chestnut-Aesculus Hippocastanum: Potential Applications in Cosmetic Skin-Care Products. *Int. J. Cosmet. Sci.* **1999**, *21*, 437–447. [CrossRef] [PubMed]
21. European Commission Cosmetic Ingredient Database Cosmetic Ingredient Database CosIng-Glossary of Ingredients. Available online: https://single-market-economy.ec.europa.eu/sectors/cosmetics/cosmetic-ingredient-database_en (accessed on 11 April 2024).
22. Cosmetic Ingredient Review CIR. Available online: https://www.cir-safety.org/ingredients (accessed on 11 April 2024).
23. Fung, E.S.; Drechsel, D.A.; Towle, K.M.; Hoang, M.T.; Novick, R.M.; Poteete, C.; Paustenbach, D.J.; Monnot, A.D. Screening-Level Safety Assessment of Personal Care Product Constituents Using Publicly Available Data. *Cosmetics* **2018**, *5*, 38. [CrossRef]
24. COSMILE Europe. Available online: https://cosmileeurope.eu/inci/ (accessed on 11 April 2024).
25. Burnett, C.L.; Bergfeld, W.F.; Belsito, D.V.; Klaassen, C.D.; Marks, J.G.; Shank, R.C.; Slaga, T.J.; Snyder, P.W.; Andersen, F.A. Final Report on the Safety Assessment of Cocos Nucifera (Coconut) Oil and Related Ingredients. *Int. J. Toxicol.* **2011**, *30*, 5S–16S. [CrossRef] [PubMed]
26. Becker, L.C.; Bergfeld, W.F.; Belsito, D.V.; Hill, R.A.; Klaassen, C.D.; Liebler, D.C.; Marks, J.G.; Shank, R.C.; Slaga, T.J.; Snyder, P.W.; et al. Safety Assessment of Glycerin as Used in Cosmetics. *Int. J. Toxicol.* **2019**, *38*, 6S–22S. [CrossRef] [PubMed]
27. Lilienblum, W.; Opinion of the Scientific Committee on Consumer Safety (SCCS). Final version of the opinion on Phenoxyethanol in cosmetic products. *Regul. Toxicol. Pharmacol.* **2016**, *82*, 156. [CrossRef] [PubMed]
28. Andersen, F.A. Annual Review of Cosmetic Ingredient Safety Assessments: 2007–2010. *Int. J. Toxicol.* **2011**, *30*, 73S–127S. [CrossRef] [PubMed]
29. Liebert, M.A. Final Report on the Safety Assessment of Phenoxyethanol. *J. Am. Coll. Toxicol.* **1990**, *9*, 259–277.
30. Johnson, W.; Bergfeld, W.F.; Belsito, D.V.; Hill, R.A.; Klaassen, C.D.; Liebler, D.; Marks, J.G.; Shank, R.C.; Slaga, T.J.; Snyder, P.W.; et al. Safety Assessment of Alkyl Glyceryl Ethers as Used in Cosmetics. *Int. J. Toxicol.* **2013**, *32*, 5S–21S. [CrossRef]
31. CIR. Safety Assessment of Hyaluronates as Used in Cosmetics. Available online: https://www.cir-safety.org/sites/default/files/SLR_HyaluronicAcid_092022.pdf (accessed on 21 April 2024).
32. Fiume, M.M.; Bergfeld, W.F.; Belsito, D.V.; Hill, R.A.; Klaassen, C.D.; Liebler, D.C.; Marks, J.G.; Shank, R.C.; Slaga, T.J.; Snyder, P.W.; et al. Safety Assessment of Monosaccharides, Disaccharides, and Related Ingredients as Used in Cosmetics. *Int. J. Toxicol.* **2019**, *38*, 5S–38S. [CrossRef]
33. Juncan, A.M.; Morgovan, C.; Rus, L.L.; Loghin, F. Development and Evaluation of a Novel Anti-Ageing Cream Based on Hyaluronic Acid and Other Innovative Cosmetic Actives. *Polymers* **2023**, *15*, 4134. [CrossRef] [PubMed]
34. Cosmetics Europe. Guidelines on Stability Testing of Cosmetic Products. 2004. Available online: https://www.cosmeticseurope.eu/files/5914/6407/8121/Guidelines_on_Stability_Testing_of_Cosmetics_CE-CTFA_-_2004.pdf (accessed on 23 April 2024).
35. *ISO/TR:18811*; Cosmetics-Guidelines on the Stability Testing of Cosmetic Products. ISO: Geneva, Switzerland, 2018.

36. Juncan, A.M.; Rus, L.L. Influence of Packaging and Stability Test Assessment of an Anti-Aging Cosmetic Cream. *Mater. Plast.* **2018**, *55*, 426–430. [CrossRef]
37. Juncan, A.M. Packaging Evaluation and Safety Assessment of a Cosmetic Product. *Mater. Plast.* **2018**, *55*, 644–647. [CrossRef]
38. Feregotto, T. Understanding the EU Cosmetics Regulation. *Personal Care*, 13 June 2017; pp. 21–23.
39. Eixarch, H.; Andrew, D. The Safety Assessment of Cosmetic Products. *Pers. Care* **2018**, *4*, 13–16.
40. *ISO 6658:2005*; Sensory Analysis—Methodology—General Guidance. ISO: Geneva, Switzerland, 2005.
41. *ISO 21149:2017*; Cosmetics—Microbiology—Enumeration and Detection of Aerobic Mesophilic Bacteria. ISO: Geneva, Switzerland, 2017.
42. *ISO 16212:2017*; Cosmetics—Microbiology—Enumeration of Yeast and Mould. ISO: Geneva, Switzerland, 2017.
43. *ISO 22718:2016*; Cosmetics—Microbiology—Detection of *Staphylococcus aureus*. ISO: Geneva, Switzerland, 2016.
44. *ISO 18416:2016*; Cosmetics—Microbiology—Detection of *Candida albicans*. ISO: Geneva, Switzerland, 2016.
45. *ISO 21150:2016*; Cosmetics—Microbiology—Detection of *Escherichia coli*. ISO: Geneva, Switzerland, 2016.
46. *ISO 22717:2016*; Cosmetics—Microbiology—Detection of *Pseudomonas aeruginosa*. ISO: Geneva, Switzerland, 2016.
47. *ISO 11930:2012*; Cosmetics—Microbiology—Evaluation of the Antimicrobial Protection of a Cosmetic Product. ISO: Geneva, Switzerland, 2012.
48. SpheraCosmolife: New Software Specific for Risk Assessment of Cosmetic Products. Available online: https://www.vegahub.eu/spheracosmolife-new-software-specific-for-risk-assessment-of-cosmetic-products/ (accessed on 23 April 2024).
49. SCCS. *The SCCS Notes of Guidance for the Testing of Cosmetic Ingredients and Their Safety Evaluation, 10th Revision*; SCCS: Cambs, UK, 24–25 October 2018; SCCS/1602/18.
50. Selvestrel, G.; Robino, F.; Baderna, D.; Manganelli, S.; Asturiol, D.; Manganaro, A.; Russo, M.Z.; Lavado, G.; Toma, C.; Roncaglioni, A.; et al. SpheraCosmolife: A New Tool for the Risk Assessment of Cosmetic Products. *ALTEX Altern. Anim. Exp.* **2021**, *38*, 565–579. [CrossRef]
51. Cosmetics Europe. Product Test Guidelines for the Assessment of Human Skin Compatibility. 1997. Available online: https://www.cosmeticseurope.eu/files/6014/6407/8875/Product_Test_Guidelines_for_the_Assessment_of_Human_Skin_Compatibility_-_1997.pdf (accessed on 24 April 2024).
52. Johansen, J.D.; Aalto-Korte, K.; Agner, T.; Andersen, K.E.; Bircher, A.; Bruze, M.; Cannavõ, A.; Giménez-Arnau, A.; Gonçalo, M.; Goossens, A.; et al. European Society of Contact Dermatitis Guideline for Diagnostic Patch Testing—Recommendations on Best Practice. *Contact Dermat.* **2015**, *73*, 195–221. [CrossRef]
53. Gruber, J.V.; Todurge, E. Ferment-Derived HA Complex Aids Skin Health. *Pers. Care* **2021**, *7*, 69–71.
54. Dudek-Makuch, M.; Matlawska, I. Flavonoids from the Flowers of Aesculus Hippocastanum. *Acta Pol. Pharm. Drug Res.* **2011**, *68*, 403–408.
55. Rizzi, V.; Gubitosa, J.; Fini, P.; Cosma, P. Neurocosmetics in Skincare-the Fascinating World of Skin-Brain Connection: A Review to Explore Ingredients, Commercial Products for Skin Aging, and Cosmetic Regulation. *Cosmetics* **2021**, *8*, 66. [CrossRef]
56. Rogiers, V. Animal-Free Cosmetics in Europe. In *The History of Alternative Test Methods in Toxicology*; Balls, M., Combes, R., Worth, A., Eds.; Elsevier Inc.: Amsterdam, The Netherlands, 2019; pp. 157–166, ISBN 9780128136980.
57. SCCS. *Notes of Guidance for the Testing of Cosmetic Ingredients and Their Safety Evaluation, 11th Revision*; SCCS: Cambs, UK, 30–31 March 2021; SCCS/1628/21.
58. SCCS. *The SCCS Notes of Guidance for the Testing of Cosmetic Ingredients and Their Safety Evaluation, 12th Revision*; SCCS: Cambs, UK, 21 December 2023; SCCS/1647/22.
59. Dréno, B.; Zuberbier, T.; Gelmetti, C.; Gontijo, G.; Marinovich, M. Safety Review of Phenoxyethanol When Used as a Preservative in Cosmetics. *J. Eur. Acad. Dermatol. Venereol.* **2019**, *33*, 15–24. [CrossRef]
60. Gilmour, N.; Kern, P.S.; Alépée, N.; Boislève, F.; Bury, D.; Clouet, E.; Hirota, M.; Hoffmann, S.; Kühnl, J.; Lalko, J.F.; et al. Development of a next Generation Risk Assessment Framework for the Evaluation of Skin Sensitisation of Cosmetic Ingredients. *Regul. Toxicol. Pharmacol.* **2020**, *116*, 104721. [CrossRef] [PubMed]
61. Silva, R.J.; Tamburic, S. A State-of-the-Art Review on the Alternatives to Animal Testing for the Safety Assessment of Cosmetics. *Cosmetics* **2022**, *9*, 90. [CrossRef]
62. Goebel, C.; Aeby, P.; Ade, N.; Alépée, N.; Aptula, A.; Araki, D.; Dufour, E.; Gilmour, N.; Hibatallah, J.; Keller, D.; et al. Guiding Principles for the Implementation of Non-Animal Safety Assessment Approaches for Cosmetics: Skin Sensitisation. *Regul. Toxicol. Pharmacol.* **2012**, *63*, 40–52. [CrossRef] [PubMed]
63. Kalicińska, J.; Wiśniowska, B.; Polak, S.; Spiewak, R. Artificial Intelligence That Predicts Sensitizing Potential of Cosmetic Ingredients with Accuracy Comparable to Animal and In Vitro Tests—How Does the Infotechnomics Compare to Other "Omics" in the Cosmetics Safety Assessment? *Int. J. Mol. Sci.* **2023**, *24*, 6801. [CrossRef] [PubMed]

toxics

MDPI

Article

An In Vitro Human Skin Test for Predicting Skin Sensitization and Adverse Immune Reactions to Biologics

Shaheda Sameena Ahmed [1], Mohammed Mahid Ahmed [1], Abbas Ishaq [1], Matthew Freer [1], Richard Stebbings [2] and Anne Mary Dickinson [1,3,*]

[1] Alcyomics Ltd., The Biosphere, Draymans Way, Newcastle Helix, Newcastle Upon Tyne NE4 5BX, UK; shaheda.ahmed@hotmail.com (S.S.A.); mahid.ahmed9@gmail.com (M.M.A.); abbas.ishaq@alcyomics.com (A.I.); matthew.freer@alcyomics.com (M.F.)

[2] National Institute for Biological Standards and Control, Blanche Lane, South Mimms, Potters Bar, Hertfordshire EN6 3QG, UK; richard.stebbings@astrazeneca.com

[3] Translational and Clinical Research Institute Faculty of Medical Sciences, Newcastle University, Newcastle-upon-Tyne NE2 4HH, UK

* Correspondence: anne.m.dickinson@alcyomics.com

Abstract: Biologics, including monoclonal antibodies (mAb), have proved to be effective and successful therapeutic agents, particularly in the treatment of cancer and immune-inflammatory conditions, as well as allergies and infections. However, their use carries an inherent risk of an immune-mediated adverse drug reaction. In this study, we describe the use of a novel pre-clinical human in vitro skin explant test for predicting skin sensitization and adverse immune reactions. The skin explant test was used to investigate the effects of therapeutic antibodies, which are known to cause a limited reaction in a small number of patients or more severe reactions. Material and Methods: Immune responses were determined by T cell proliferation and multiplex cytokine analysis, as well as histopathological analysis of skin damage (grades I–IV in increasing severity), predicting a negative (grade I) or positive (grade $\geq$ II) response for an adverse skin sensitization effect. Results: T cell proliferation responses were significantly increased in the positive group ($p < 0.004$). Multiplex cytokine analysis showed significantly increased levels of IFNγ, TNFα, IL-10, IL-12, IL-13, IL-1β, and IL-4 in the positive response group compared with the negative response group ($p < 0.0001$, $p < 0.0001$, $p < 0.002$, $p < 0.01$, $p < 0.04$, $p < 0.006$, and $p < 0.004$, respectively). Conclusions: Overall, the skin explant test correctly predicted the clinical outcome of 13 out of 16 therapeutic monoclonal antibodies with a correlation coefficient of 0.770 ($p = 0.0001$). This assay therefore provides a valuable pre-clinical test for predicting adverse immune reactions, including T cell proliferation and cytokine release, both associated with skin sensitization to monoclonal antibodies.

Keywords: skin explant; adverse event; T cell proliferation; monoclonal antibodies; cytokine release

1. Introduction

Biologics, including monoclonal antibodies (mAbs), are widely used as therapeutic agents for the treatment of disease. The development of therapeutic antibodies has advanced over the last few decades. The transition from mouse monoclonal antibodies [1] to humanized forms [2,3] has resulted in reduced immunotoxicity and improved safety and efficacy. However, a low number of adverse events still occur, halting clinical trials, despite the improvement in the use of humanized monoclonal antibodies. Common adverse events include, but are not limited to, skin sensitization, infusion reactions, anaphylaxis, rash, anti-drug antibody (ADA) development, and cytokine release syndrome, as well as other complications [4].

Hypersensitivity involves cytokine release and symptoms, which can range from skin sensitization and rashes to anaphylaxis and organ failure and are classified from type I to IV [5]. Type I immediate adverse response is associated with IgE antibody release and anaphylaxis [6]. A type II response is an antibody-dependent cytotoxic response [7], and a type IV response is a delayed T cell-mediated hypersensitivity response [8]. Marketed monoclonal antibody therapeutics have been associated with hypersensitivity reactions [9,10] involving elevated levels of circulating cytokines (Cytokine Release Syndrome (CRS)) and subsequent serious adverse reactions, including anaphylaxis [11].

Current practices for preclinical safety evaluation of mAbs or biopharmaceuticals involve the use of in vivo animal studies, followed by standard in vitro toxicity studies [12]. The main aim of these pre-clinical studies is to determine the immunotoxic properties of mAbs and determine a safe first-in-human dose to conduct Phase I clinical trials. However, the success of the use of these methods is limited to the selection of relevant animal species and the correct interpretation of animal study data for humans [13,14]. However, this approach can lead to misleading data, as in the case of TeGenero in 2006 [15]. In this trial, cynomolgus and rhesus monkeys were chosen to perform preclinical in vivo studies because of the similarity in their CD28 receptor to the human CD28 receptor and their binding affinity to the TGN1412 antibody [16]. A repeat dose study was performed, and no adverse reactions or toxicity were observed. The immunotoxicity of the TGN1412 antibody, however, did not become apparent until administration to Phase I volunteers, resulting in severe cytokine release syndrome (CRS) and organ failure. There is a pressing need to develop preclinical human in vitro tests that can predict the immunotoxic properties of biologics. To this end, we have developed a novel human in vitro skin explant test to predict cytokine-associated adverse responses, skin sensitization, or immunotoxic properties of biologics as a first-in-man test. The test has been used to predict adverse immune reactions to sensitizers and non-sensitizers [17] as well as hypersensitivity responses to low molecular weight (LMW) drugs [18]. In the current study, the skin explant test was further modified for the safety assessment of biologics. The test mimics the in vivo immunological response upon exposure to a mAb, with the elicitation of an immune response in the form of histopathological skin damage (skin sensitization reactions). The predictive endpoint of the test is based on the observed histopathological tissue damage, which is scored (grades I–IV) in order of increasing severity of damage in the skin [19]. Furthermore, the grading of damage in the skin allows for the determination of dose response. This novel endpoint has not been previously described for any tests predicting cytokine-associated adverse events or direct skin sensitization.

In this study, we investigated the effects of 16 known therapeutic mAbs, each with the potential to stimulate a different arm of the immune system (Fab or Fc).

2. Materials and Methods

2.1. Healthy Volunteers

This study involved the use of blood (60 mL) and skin biopsies (two 4 mm punch biopsies) taken from 10 healthy volunteers with informed consent at a dermatology clinic by a National Health Service research nurse who also conducted a questionnaire to ensure the volunteers had no underlying condition and were fit and healthy. Each set of blood and skin biopsies was used to test each mAb. This research was approved by the Local Research Ethics Committee (LREC).

2.2. Antibodies

Sixteen test antibodies were provided by the National Institute of Biological Standards and Control (NIBSC). Antibodies were selected based on the known ability of the antibody, as defined by the box warnings, to give rise to responses ranging from extreme to weak in the form of a subcutaneous rash or a systemic response causing organ toxicity (Table 1). These included 11 antibodies of the IgG1 subclass: campath-1H (alemtuzumab), mabthera (rituximab), remicade (infliximab), remsima (infliximab), inflectra (infliximab), embrel (etanercept), humira (adalimumab), simulect (basliximab), erbitux (cetuximab), arzerra (ofatumumab), simponi (golimumab), 2 antibodies of the IgG2 subclass orthoclon-OKT3 (muromonab) and vectibix (panitumumab), and 2 antibodies of the IgG4 subclass: tysabri (natalizumab) and a TGN1412 homolog. Finally, the biological PEGylated fragment of the monoclonal antibody cimzia (certolizumab) was also tested. All antibodies were tested at 1 μg/mL. This dose was recommended by colleagues at the NIBSC and was also shown to be within the optimal range in assays for assessment of monoclonal antibodies as published by Stebbings et al. [20], but it was also the minimum dose in pilot studies that produced negative skin explant grades (Grades 0–I) with the weak positive control antibody tysabri and positive grades II–III with the positive antibody OKT3. Human isotype controls were used in all assays; this included IgG1 (A50183H, AMS Biotechnology, Abingdon, UK) for 11 biologics, IgG2 (HCA193, Bio-Rad, formerly AbD Serotec, Hercules, CA, USA) for panitumumab and muromonab, and finally, IgG4 (A01241H, AMS Biotechnology) for natalizumab and TGN1412.

2.3. Skin Explant Test

The skin explant tests were performed on a minimum of 10 healthy volunteers. Whole blood (60 mL) was used to isolate peripheral blood mononuclear cells (PBMCs) using the Lymphoprep™ (Gibco, Grand Island, NY, USA) method and density gradient centrifugation. Five ml of whole blood was used to collect serum, which was heat inactivated (56 °C for 30 min) prior to use. Two 4 mm skin biopsies were collected in X-Vivo-10 medium (Lonza, Basel, Switzerland) and were cut into 12 equal sections prior to use. The antibody of interest was co-cultured at 1 μg/mL concentration with or without PBMCs (1×10^6) and autologous skin sections in RPMI 1640 (Gibco, London, UK) containing 100 IU/mL penicillin, 100 μg/mL streptomycin (Gibco UK), and 2 mM L-glutamine (Gibco UK) supplemented with 20% heat-inactivated autologous serum for 3 days. Routinely, all PBMC showed greater than 95% viability with the addition of PBMC to the assay. Background control skin was cultured with medium alone [21]. To determine non-specific binding, the antibody isotype was used as a negative control to assess background histopathological damage. Following the 3-day incubation, supernatants were collected for cytokine analysis, and the skin was paraffin embedded, sectioned, and stained with hematoxylin and eosin. Skin explants were examined blindly by two independent observers for histopathological damage. Using the Lerner histological grading system [19], the damage observed was assigned a grade (I–IV) relative to the severity of the damage observed (Figure 1). The grading system was as previously described [21] and as follows: a grade I response is considered a negative reaction with normal skin pathology but may show some mild vacuolization of cells. Grades II–IV are considered positive reactions. Grade II is like a skin sensitization rash and shows damage in the form of vacuolization of keratinocytes and dyskeratotic bodies. Grade III is analogous to blistering and damage at the dermal/epidermal junction with cleft formation. Grade IV is similar to skin peeling and displays complete separation of the dermis from the epidermis.

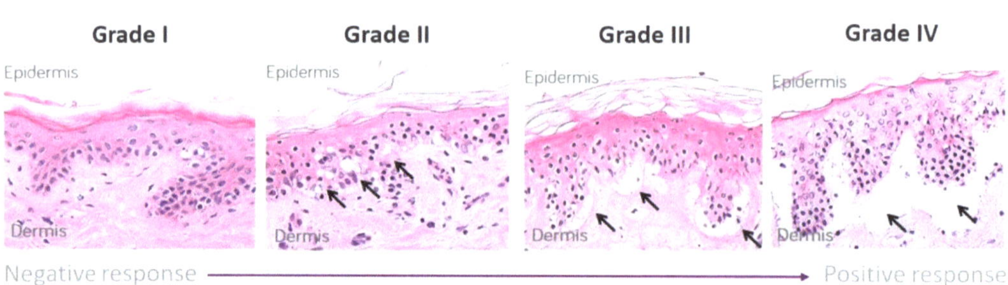

Figure 1. Lerner grading scheme for sub-epidermal lesions in the skin explant assay. The Grade II arrows indicate epidermal skin damage with vascularization and dyskeratosis; the Grade III arrows show Grade II damage as well as sub-epidermal cleft formation; Grade IV arrows show extensive sub-epidermal damage with separation of the dermis and epidermis.

Table 1. Therapeutic antibodies tested in the study. Antibodies were selected for their known clinical outcome for adverse events. The antibody name, type, target, mechanism, clinical class of response, positive (extreme, strong, or weak) or negative, box warnings occurrence is given. The total number of tests that gave a positive histopathological response (grade II and above) over the total number of tests performed (n = 10) is shown and expressed as a percentage, and the average skin explant grade (converted to grades 1–4) (rather than expressed as roman numerals) from the results of the ten donors is also given. A √ tick indicates a correct correlation with expected clinical responses. Responses are shown as weak positives 10–30%, positives 31–70%, and extreme responses >71%.

Antibodies	Target	Mechanism of Action (MOA)	Box Warning	Clinical Classification	Most Common Adverse Reaction	% Positive Skin Explant Test	Skin Explant Test Classification	Percentage Group	Average Skin Explant Test Grade (10 Donors)	Aggregated Score (Grade * Percentage)
OKT3 (Muromonab)	CD3	Blocks T cell activation	Anaphylaxis, Cytokine storm	Extreme (>10%)	Cytokine release syndrome, Pyrexia	96% (√)	Strongly Positive	71–100	2.8	2.7
NIB1412 (TGN1412)	CD28	Activation of regulatory T cells	Caused severe systemic immune reaction	Extreme (>10%)	Cytokine storm	90% (√)	Strongly Positive	71–100	2.3	2.1
Campath® (Alem-tuzumab)	CD52	Binds CD52 protein on lymphocytes	Systemic immunogenicity, Rash, urticaria, erythema.	Common (6%)	Infusion reactions, fever	70% (√)	Positive	71–100	2.4	1.6
Humira® (Adali-mumab)*	TNFα	Prevents TNF receptor activation, down regulates inflammation	Skin reactivity and hypersensitivity	Common (7%)	Injection site reactions, Rash	70% (√)	Positive	71–100	2.5	1.8
Simulect® (Basiliximab)	IL-2	Saturates IL-2 receptors, prevents T cell + B cell activation	Immunogenicity, hypersensitivity, Rash	Uncommon 1/1000	Fever, acne. Rash, skin ulceration	54% (X)	Positive	31–70	2.0	1.1
Mabthera® (Rituximab)	B cell	Destroys normal & malignant CD20+ B cells	Fatal IRs, Hypersensitivity, anaphylaxis, Cutaneous SJS, TEN	Common (10%)	Fever, urticaria/rash	50% (√)	Positive	31–70	2.0	1.0
Arzerra® (Ofatu-mumab)	CD-20	Inhibits early B cell activation	Cutaneous: Rash, urticaria,	Common (10%)	Rash, pyrexia	50% (√)	Positive	31–70	1.7	0.9
Erbitux® (Cetuximab)	EGFR	Targets EGFR expressing cells	Serious IR, dermatologic toxicity, rash, xeroderma, inflammation	Common (10%)	Urticaria, Rash, Pyrexia	50% (√)	Positive	31–70	1.7	0.8

Table 1. *Cont.*

Antibodies	Target	Mechanism of Action (MOA)	Box Warning	Clinical Classification	Most Common Adverse Reaction	% Positive Skin Explant Test	Skin Explant Test Classification	Percentage Group	Average Skin Explant Test Grade (10 Donors)	Aggregated Score (Grade * Percentage)
Avastin® (Bevacizumab)	VEGF	Binds to and inhibits VEGF	Hypersensitivity and infusion reactions: Rash, urticaria.	Common (8.4%)	Dry skin, exfoliative dermatitis	50% (√)	Positive	31–70	N/A	N/A
Inflectra® (Infliximab)	TNFα	Neutralizing bind to TNFα	Anaphylaxis, IRs, vasculitis; SJS; EM; psoriasis	Common (3%)	Fever, chills, Anaphalaxis	40% (√)	Positive	31–70	1.6	0.6
Simponi® (Golimumab)	TNFα	Targets soluble and transmem-brane TNFα	Skin exfoliation, rash	Common (2%)	Injection site reactions, erythema, urticaria, induration, Psoriasis, Palmar/Plantar	40% (√)	Positive	31–70	1.7	0.7
Remicade® (Infliximab)	TNFα	Neutralizing bind to TNFα	Anaphylaxis, IRs, vasculitis, SJS; EM; psoriasis	Common (3%)	Urticaria, anaphylaxis erythematous rash	35% (√)	Positive	31–70	1.5	0.6
Vectibix® (Panitu-mumab)	EGFR	Blocks signals on EGFR expressing cancer cells	Dermatologic toxicity, IR	Uncommon <1%	Severe skin toxicity (16%)	30% (√)	Weak Positive	10–30	1.3	0.4
Cimzia® (Cer-tolizumab)	TNFα	Binds free TNFα preventing its action	Allergic reactions, Dermatitis	Common (1 to 2.5%)	Pyrexia, Rash (0.3%)	30% (X)	Weak Positive	10–30	1.4	0.4
Tysabri® (Natal-izumab)	α4-integrin	Prevents immune cells crossing blood vessel walls	Hypersensitivity, Rash	Uncommon <1%	Hypersensitivity, anaphylaxis	20% (√)	Weak positive	10–30	1.6	0.3
Enbrel® (Etanercept)	TNFα	Binds TNF prevents its binding to WBCs	Psoriasis, allergic reactions, autoimmune reactions	Common (4%)	Rash, Pyrexia	10% (X)	Weak positive	10–30	1.8	0.2

2.4. [3H]-Thymidine T Cell Proliferation Measurement

T cell proliferation responses were measured in the same donors for the skin explant tests following exposure of PBMC's to the biologics at 1 µg/mL concentration. T cell responses were correlated to a negative response (grade I) or a positive response (≥grade II) in the skin explant tests for each donor (n = 10) and each test antibody. PBMC's were incubated with the test antibody isotype or control antibody (1 µg/mL) for 5 days in triplicate in RPMI 1640 (Gibco, UK) containing 100 IU/mL penicillin, 100 µg/mL streptomycin (Gibco UK), and 2 mM L-glutamine (Gibco UK) supplemented with 10% heat-inactivated fetal calf serum. Cells were harvested and [3H] Thymidine added for the last 16–18 h. [3H]-thymidine primary stock was stored at 37 MBq and used at 3.7 MBq (1/10 dilution), and subsequent uptake was measured using a microbeta-scintillation counter in counts per minute (cpm). Data were interpreted using Prism GraphPad software (V5, San Diego, CA, USA).

2.5. Analysis of Cytokine Concentration in Cell Culture Supernatants

Supernatants collected from the skin explant cultures were measured according to the manufacturer's instructions by Meso Scale Discovery (MSD) V-PLEX Proinflammatory Panel 1 for human IFNγ, TNFα, IL-10, IL-12, IL-13, IL-1β, IL-2, IL-4, and IL-6.

2.6. Statistical Analysis

(1) Skin Explant Data

Skin explant results were assigned grading scores for histopathological damage, and this positive or negative score was then compared with clinical responses and box warnings as reported for each test antibody. The classification of the data was as follows, with the clinical comparison given in brackets. An antibody was a strong positive if 71% or more of the total tests performed gave a grade ≥II response (frequent > 10%), a positive if 31–70% of the total tests performed gave a ≥grade II (common 1–10%), and a weak positive if 10–30% of the total tests performed gave a ≥grade II (uncommon event 0.1–1%). Pearson's correlation coefficient was used to correlate skin explant grades with clinical outcome data, i.e., frequent, common, uncommon, or rare clinical events. Additionally, average grades over 10 donors for each tested compound were calculated. An aggregate score was then derived by multiplying the average grades by the percentage of positive donors. The classifications and aggregate scores are in-set predictions. The aggregate score was then compared against the clinical frequency classification above using Pearson's correlation.

Where relevant, p-values were adjusted for the false discovery rate using a Benjamini-Hochberg correction. The adjusted values are indicated on the individual graphs.

(2) T Cell Proliferation Responses

T cell proliferation data was obtained by determining the mean cpm from triplicate culture wells of [3H] thymidine incorporation, converted to a log-fold increase to allow for equal distribution of data. T cell proliferation stimulation indices were matched to the graded response observed in the skin explant tests for each donor. Responses were grouped as negative (grade I) or positive (grade ≥ II). T cell proliferation responses were compared between the two groups, and statistical analysis was performed with an unpaired t test.

(3) Cytokine Levels in Cell Culture Supernatants

Cytokine levels in cell culture supernatants were measured as concentrations of pg/mL. Cytokine levels were compared between the two groups, and statistical analysis was performed with an unpaired t test. A p-value of <0.05 was statistically significant. Correlation coefficients were determined using Statistical Package for Social Sciences (SPSS) Statistics 7. Unpaired t-tests were replicated in R (R Studio Build 402), and false discovery rate correction was applied using the Benjamini-Hochberg method. The q-values did not significantly modify the t-test p-values.

3. Results

3.1. Skin Explant Test Demonstrates a Positive Correlation with Clinical Responses in 14/16 Therapeutic Antibodies

Results for each test biologic are summarized in Table 1, and representative images of the biologics showing positive and negative responses and controls are shown in Figure 2. The three Infliximab antibodies are represented as one image, as they all showed the same average grades. Positive reactions presented histopathological damage in the skin consisting of vacuolization of basal cells with some dyskeratotic bodies (grade II) (Figure 2E–I) or sub-epidermal cleft formation (grade III) (Figure 2K–N). Negative reactions showed a grade I response in the skin (Figure 2A–D and J). Isotype antibodies or skin cultured in medium alone were used as negative controls and gave a negative response (grade I) in all tests (Figure 2O–Q).

13 out of the total 16 tested monoclonal antibodies were correctly categorized by the skin explant test according to the box warnings and clinically relevant classifications (frequent, common, uncommon, and rare) observed in clinical trials for immunological reactivity (Table 1). The skin explant categorization was an in-set prediction.

Biologics that were correctly predicted positive were TGN1412 homolog (9 of 10 tests showed a grade $\geq$ II positive response), OKT3 was positive in 9 of 10 initial experiments and was then subsequently used as a positive control in all tests of the therapeutic antibodies. OKT3 was positive in 29 of 30 tests with a grade >II response. Alemtuzumab showed a grade $\geq$ II positive response in 7 of 10 tests. Rituximab, cetuximab, ofatumumab, and bevacizumab also showed a positive response (5 of 10 tests were positive, respectively), and Simponi (4 of 10 tests showed a $\geq$grade II). Panitumumab, infliximab, and certolizumab (3 of 10 tests positive) were also categorized as positive; however, the grading was generally weaker.

Natalizumab (8 of 10 tests showed a grade I negative response) and etanercept (9 of 10 tests showed a grade I negative response) were categorized as weakly positive.

The three incorrectly categorized compounds were simulect and panitumumab, which were overclassified, and etanercept, which was under classified, respectively.

A strong Pearson's correlation coefficient of r = 0.834 and a p value of <0.0001 were observed between the two comparatives (skin explant and clinical outcome data). Furthermore, the percentage of donors positive for a skin explant test reaction strongly correlated with the average grade across all 10 donors (r = 0.896).

3.2. T Cell and IGNɣ Responses Correlated with Responses Induced by Therapeutic Biologics and Histopathological Damage Observed in Skin Explants

The mean T cell proliferation observed in the positive group was 0.79 ((Log_2) $\pm$ 1.28 SEM) and in the negative group -4.15 ((Log_2) $\pm$ 1.23 SEM). A significant increase in T cell proliferation was observed in the positive response group ($p < 0.004$) compared with the negative response group. Results showed a correlation coefficient of 0.61 ($p < 0.0001$) with IFNɣ levels and skin explant grades (Figure 3).

3.3. Cytokines were Increased in Cell Culture Supernatants from Positive Skin Explant Tests

Cytokines, IFNγ, TNFα, IL-10, IL-12, IL-13, IL-1β, IL-6, IL-4, and IL-2 were measured (Figure 4) in cell culture supernatants collected from skin explant assays. Cytokine levels were correlated to negative (grade I) or positive (grade $\geq$ II) responses in the skin explant tests. Significantly increased levels of IFNγ, TNFα, IL-10, IL-12, IL-13, IL-1β, and IL-4 were observed in the positive response group compared with the negative response ($p < 0.0001$, $p < 0.0001$, $p < 0.002$, $p < 0.01$, $p < 0.04$, $p < 0.006$, and $p < 0.004$, respectively). No differences were observed in IL-2 and IL-6 levels between the positive and negative groups. In the later case, this was due to increased IL-6 levels being produced in normal skin cultures in medium alone (Figure 4).

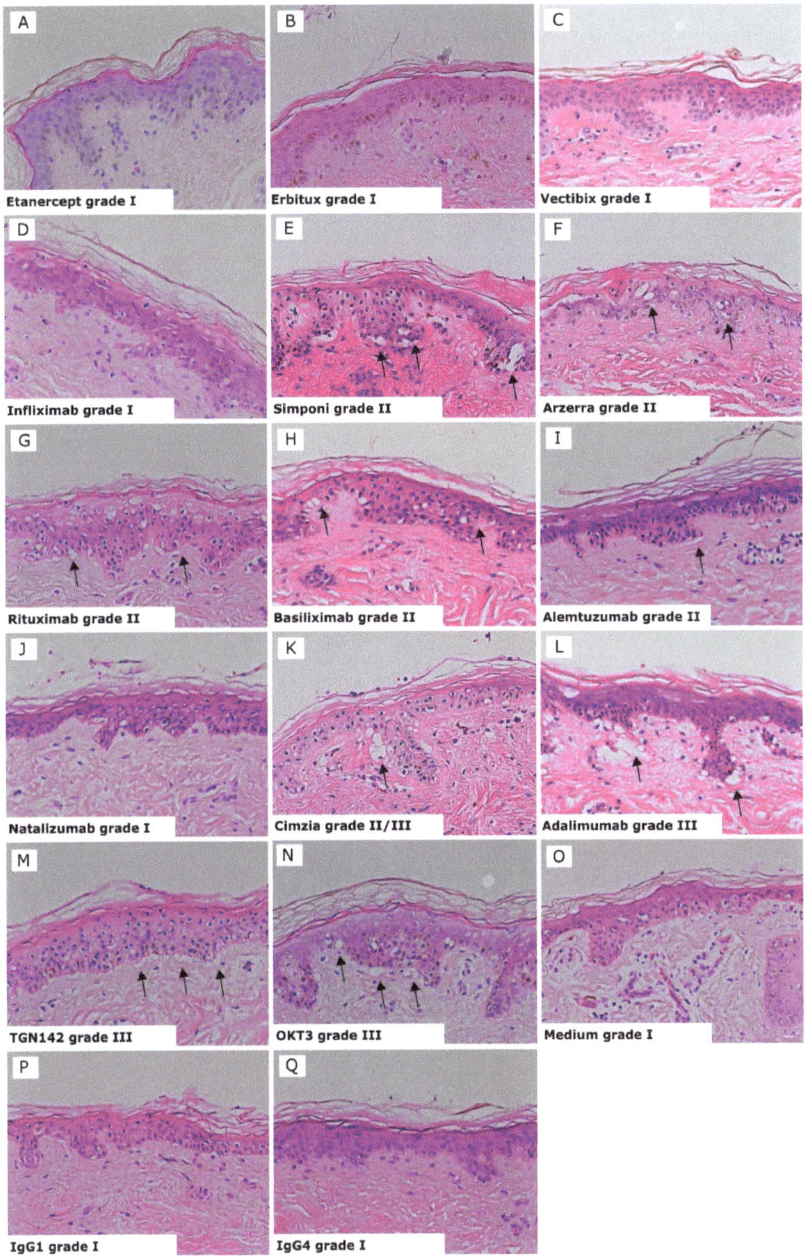

Figure 2. Skin explant results gave either a negative response (grade I) or positive response indicated by black arrows (grades II–III). (**A**) Etanercept grade I; (**B**) Erbitux grade I; (**C**) Vectibix grade I; (**D**) Infliximab, all brands, grade I; (**E**) Simponi grade II; (**F**) Arzerra grade II; (**G**) Rituximab grade II; (**H**) Basiliximab grade II; (**I**) Alemtuzumab grade II with vacuolarization and dyskeratosis of the epidermis; (**J**) Natalizumab grade I; (**K**) Cimzia grade II/III; (**L**) Adalimumab grade III; (**M**) TGN1412 grade III with separation of the dermis and epidermis; (**N**) OKT3 grade III; with vaculorization and dyskeratosis of the epidermis; (**O**) Media grade I; (**P**) IgG1 isotype control grade I; (**Q**) IgG4 isotype control grade I.

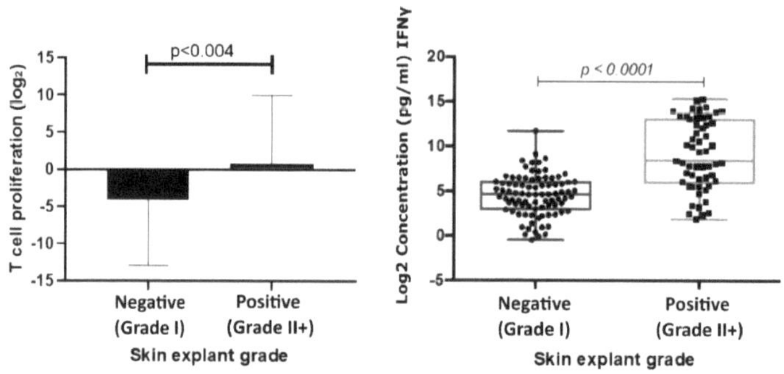

Figure 3. T cell proliferation responses (**left**) correlated with responses observed in skin explant tests. T cell proliferation responses were measured following incubation of PBMCs with test biologics. The graph shows a comparison of Log2 stimulation indices of T cell proliferation, calculated from the base line T cell proliferation responses, to responses observed in the skin explant tests. Data show a significant increase ($p < 0.004$) in T cell proliferation in the positive (grade >II; square symbols) skin explant group compared with the negative (grade I; round symbols) skin explant group. IFNγ responses (**right**) were measured in cell culture supernatants from the skin explant tests. The graph shows Log2 IFNγ levels (pg/mL). IFNγ responses showed a significant increase ($p < 0.0001$) in the positive skin explant group (grade $\geq$ II) compared with the negative skin explant group (grade I). The *p*-values are FDR-adjusted after a Benjamini-Hochberg correction.

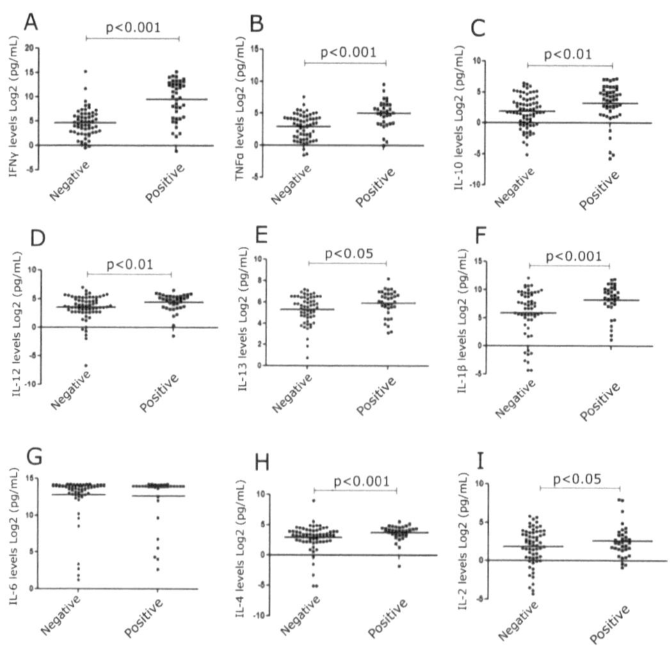

Figure 4. Levels of cytokines in skin explant cell culture supernatants following exposure to mAbs. Cytokine levels were determined in the skin explant test cell culture supernatants. Graphs show cytokine levels (Log2 pg/mL). A significant increase was observed for (**A**) IFNγ ($p < 0.0001$), (**B**) TNFα ($p < 0.0001$), (**C**) IL-10 ($p < 0.01$), (**D**) IL-12 ($p < 0.01$), (**E**) IL-13 ($p < 0.05$), (**F**) IL-1β ($p < 0.001$) (**G**) IL-6, (**H**) IL-4 ($p < 0.001$), and (**I**) IL-2 ($p < 0.05$). The *p*-values are FDR-adjusted after Benjamini-Hochberg correction. The square symbols represent the positive group and the round symbols represent the negative group.

3.4. TGN1412 Showed a Delayed Time Response In Vitro

TGN1412 was co-cultured with healthy skin and autologous PBMC over a period of 3 days. A skin sample was removed every 24 h, and histopathological analysis was performed. Data (n = 4) clearly showed TGN1412 exposure did not cause any damage to skin until day 3 following antibody co-incubation with skin and autologous cells. Negative controls (medium alone incubated with skin, autologous T cells incubated with skin, or the IgG4 isotype antibody incubated with skin in the absence of autologous cells) showed a grade I negative response over the 3 days. TGN1412 exerted a significant immune response, showing an increase in histopathological damage from a grade I negative response at 24 h to a grade III positive response by 72 h (Figure 5), in 2/3 tests, indicating a time-dependent response relationship.

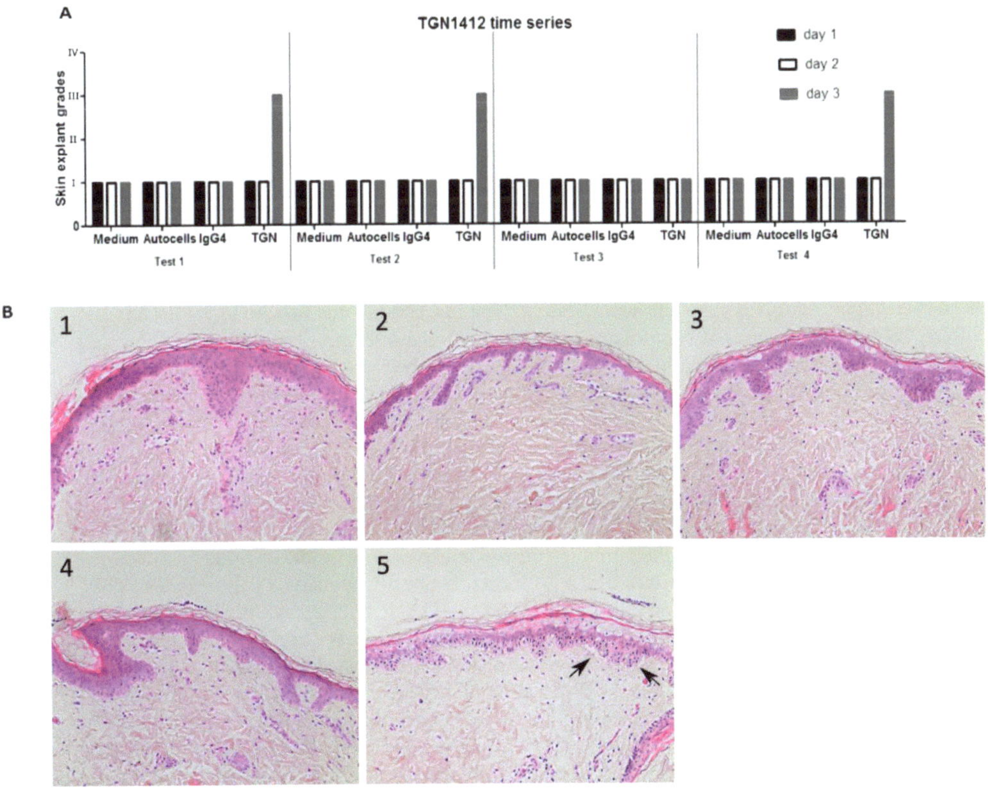

Figure 5. (**A**) In vitro responses (n = 4) of TGN1412 over a 3-day time period, day 1 (black bar), day 2 (white bar), and day 3 (grey bar). (**B**) Skin incubated in medium control or with autologous cells or an IgG4 isotype antibody gave a grade I negative response in the skin explant tests on days 1, 2, and 3. TGN1412 gave a grade I negative response on days 1 and 2. On day 3, a grade III positive response was observed in 2/3 tests (as shown by the arrows) and a grade I negative response in the day 1 test. (**B**) Representative images of (1) medium day 3, (2) autologous cells day 3, (3) IgG4 day 3, (4) TGN1412 day 1, (5) TGN1412 a day 3.

4. Discussion

Adverse reactions in response to biological drugs can present in several ways. To further complicate this matter, the true incidence of occurrence is largely unknown. Reaction incidences are reported as variable, influenced by drug dosage, multiple drug administrations [22,23], and even geographical location [24]. Additionally, studies are often reported in small patient groups. In vitro tests such as patch tests, skin prick tests, and intradermal tests are sometimes used; however, some patients still present with adverse immune reactions, including skin sensitization, despite obtaining a negative result in in vitro diagnostic tests [25–27]. Steroid drug administration is a frequently used strategy to dampen adverse effects of drugs; for example, adalimumab has been shown to cause an adverse reaction in up to 12% of rheumatoid arthritis (RA) patients [28] when administered alone, and this has been seen to be reduced to 1% when co-administered with methotrexate [29,30].

During the TeGenero (TGN1412) clinical trial, the adverse events in response to TGN1412 started to present as early as 3 h post-infusion, with multiple organ failure occurring by 12 h [21]. We used the skin explant assay to determine if the response period of the immune system in vitro was similar to that observed in vivo. The TGN1412 homolog antibody was tested at 1 μg/mL, the estimated concentration to be present in the blood of individuals during the Northwick Park trial [22]. An interesting skin explant assay result was the delay in the response in vitro to the TGN1412 mAb in comparison to the early response reported in vivo. This delayed response may well not be specific to that of TGN1412 and may be related to the fact that autologous PBMCs, which are immature, were used and required a period following activation to become effectors and cause damage.

Following the TGN1412 Northwick Park trial, the need and usefulness of human in vitro studies to determine human safety data and dose ranges have become increasingly recognized. This is illustrated by the work of Stebbings et al. (2007) [20], where they compared several protocols to assess cytokine release, with the most successful being when TGN1412 was immobilized on endothelial cells and demonstrated that the simple interaction of TGN1412 with peripheral blood lymphocytes was not sufficient to cause cytokine release as shown in vivo. The need for improved in vitro assays to predict adverse immune reactions to biologics occurring in vivo has also been debated for some time [31]. This was recently highlighted last year by the FDA in their new Modernization Act to reduce reliance on animal model systems and develop new ways to de-risk novel therapeutics. The key aim of our study was to develop and evaluate a suitable pre-clinical human in vitro test that could predict cytokine release associated with immunotoxicity and other immunological reactions in humans as a first-in-man test to bridge the gap between animal studies and human studies and to avoid a similar occurrence as observed in the TeGenero trial, or skin sensitization or injection site reactions that could halt a clinical trial.

We have shown that the human in vitro skin explant test is able to detect an unwanted immune response to a therapeutic monoclonal antibody. By using an autologous PBMC population and co-culturing these with autologous skin in the presence of the test biologics, we observed histopathological damage in the skin if the therapeutic had caused an adverse immune response in the clinic. As a result, the test distinguished between extreme positive or limited adverse responses to mAbs. In conjunction with other in vitro tests, this test could prove to be valuable during both the early and preclinical development stages of mAbs. It can offer the opportunity to further test modified compounds, for example, protein modification, to decrease any immunotoxic potential and enable compound elimination early in the drug development pipeline.

The skin explant test is dissimilar to alternative tests available for immunotoxicity testing in that it offers observation of histological changes in the skin tissue and observation of skin sensitization as a direct result of immune cell activation and cytokine release. The use of autologous immune cells in this model allows the model to closely mimic an in vivo immune response in vitro.

Cytokine release assays are recognized as a valuable alternative to determining a risk-based evaluation [32] and as early indicators of cytokine release syndrome. Cytokine release observed in the positive Mab skin explants corresponded to Th1 (IFNγ and TNFα) and Th2 (IL-13, IL-10, and IL-4) type responses. Our results were able to replicate previous reports of increased IFNγ [20] or TNFα [31] levels, but not IL-2. T cell proliferation upon exposure to positive mAbs gave positive responses, as previously reported [30].

In an attempt to develop safer biologics, predictive modeling has attracted much interest as a tool for the identification of potential health hazards. A large number of these models are based primarily on the use of drug molecular structure and its activity in a quantifiable approach known as quantitative structure-activity relationship (QSAR) models [33]. They do, however, have limitations. [34]. Another approach involves the development of algorithms allowing T cell epitope mapping to predict or identify immunotoxic epitopes [35,36]. In comparison to in vivo studies, which are time-consuming and expensive, the speed and low cost of these protein immunotoxicology tools make them very attractive. However, they do not provide any information with regard to T cell activation or the prediction of cytokine release.

Recent efforts have also combined the use of in silico and in vitro prediction tools to develop better models to predict adverse reactions, such as the Epibase® in silico and in vitro tools. In vivo animal studies also face much criticism for the pre-clinical assessment of drugs, as animal models can often fail to predict the human immune response, as highlighted by the 2006 TGN1412 clinical trial [37]. There are regulatory and ethical issues around the use of animals in pre-clinical research, as well as the interpretation of the cross-species barrier, and the effectiveness of these models is an important topic of debate [38].

The skin explant test, which effectively distinguishes between monoclonal antibodies showing extreme positive or limited responses, could be used as a valuable preclinical tool in the safety assessment of biologics.

Three compounds were incorrectly categorized, namely simulect and panitumumab, overestimated and etanercept underestimated. Panitumumab carries a box warning for dermatological toxicities and induces a severe skin reaction in approximately 16% of patients. It is therefore likely that the misclassification of panitumumab from uncommon (clinical) to common (skin explant test) was due to the skin-related pathology rather than predicting a more generally immune-mediated reaction. Similarly, Simulect carries a box warning for skin ulceration [39].

Skin-specific reactions are not uncommon for the majority of the biologics tested in this study, including injection site reactions, rash, and inflammation, which are commonly observed [39].

In an exploratory approach to further stratify the skin explant test classification (Table 1), the grading over 10 donors was averaged, using 1–4 rather than the I–IV classification. The grades were grouped as <1.2 for negative reactions, 1.3–2.0 for weak positives, and 2.0–3.0 for strong positives. The percentages were grouped as <10% for negative reactions, 10–30 for weak positives, 31–70 for positives, and 71–100 for extreme reactions. Hypothetically, the product of percentage positive donors and average grade would produce an aggregated score that is more comprehensive than either parameter alone. Due to the strong correlation between both groups ($r = 0.770$), the aggregated score strongly reflects the percentage group. However, the percentage group parameter does not account for edge cases where 1 or 2 out of 10 donors may be grade 3, or 10 out of 10 donors may be grade 2, whereas the aggregated score rates the former as 1.5 and the latter as a 2. The aggregate score would represent severity and occurrence, where high occurrences of low grades are as noteworthy as low occurrences of high grades. The classifications were derived in-set, and the strong Pearson's correlation between the various parameters could partially be due to overfitting of the predictions. Further ongoing testing of the skin explant test assay using a targeted selection of novel therapeutics, graded by blinded observers and correlated against clinical trial data, has indicated that the skin explant test assay is predictive. The classification and correlation analysis performed for the 16 therapeutic antibodies analyzed in this study will be repeated with other therapeutics in on-going studies.

Another point to consider is that testing was performed in a healthy donor population, and responses in clinical patients may differ. As a large proportion of monoclonal antibodies are targeted towards immune-related disorders, patients with these disorders may have altered or enhanced immune responses. This could be one reason why commonly positive biologics such as etanercept were classified as weak positives in this test.

As of 2023, the FDA no longer requires animal testing for new therapeutics. It is anticipated that ex vivo human disease models, instead of murine models, will gradually be seen as more relevant models of human diseases. The skin explant assay is a patented technology (traded as Skimune®), and additional validation data is ongoing through the testing of blinded compounds from industry where clinical outcomes are known. As demonstrated, the classification criteria are strongly correlated to the histopathology grades. Thus, for underpowered studies with fewer than 10 donors, the average grades are a good indicator of a positive or negative response in the assay and can be used as a pre-clinical screen with the understanding that no mAb is completely negative and that a very small population may show weak positivity with regard to adverse immune events. This, however, does not preclude them from being marketable products.

Further work is being performed to stratify the skin explant assay grading between subepidermal damage and epidermal damage, such as toxic epidermal necrolysis [40,41], and to include acantholysis as demonstrated in palmar-plantar erythrodysesthesia.

5. Conclusions

Overall, the results suggest that the skin explant test could be an effective tool in bridging the gap between animal studies and first-in-human clinical trials. The skin explant assay involves the skin sensitization pathway, a molecular initiating event, an inflammatory response and release of cytokines, and an adverse event in the form of keratinocyte activation culminating in skin damage. The assay, therefore, significantly mimics the adverse event pathway of skin sensitization. This assay accurately predicts the adverse reaction incidence rate for a variety of biological therapeutics with varying mechanisms of action and provides a valuable and ethical alternative to animal studies for immunological activity evaluation. The test could also reduce the time and cost of the drug development process by identifying hazardous biological candidates likely to fail in Phase I clinical trials and identifying early skin sensitization reactions implicated in their failure.

6. Patents

Alcyomics holds patents on the skin explant assay (trading as Skimune®). See the table below.

Case Code	Applicant	Country	Status	Official Number	Title
P131663CH	Alcyomics Limited	Switzerland	Granted	'EP2524227	Skin Explant Assay
P131663DE	Alcyomics Limited	Germany	Granted	'602011004834	Skin Explant Assay
P131663DK	Alcyomics Limited	Denmark	Granted	'EP2524227	Skin Explant Assay
P131663ES	Alcyomics Limited	Spain	Granted	'EP2524227	Skin Explant Assay
P131663FR	Alcyomics Limited	France	Granted	'EP2524227	Skin Explant Assay
P131663GB1	Alcyomics Limited	United Kingdom	Granted	'EP2524227	Skin Explant Assay
P131663IT	Alcyomics Limited	Italy	Granted	'502014902254005	Skin Explant Assay
P131663US	Alcyomics Limited	USA	Granted	'9,651,544	Skin Explant Assay
P131663USD1	Alcyomics Limited	USA	Granted	'10,073,084	Skin Explant Assay

Author Contributions: S.S.A. and M.M.A., performed experiments and result analysis. A.M.D. performed histopathological analysis. S.S.A. and M.F. performed statistical analysis. A.I., M.F. and A.M.D. wrote the paper. A.M.D. and R.S. designed the research. All authors have read and agreed to the published version of the manuscript.

Funding: This study was supported by a Technology Strategy Board Grant No. 720110/775. This work was also partly funded by the UK Government Advanced Manufacturing Supply Chain Initiative as part of the Biostreamline Project, which exists to develop new approaches for the more efficient development and manufacture of next-generation biologics.

Institutional Review Board Statement: The study was conducted in accordance with the Declaration of Helsinki, and approved by the Ethics Committee of Newcastle and North Tyneside (UK) (reference no. 10/H0906/58 and 2 October 2020) for studies involving humans.

Informed Consent Statement: Informed consent was obtained from all subjects involved in the study.

Data Availability Statement: The original data presented in the study are included in the article; further inquiries can be directed to the corresponding author.

Acknowledgments: The authors thank Louis Bibby, Elizabeth Douglas (in memorandum), the National Institute of Biologics and Control (Robin Thorpe), and Richard Stebbings (currently employed at Astra Zeneca) for their assistance and contribution to this study.

Conflicts of Interest: Ahmed, S.S., Ahmed, M.M., Ishaq, A., Freer, M., Stebbings, R., Dickinson, A.M., all are employed by Alcyomcis Ltd., except R. Stebbings, who is employed by AstraZeneca. None of the authors has received or expects to receive research grants from funding. None of the authors has received support from commercial sources of funding by companies that sell drugs, medical devices, or provide medical services. None of the authors have received honoraria for speaking at symposia. None of the authors holds a position on the advisory boards associated with the publication.

References

1. Kohler, G.; Milstein, C. Continuous cultures of fused cells secreting antibody of predefined specificity. *Nature* **1975**, *256*, 495–497. [CrossRef]
2. Almagro, J.C.; Fransson, J. Humanization of antibodies. *Front. Biosci.* **2008**, *13*, 1619–1633. [PubMed]
3. Lonberg, N. Fully human antibodies from transgenic mouse and phage display platforms. *Curr. Opin. Immunol.* **2008**, *20*, 450–459. [CrossRef] [PubMed]
4. Hansel, T.T.; Kropshofer, H.; Singer, T.; Mitchell, J.A.; George, A.J.T. The safety and side effects of monoclonal antibodies. *Nat. Rev. Drug Discov.* **2010**, *9*, 325–338. [CrossRef] [PubMed]
5. Gell, P.G.H.; Coombs, R.R.A. *Clinical Aspects of Immunology*, 1st ed.; Blackwell: Oxford, UK, 1963; p. 883. [CrossRef]
6. Pepys, J. IgE-mediated hypersensitivity. *Clin. Sci.* **1971**, *41*, 1P–2P. [CrossRef]
7. Ring, J. Applied allergology. 12. Allergic blood disorders: Type II cytotoxic reactions. *MMW Munch. Med. Wochenschr.* **1982**, *124*, 78–82. [PubMed]
8. Weston, W.L. Type IV reactions in the skin. *Ann. Allergy* **1976**, *37*, 346–352.
9. Jankowitz, R.; Joyce, J.; Jacobs, S.A. Anaphylaxis after administration of ibritumomab tiuxetan for follicular non-hodgkin lymphoma. *Clin. Nucl. Med.* **2008**, *33*, 94–96. [CrossRef] [PubMed]
10. Chung, C.H.; Mirakhur, B.; Chan, E.; Le, Q.-T.; Berlin, J.; Morse, M.; Murphy, B.A.; Satinover, S.M.; Hosen, J.; Mauro, D.; et al. Cetuximab-induced anaphylaxis and IgE specific for galactose-alpha-1,3-galactose. *N. Engl. J. Med.* **2008**, *358*, 1109–1117. [CrossRef]
11. Bugelski, P.J.; Achuthanandam, R.; Capocasale, R.J.; Treacy, G.; Bouman-Thio, E. Monoclonal antibody-induced cytokine-release syndrome. *Expert. Rev. Clin. Immunol.* **2009**, *5*, 499–521. [CrossRef]
12. Lynch, C.M.; Grewal, I.S. Preclinical safety evaluation of monoclonal antibodies. *Handb. Exp. Pharmacol.* **2008**, *181*, 19–44. [CrossRef] [PubMed]
13. Pound, P.; Ebrahim, S.; Sandercock, P.; Bracken, M.B.; Roberts, I. Where is the evidence that animal research benefits humans? *BMJ* **2004**, *328*, 514–517. [CrossRef] [PubMed]
14. van der Worp, H.B.; Howells, D.W.; Sena, E.S.; Porritt, M.J.; Rewell, S.; O'Collins, V.; Macleod, M.R. Can animal models of disease reliably inform human studies? *PLoS Med.* **2010**, *7*, e1000245. [CrossRef] [PubMed]
15. Dowsing, T.; Kendall, M.J. The Northwick Park tragedy--protecting healthy volunteers in future first-in-man trials. *J. Clin. Pharm. Ther.* **2007**, *32*, 203–207. [CrossRef] [PubMed]
16. Hanke, T. Lessons from TGN1412. *Lancet* **2006**, *368*, 1569–1570. [CrossRef]
17. Ahmed, S.S.; Wang, X.N.; Fielding, M.; Kerry, A.; Dickinson, I.; Munuswamy, R.; Kimber, I.; Dickinson, A.M. An in vitro human skin test for assessing sensitization potential. *J. Appl. Toxicol.* **2016**, *36*, 669–684. [CrossRef] [PubMed]
18. Ahmed, S.S.; Whritenour, J.; Ahmed, M.; Bibby, L.; Darby, L.; Wang, X.; Watson, J.; Dickinson, A. Evaluation of a human in vitro skin test for predicting drug hypersensitivity reactions. *Toxicol. Appl. Pharmacol.* **2019**, *369*, 39–48. [CrossRef] [PubMed]
19. Glucksberg, H.; Storb, R.; Fefer, A.; Buckner, C.D.; Neiman, P.E.; Clift, R.A.; Lerner, K.G.; Thomas, E.D. Clinical manifestations of graft-versus-host disease in human recipients of marrow from HLA-matched sibling donors. *Transplantation* **1974**, *18*, 295–304. [CrossRef]
20. Stebbings, R.; Findlay, L.; Edwards, C.; Eastwood, D.; Bird, C.; North, D.; Mistry, Y.; Dilger, P.; Liefooghe, E.; Cludts, I.; et al. "Cytokine storm" in the phase I trial of monoclonal antibody TGN1412: Better understanding the causes to improve preclinical testing of immunotherapeutics. *J. Immunol.* **2007**, *179*, 3325–3331. [CrossRef]
21. Dickinson, A.M.; Sviland, L.; Carey, P.; Reid, M.M.; Hamilton, P.J.; Pearson, A.J.; Proctor, S.J. Skin explant culture as a model for cutaneous graft-versus-host disease in humans. *Bone Marrow Transplant.* **1988**, *3*, 323–329.
22. Suntharalingam, G.; Perry, M.R.; Ward, S.; Brett, S.J.; Castello-Cortes, A.; Brunner, M.D.; Panoskaltsis, N. Cytokine storm in a phase 1 trial of the anti-CD28 monoclonal antibody TGN1412. *N. Engl. J. Med.* **2006**, *355*, 1018–1028. [CrossRef] [PubMed]

23. Waibler, Z.; Sender, L.Y.; Kamp, C.; Müller-Berghaus, J.; Liedert, B.; Schneider, C.K.; Löwer, J.; Kalinke, U. Toward experimental assessment of receptor occupancy: TGN1412 revisited. *J. Allergy Clin. Immunol.* **2008**, *122*, 890–892. [CrossRef] [PubMed]
24. Brennan, P.J.; Bouza, T.R.; Hsu, F.I.; Sloane, D.E.; Castells, M.C. Hypersensitivity reactions to mAbs: 105 desensitizations in 23 patients, from evaluation to treatment. *J. Allergy Clin. Immunol.* **2009**, *124*, 1259–1266. [CrossRef]
25. O'Neil, B.H.; Allen, R.; Spigel, D.R.; Stinchcombe, T.E.; Moore, D.T.; Berlin, J.D.; Goldberg, R.M. High incidence of cetuximab-related infusion reactions in Tennessee and North Carolina and the association with atopic history. *J. Clin. Oncol.* **2007**, *25*, 3644–3648. [CrossRef]
26. Torres, M.J.; Romano, A.; Mayorga, C.; Carmen, M.; Guzman, A.E.; Reche, M.; Juarez, C.; Blanca, M. Diagnostic evaluation of a large group of patients with immediate allergy to penicillins: The role of skin testing. *Allergy* **2001**, *56*, 850–856. [CrossRef] [PubMed]
27. Torres, M.J.; Mayorga, C.; Cornejo-García, J.A.; Romano, A.; Blanca, M. IgE antibodies to penicillin in skin test negative patients. *Allergy* **2002**, *57*, 965. [CrossRef] [PubMed]
28. Hjortlund, J.; Mortz, C.G.; Skov, P.S.; Bindslev-Jensen, C. Diagnosis of penicillin allergy revisited: The value of case history, skin testing, specific IgE and prolonged challenge. *Allergy* **2013**, *68*, 1057–1064. [CrossRef]
29. van de Putte, L.B.; Atkins, C.; Malaise, M.; Sany, J.; Russell, A.S.; van Riel, P.L.C.M.; Settas, L.; Bijlsma, J.W.; Todesco, S.; Dougados, M.; et al. Efficacy and safety of adalimumab as monotherapy in patients with rheumatoid arthritis for whom previous disease modifying antirheumatic drug treatment has failed. *Ann. Rheum. Dis.* **2004**, *63*, 508–516. [CrossRef] [PubMed]
30. Weinblatt, M.E.; Keystone, E.C.; Furst, D.E.; Moreland, L.W.; Weisman, M.H.; Birbara, C.A.; Teoh, L.A.; Fischkoff, S.A.; Chartash, E.K. Adalimumab, a fully human anti-tumor necrosis factor alpha monoclonal antibody, for the treatment of rheumatoid arthritis in patients taking concomitant methotrexate: The ARMADA trial. *Arthritis Rheum.* **2003**, *48*, 35–45. [CrossRef]
31. Stebbings, R.; Eastwood, D.; Poole, S.; Thorpe, R. After TGN1412: Recent developments in cytokine release assays. *J. Immunotoxicol.* **2013**, *10*, 75–82. [CrossRef]
32. Vidal, J.M.; Kawabata, T.T.; Thorpe, R.; Silva-Lima, B.; Cederbrant, K.; Poole, S.; Mueller-Berghaus, J.; Pallardy, M.; Van der Laan, J.-W. In vitro cytokine release assays for predicting cytokine release syndrome: The current state-of-the-science. Report of a European Medicines Agency Workshop. *Cytokine* **2010**, *51*, 213–215. [CrossRef]
33. Nantasenamat, C.; Isarankura-Na-Ayudhya, C.; Prachayasittikul, V. Advances in computational methods to predict the biological activity of compounds. *Expert. Opin. Drug Discov.* **2010**, *5*, 633–654. [CrossRef]
34. Valerio, L.G., Jr. In silico toxicology for the pharmaceutical sciences. *Toxicol. Appl. Pharmacol.* **2009**, *241*, 356–370. [CrossRef] [PubMed]
35. Koren, E.; De Groot, A.; Jawa, V.; Beck, K.; Boone, T.; Rivera, D.; Li, L.; Mytych, D.; Koscec, M.; Weeraratne, D.; et al. Clinical validation of the "in silico" prediction of immunogenicity of a human recombinant therapeutic protein. *Clin. Immunol.* **2007**, *124*, 26–32. [CrossRef]
36. Bian, H.; Hammer, J. Discovery of promiscuous HLA-II-restricted T cell epitopes with TEPITOPE. *Methods* **2004**, *34*, 468–475. [CrossRef]
37. Mak, I.W.; Evaniew, N.; Ghert, M. Lost in translation: Animal models and clinical trials in cancer treatment. *Am. J. Transl. Res.* **2014**, *6*, 114–118. [PubMed]
38. Bugelski, P.J.; Treacy, G. Predictive power of preclinical studies in animals for the immunogenicity of recombinant therapeutic proteins in humans. *Curr. Opin. Mol. Ther.* **2004**, *6*, 10–16. [PubMed]
39. Baldo, B.A. Immune- and Non-Immune-Mediated Adverse Effects of Monoclonal Antibody Therapy: A Survey of 110 Approved Antibodies. *Antibodies* **2022**, *11*, 17. [CrossRef]
40. Nwikue, G.; Olsson-Brown, A.; Aboheimed, N.; Yip, V.; Jolly, C.; Luchian, A.; Ressel, L.; Sharma, A.; Bergfeld, W.; Ahmed, S.; et al. TNF-α induced extracellular release of keratinocyte high-mobility group box 1 in Stevens-Johnson syndrome/toxic epidermal necrolysis: Biomarker and putative mechanism of pathogenesis. *J. Dermatol.* **2023**, *50*, 1129–1139. [CrossRef]
41. Olsson-Brown, A.; Yip, V.; Ogiji, E.D.; Jolly, C.; Ressel, L.; Sharma, A.; Bergfeld, W.; Liu, X.; Khirwadkar, N.; Bellon, T.; et al. TNF-α-Mediated Keratinocyte Expression and Release of Matrix Metalloproteinase 9: Putative Mechanism of Pathogenesis in Stevens-Johnson Syndrome/Toxic Epidermal Necrolysis. *J. Investig. Dermatol.* **2023**, *143*, 1023–1030.e7. [CrossRef]

toxics

MDPI

Article

A Human Skin Explant Test as a Novel In Vitro Assay for the Detection of Skin Sensitization to Aggregated Monoclonal Antibodies

Ana Martins-Ribeiro [1,2], Arathi Kizhedath [3], Shaheda Sameena Ahmed [1], Jarka Glassey [3], Abbas Ishaq [1], Matthew Freer [1] and Anne Mary Dickinson [1,2,*]

1 Alcyomics Ltd., The Biosphere, Draymans Way, Newcastle Helix, Newcastle Upon Tyne NE4 5BX, UK; anappmribeiro@gmail.com (A.M.-R.); matthew.freer@alcyomics.com (M.F.)
2 Translational and Clinical Research Institute Faculty of Medical Sciences, Newcastle University, Newcastle Upon Tyne NE2 4HH, UK
3 Chemical Engineering and Advanced Materials, Newcastle University, Newcastle Upon Tyne NE1 7RU, UK
* Correspondence: anne.m.dickinson@alcyomics.com

Abstract: Introduction: Monoclonal antibodies (mAbs) are important therapeutics. However, the enhanced potential for aggregation has become a critical quality parameter during the production of mAbs. Furthermore, mAb aggregation may also present a potential health risk in a clinical setting during the administration of mAb therapeutics to patients. While the extent of immunotoxicity in patient populations is uncertain, reports show it can lead to immune responses via cell activation and cytokine release. In this study, an autologous in vitro skin test designed to predict adverse immune events, including skin sensitization, was used as a novel assay for the assessment of immunotoxicity caused by mAb aggregation. **Material and Methods**: Aggregation of mAbs was induced by a heat stress protocol, followed by characterization of protein content by analytical ultra-centrifugation and transmission electron microscopy, revealing a 4% aggregation level of total protein content. Immunotoxicity and potential skin sensitization caused by the aggregates, were then tested in a skin explant assay. **Results:** Aggregated Herceptin and Rituximab caused skin sensitization, as shown by histopathological damage (grade II–III positive response) together with positive staining for Heat Shock Protein 70 (HSP70). Changes in T cell proliferation were not observed. Cytokine analysis revealed a significant increase of IL-10 for the most extreme condition of aggregation (65 °C at pH3) and a trend for an overall increase of IFN-γ, especially in response to Rituximab. **Conclusions:** The skin explant assay demonstrated that aggregated mAbs showed adverse immune reactions, as demonstrated as skin sensitization, with histopathological grades II-III. The assay may, therefore, be a novel tool for assessing immunotoxicity and skin sensitization caused by mAb aggregation.

Keywords: mAb aggregation; skin sensitization; immunotoxicity; in vitro test; T-cell proliferation; cell death; skin explant; safety assessment; immunotoxicity

1. Introduction

mAb industrial production is a complex process where several stress factors must be taken into consideration during industrial scale-up and manufacturing of therapeutic antibody proteins. Temperature is one such critical parameter since thermal changes can affect the conformational structure and stability of the IgG mAb molecule [1]. This poses a serious cost issue due to production failure, as it can lead to epitope loss and loss of antibody efficacy together with irreversible aggregation events. At high temperatures, the disulphide bridges, binding the polypeptide chains of the antibody, start to weaken and denature. This leads to an unfolding of the overall structure and an unstable conformational state [1–4]. pH can also induce antibody aggregation via protein instability [2,5,6]. Acid-induced changes can promote loss of native structure by means of secondary structure refolding [6]. More specifically, one study showed [2] that while the Fab fragment was very sensitive to heat stress, the Fc region was more sensitive to acidic pH. Physical or mechanical stress can also potentiate antibody aggregation. This stress can be caused by agitation during manufacture [7], storage [8], shipping, or shaking during administration of the antibody [9]. Commercially available mAbs are formulated to be very stable endpoint products, but exposure to temperature, pH, or stirring stress factors can compromise mAb stability, ultimately potentially having effects on clinical safety.

Considering the size of protein aggregates (ranging from 100 to 1000 kDa), their characterization becomes quite complex. Therefore, multi-technique analysis or a combination approach provides more reliable information. New analytical techniques, include, analytical ultracentrifugation (AUC) [10], size exclusion chromatography-multi angle light scattering (SEC-MALS) [11], and asymmetric flow field-flow fractionation multi-angle light scattering [12].

Asymmetric flow field-flow fractionation (AF4) is a technique that offers the possibility to physically separate biomolecules across a wide size range. AF4 can be used in combination with different types of detectors depending on the information required and becomes especially powerful when combined with multi-angle light scattering (MALS) [13] and (AF4-MALS) [14] are the current most reliable methods for quantifying low levels of protein aggregates. AUC allows for the quantification of protein aggregation and the formulation of large supramolecular complexes based on their sedimentation properties [15]. Sedimentation velocity (SV)-AUC is a hydrodynamic approach that provides details of particle mass and shape and is particularly useful for studying multicomponent irreversible and reversible mixtures of species [15]. In this study, SV-AUC was used to explore the formation of stable and metastable oligomers, irreversible aggregates, and degradation products due to storage and thermal stress.

There is a concern that protein and antibody aggregation may represent a potentially serious cause of adverse immunotoxicity in humans when administered therapeutically. For example, one study investigating the instability of antibodies during aerosolization of lung mucosa in mice found the development of aggregates, which led to pro-inflammatory and cytotoxic effects [16]. The Clinical Oncology Society of Australia Position Statement [17] also reviews the safe handling of monoclonal antibodies in the health care setting where distinct exposure mechanisms, dermal, mucosal, inhalation, and oral, were discussed with regard to health risk, e.g., dermatitis and skin sensitization.

Indeed, the aggregation of proteins is a complex multifactorial process that includes conformational change, nucleation, and the growth of protein particles. Protein-based vaccines, for example, are a demonstration of when immunogenicity is desirable as they are more effective at eliciting an immune response as aggregated in an adjuvant particle. However, in the case of immunogenicity, the aggregation of proteins can be undesirable and cause an adverse immune response [18]. In this instance, immunotoxicity can be defined as an adverse immune activation event, including T cell proliferation and cytokine stimulation caused by aggregated mAb therapeutics. However, little is known about mAb aggregation following patient administration. Indeed, previous studies have shown that there is a significant chance that mAb preparations alone will cause some degree of immunotoxicity [19–21]. Other reports have shown that aggregated mAb have been found in human biological fluids, including serum [22,23] and plasma [22]. mAb aggregation can, therefore, appear to increase the likelihood of an adverse immune response.

In vitro, cell-based assays that mimic the normal immune response have demonstrated that exposure to aggregated mAbs can induce cell activation, proliferation, and cytokine secretion of immune system components through the activation of dendritic cells [24] and T cells [25–28]. In vivo, responses to mAb aggregates from animal-based studies showed that heat-aggregated IgG immune complexes lead to cellular activation and immune complex formation [29,30].

In this study, we have used a very sensitive human in vitro skin assay, previously described for predicting adverse immune events to chemicals [31] and small molecule drugs [32], to assess the immunotoxicity and skin sensitization caused by mAb aggregation. This assay is sensitive as it includes dermal exposure of the aggregated monoclonal antibody, T cell activation, cytokine release, and histopathological skin damage as parameters for the detection of adverse immune responses and skin sensitization. In this current study, two commercially available monoclonal antibodies, Herceptin and Rituximab, were subjected to thermal stress conditions to induce aggregation, followed by characterization of protein aggregates. Immune activation was assessed by testing the heat-stressed mAb samples in both T cell proliferation and cytokine release assays alongside a human in vitro skin explant assay [31,32].

2. Materials and Methods

2.1. Degradation of mAb Samples

Herceptin (Trastuzumab) and Rituximab (Mabthera) monoclonal antibodies (IgG1 class) were used at a concentration of 1 mg/mL and exposed to three different heat stress conditions: 4 °C (fridge), 37 °C (incubator) and 40 °C (water bath) for the following time points: 0, 3, 6, 12, 24 and 48 h. A positive control for aggregation was also prepared for each mAb by leaving the samples at 65 °C for one hour in an acidic buffer (pH3). Samples were given a 3-digit code name for identification, with the first digit representing the mAb tested (Rituximab, R or Herceptin, H), the second digit representing the temperature exposed (4, 37, or 40 °C), and the third digit representing the time of exposure (0, 3, 6, 12, 24 or 48 h). Samples were stored immediately after the heat stress protocol at 4 °C until further analysis.

2.2. Protein Analysis of Stressed mAb Samples

The protein content of the heat-stressed mAb samples was quantified by sedimentation velocity-analytic ultra-centrifugation (SV-AUC). Sedimentation analysis was carried out using a ProteomeLab XL-I analytical centrifuge (Beckman Coulter, Palo Alto, CA, USA). The following conditions were used in all centrifugation runs: 40,000-rpm angular velocity, 20 °C rotor temperature, and 280-nm absorbance scanned. Absorbance and interference data were collected for each experiment with a minimum of 65 scans. Protein quantification (percentage of monomers, dimers, and larger molecules) was calculated using SEDNTERP software (2.01). Sedimentation velocity profiles were treated using size-distribution models and refined with Bayesian statistics. Root mean square deviation (RMSD) was also calculated as the sample standard deviation of the differences between predicted and observed values. This quantification was carried out using Newcastle University Protein and Proteome Analysis (NUPPA) service facilities.

Protein characterization was analyzed using transmission electron microscopy (TEM). 10 μL of the mAb stressed samples were deposited on carbon-coated TEM grids, after which the grids were stained with uranyl acetate for negative staining. The grid was air-dried, and the excessive staining was removed from the grid specimens. The grids were then analyzed under an acceleration voltage of 100 kV under a Philips CM100 TEM.

2.3. Human Blood and Skin Samples

This study was approved by the Local Research Ethics Committee (LREC). A total of five donor human blood and skin biopsy samples were obtained from healthy volunteers after informed consent by a research nurse at a National Health Service dermatology clinic. Each volunteer donated 60 mL of peripheral blood, collected in heparin (455051, Greiner bio-one), and two 4-mm skin biopsies, which were used for the in vitro skin explant assay. Peripheral blood mononuclear cells (PBMCs) were isolated by density-gradient centrifugation from fresh blood. Skin biopsies (abdominal area) were collected fresh in X-Vivo™ medium (Lonza, Basel, Switzerland) and processed by washing in PBS (Sigma, Burlington, MA, USA), then trimmed of excess fat and dissected into small sections for use in the skin explant assay.

2.4. Detection of Aggregated mAb-Induced Immune Activation by T-Cell Proliferation Assay

A 2×10^5 PBMC/well was incubated with aggregated mAb samples at 1 and 10 μg/mL for 3 days. A negative isotype control (IgG1) and a positive control for immune activation (OKT3) [33] were used in each test. Cells were harvested, and [3H] Thymidine was added for the last 16–18 h. [3H] Thymidine primary stock was stored at 37 MBq and used at 3.7 MBq (1/10 dilution), and subsequent uptake was measured using a Micro Beta β-scintillation counter in counts per minute (cpm). Data were interpreted using Prism GraphPad software (V5, Fremont, CA, USA).

2.5. Aggregated mAb-Induced Immunotoxicity, including Skin Sensitization Using the Skin Explant Assay

The 4 mm skin biopsies, obtained from volunteers, were cut into multiple smaller sections of approximately 1 mm^2 for analysis. Each test condition utilized 1 piece of 1 mm^2 of healthy skin from each volunteer. Skin biopsy sections were incubated in a medium supplemented with autologous, heat-inactivated serum containing both 1×10^6 PBMCs/well and treated mAb samples at a concentration of either 1 or 10 μg/mL for 3 days at 37 °C, 5% CO_2, 95% humidity. Skin biopsies were then fixed in formalin prior to being paraffin-embedded, sectioned, and stained with haematoxylin and eosin (H&E). Negative controls consisted of co-culture of PBMCs and autologous skin alone and co-culture of the skin with the isotype antibody IgG1. The positive control was OKT3, which has previously been shown to give rise to grade II or III response in skin explant assays [34] Controls were included in all assays. The endpoint of the assay was the assessment of the histopathological damage observed in the skin tissue caused by exposure to the mAb sample. This output was measured as skin grades (grades I to IV) according to the severity of the skin tissue damage observed (Figure 1) [35]. Grade I is considered negative, with an intact upper keratinocyte layer. Grade II includes dyskeratosis and vacuolization of the epidermis and dyskeratotic bodies. Grade III shows more severe damage to the epidermal layer, with the initial separation of the epidermal and dermal layers observed as cleft formation. Grade IV refers to severe damage, with complete separation of the epidermal and dermal layers. Grade II or higher is regarded as a positive response. Grading was performed blind by 2 individual assessors.

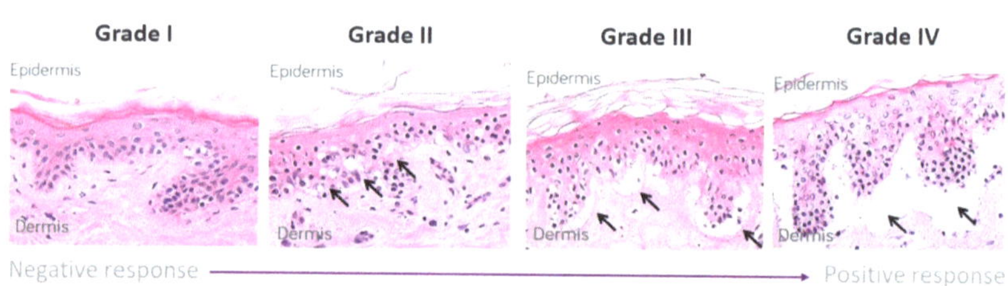

Figure 1. Lerner grading scheme for sub-epidermal lesions in the skin explant assay.

2.6. Multiplex Cytokine Analysis

Multiplex cytokine analysis was performed on the skin explant assay supernatants using MSD V-Plex kits (Meso Scale Diagnostics, Rockville, MD, USA) following the manufacturer's instructions. The biomarkers included in this panel were IFN-γ, IL-10, and TNF-α.

2.7. Immunofluorescence Labelling

Anti-Hsp70 (Heat-Shock protein 70) (1:20 dilution) antibody (Abcam, Cambridge, UK) was used to detect apoptosis in the skin explant biopsies. Heat shock proteins (HSPs) have a wide array of functions in apoptosis, leading to the suppression of apoptotic pathways. However, stress signals that trigger apoptosis also stimulate the expression and release of HSPs [36]. In this study, we investigated the expression of HsP70, which has been shown in previous studies of skin explants to correlate with increased immune responses [37].

Alexa Fluor 488 (A488, Life Technologies, Carlsbad, CA, USA) was used as the secondary antibody at a concentration of 1 μg/mL. 4′,6-diamidino-2-phenylindole (DAPI) staining was used at a concentration of 1.5 μg/mL, with VECTASHIELD® Mounting Medium (Vector Laboratories, Newark, CA, USA).

2.8. Statistical Analysis

Skin explant, T cell proliferation assays, and cytokine assessment were performed using samples from five healthy donors ($n = 5$). The T cell proliferation assay endpoint was a log fold-increase together with Stimulation Indices (SI) in response to exposure to the test compound. SI was measured using the ratio between the test conditions (mAb exposure) and the untreated control (no mAb exposure). The cut-off SI value for a positive response was a log 2-fold increase [32]. Statistical analysis was carried out using repeated measures one-way ANOVA.

For the skin explant test, using the histopathological grading classification, statistical analysis was performed to firstly to compare the mAb concentration (1 and 10 μg/mL) in the same donor by a repeated measures two-way ANOVA with a Bonferroni correction (post-test) and then to see the effect of the different mAb temperature conditions in the same donor by a one-way repeated measures ANOVA with a Bonferroni correction post-test.

Cytokine assessment was analyzed using the MSD Workbench software, then statistically analyzed using a one-way repeated measures ANOVA with a Bonferroni correction post-test.

All statistical tests were carried out using Prism GraphPad software (version 5). Statistical differences were considered significant if p-value < 0.05 = *, <0.005 = **.

3. Results

In this study, two therapeutic antibodies—Rituximab (chimeric IgG1 mAb) and Herceptin (humanized IgG1 mAb), were exposed to a heat stress protocol. The testing temperatures were intended to mimic storage conditions at 4 °C, 37 °C (normal physiological temperature), and 40 °C (elevated body temperature during an infectious episode). The selected heat stress conditions ensured that mAb degradation mimicked storage or in vivo conditions rather than industrial manufacturing.

3.1. Protein Analysis of the Stressed mAb Samples

The aggregated samples were analyzed by SV-AUC to determine the protein content of the heat-stressed mAb samples. Only the control and extreme testing conditions were analyzed at the following time points: 4 °C for 0 h, 37 °C for 48 h, 40 °C for 48 h, and 65 °C for 1 h (positive control for aggregation) (Table 1). Overall, the loss of monomers by heat stress at 4 °C, 37 °C and 40 °C was found to be very low, with up to <6% aggregation of the total protein content, including dimers and larger molecules at 4 °C, 37 °C and 40 °C (Table 1). This result is in concordance with previously reported studies, where a low level (<3%) of total protein content was reported as aggregated [38]. The positive controls (65 °C), however, showed a high loss of monomers of >97% in both instances.

The 4 °C for 0 h was considered as the baseline condition and showed more than 97% of the protein content in the monomer form in both heat-stressed Rituximab and Herceptin. Rituximab showed little variation in the monomer content throughout the different heat stress conditions, with 97.38% monomer content at 37 °C for 48 h and 97.73% at 40 °C for 48 h. Dimer content at baseline was 1.14% of the overall protein content, with a small increase to 1.61% after 37 °C and 1.15% after 40 °C. For the larger molecules, e.g., trimers, tetramers, and heavier molecules, Rituximab demonstrated a small increase from 0.93% at baseline to 1.48% at 37 °C and 1.44% at 40 °C. The decrease in monomer content for Herceptin was more evident. At baseline, the monomer content was 97%, which was reduced to 94.8% at 37 °C and to 95.51% at 40 °C. Correspondingly, the dimer content increased from 1.701% at baseline to 2.854% after 37 °C and 2.611% at 40 °C. There was also an increase in larger molecules, ranging from 0.468% at baseline to 2.653% at 37 °C and to 1.929% at 40 °C.

Table 1. Quantification of the aggregated content of heat-stressed mAb samples by analytical ultracentrifugation (SV-AUC).

			Monomer			Dimer			Larger Molecules	RMSD
	Temp (°C)	Time (h)	Sedimentation (S)	Mass (kDa)	%	Sedimentation (S)	Mass (kDa)	%	%	
Rituximab	4	0	6.412	152.1	97.92	9.25	263.653	1.14	0.93	0.015
		48	6.415	138.3	97.72	9.715	257.707	1.7	0.58	0.017
	37	0	6.411	153.4	96.16	9.089	258.781	2.69	1.15	0.014
		3	6.411	149.9	97.41	9.27	274.93	1.68	0.91	0.013
		6	6.408	152.5	96.88	9.19	261.938	2.07	1.05	0.011
		12	6.413	149.3	96.69	9.522	270.064	2.28	1.11	0.010
		24	6.426	146.8	99	-	-	-	1	0.018
		48	6.424	148.3	97.38	8.34	229.272	1.61	1.48	0.014
	40	0	6.415	153	97.26	9.042	262.435	2.23	1.24	0.016
		3	6.407	147.4	98.34	8.997	262.004	1.18	1.3	0.017
		6	6.399	149.5	98.16	8.768	249.582	1.15	1.04	0.017
		12	6.397	155.3	98.40	-	-	-	1.60	0.017
		24	6.425	150.3	97.38	9.461	281.061	1.32	1.42	0.018
		48	6.42	147.6	97.73	9.649	281.75	1.15	1.44	0.016
	65	1	1.587	13.11	0.706	6.503	108.75	2.544	73.01	-
Herceptin	4	0 *	6.441	149	97.96	9.432	263.977	1.701	0.468	0.010
	37	48	6.427	148.1	94.8	9.109	249.839	2.854	2.653	0.010
	40	48	6.421	148.1	95.51	9.235	255.364	2.611	1.929	0.011
	65	1	6.346	132.6	2.661	9.862	256.928	6.107	69.246	0.010

Quantification of monomers, dimers, and larger molecules in Rituximab and Herceptin monoclonal antibodies (at 1 mg/mL) after exposure to a heat-stress protocol (4, 37, and 40 °C for 0, 3, 6, 12, 24, and 48 h). Time 0 * for Herceptin was identical for the temperature ranges (4–40 °C). Quantification by sedimentation velocity (S), mass (kDa), and percentage in overall mAb sample (%) based on absorbance data. Standard deviation is calculated as Root Mean Square Deviation (RMSD). Missing values (-).

Size distribution of the sedimentation velocity of aggregated Rituximab (Figure 2A) and Herceptin (Figure 2B) showed a good distribution of the monomer, dimer, and larger molecule forms across the different temperature ranges. Size distribution showed that the majority of monomeric species sedimented at 6.3S with a molecular weight of 152 kDa when present in the solution, alongside some dimeric species. The dimeric species had different stoichiometry with both elongated configuration and sedimentation close to 8S and more globular configuration with sedimentation just below 10S. At 37 °C, even larger species were observed with sedimentation up to 15S. This change in protein size reflected the heat stress, causing the mAb to become heavier as it aggregated. As a result, the sedimentation velocity increased when compared to non-aggregated samples.

The internal positive control for aggregation (65 °C incubation for one hour in acidic conditions) aggregated very easily, as expected, with 73% content of larger molecules for Rituximab and 69% for Herceptin. Size-distribution showed a lower number of monomer content and an increased number of trimers, tetramers, and larger molecules (up to 10–15 mers), demonstrating a more aggregated state of this sample.

Overall, the results from the SV-AUC showed that Rituximab was less susceptible to thermodynamic changes when compared to Herceptin, since only Herceptin showed variation in content of monomer, dimer, and large structure molecules throughout the heat stress protocol. These results correlate with the outcome predicted by the Bayesian model.

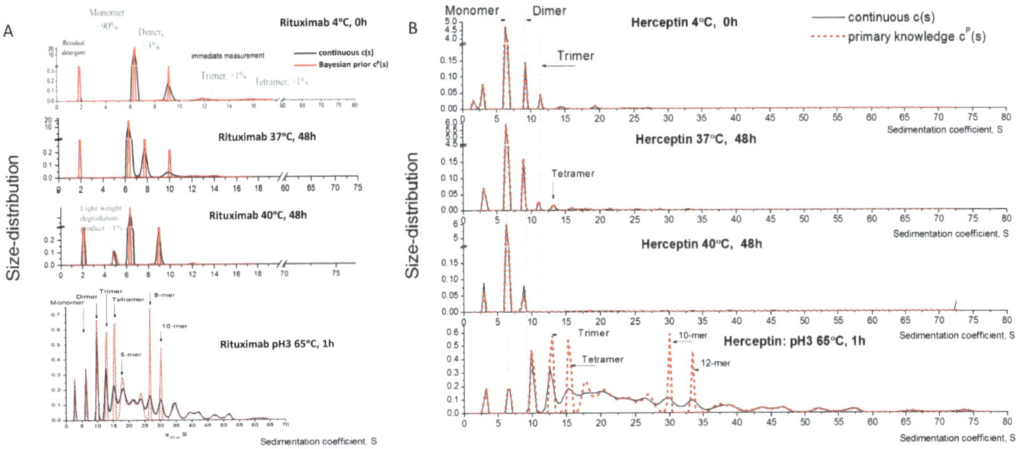

Figure 2. Size distribution based on sedimentation coefficient of the heat-stressed aggregated samples of Rituximab and Herceptin. Quantification of the aggregation state of heat-stressed (**A**) Rituximab and (**B**) Herceptin samples at 4 °C for 0 h, 37 °C for 48 h, 40 °C for 48 h, and 65 °C for 1 h (black line). The results fit the Bayesian prediction model (red dotted line). Monomer species are at around 6S, with elongated dimers appearing at 8S and globular dimers appearing at 10S.

3.2. Visual Characterization of Aggregates

Transmission Electron Microscopy (TEM) was performed to allow visual characterization of the aggregated heat-stressed mAb samples and provide a more qualitative analysis of aggregation. As depicted in Figure 3, the unstressed mAb samples at 4 °C contained almost no microscopic particles, whereas the heat-stressed samples at 40 °C showed small aggregates (black arrows). The number of detectable aggregates was even greater in the positive control for aggregation (65 °C). These results confirm the earlier results reported for protein analysis. Together, the two protein quantification methods indicated that heat stress modified the secondary structure of the mAb and promoted low levels of aggregation. Furthermore, the intensity of the heat stress protocol, as in the case of the positive control at 65 °C, exacerbated the level of aggregation.

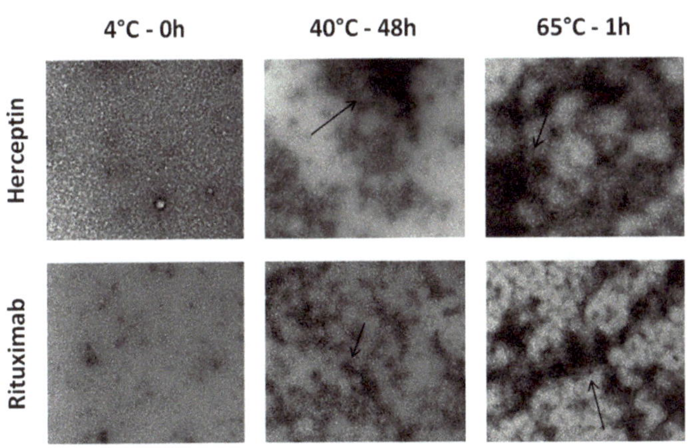

Figure 3. Visual characterization of two heat-stressed mAb samples by Transmission Electron Microscopy (TEM). Comparison of aggregation status of two mAbs, Rituximab and Herceptin, after a heat stress protocol. Increased aggregation highlighted by arrows is shown in heat treated conditions from 4 to 65 °C.

When comparing the protein content observed between heat stressed Herceptin and Rituximab, Herceptin showed a slightly increased aggregation level compared to Rituximab. Differences in the primary sequence of these mAbs may explain the differences in the degree of aggregation since Rituximab is a chimeric mAb and Herceptin is a humanized mAb [37].

3.3. Aggregated mAb Samples Induce Immune Activation and Skin Sensitization

In view of the observations that heat stress can alter mAb structure and induce aggregation, it was important to understand if this aggregation could lead to immunotoxic responses and potential skin sensitization. The effect of the heat stressed mAbs on immune activation was assessed by a human skin explant assay, T cell proliferation and cytokine release assays. PBMCs and skin biopsies from five healthy donors were stimulated with heat stressed aggregated samples of Rituximab and Herceptin (at 1 and 10 µg/mL) and assessed for increased immune activation.

Responses in the skin explant assay showed tissue damage in response to both aggregated Rituximab and Herceptin samples (Figures 4 and 5). Using Lerner's skin damage classification, skin damage classified as a grade II or higher response is indicative of an adverse immune reaction (Figure 1). At a concentration of 1 µg/mL, Herceptin caused significant tissue damage ($p < 0.005$) with grade III damage in three out of five tests at the 40 °C/48 h testing condition when compared to the negative control (Figure 4).

Herceptin 1 µg/mL tested at 4 °C and 37 °C, caused either a grade I or grade II skin damage response in all tests, with the exception to one test which showed a grade III positive response at 37 °C/48 h. Responses observed at 4 °C and 37 °C were not statistically different from the negative control. Similarly, at 10 µg/mL concentration, Herceptin caused significant tissue damage at 40 °C/48 h ($p < 0.05$), with grade III skin damage observed in four tests and a grade II skin damage response observed in one out of five tests. OKT3, as the positive control for skin damage, caused significant grade III skin damage at both 1 and 10 µg/mL ($p < 0.05$). The positive control for aggregation at 60 °C (pH = 3) caused grade II damage in three out of five tests at both 1 µg/mL and 10 µg/mL and a grade III response in one out of five tests at 10 µg/mL While still considered positive for skin damage, it was, however, not statistically significant from the negative control. These differences in the responses may be related to donor variability.

At 1 µg/mL, Rituximab did not cause any significant skin damage at the testing condition of 4 °C/0 h, where a grade I response was observed in all tests, whereas, at 4 °C/48 h, three of the five tests showed grade II skin damage. At 37 °C/0 h, three tests showed grade II skin damage, and one showed grade III skin damage, whereas, at 37 °C/48 h, only two tests of the five showed grade II skin damage response. At 40 °C/0 h, one test showed a grade II and one test showed a grade III skin damage, whereas, at 40 °C/48 h, only one test showed a positive response (grade II). OKT3 caused significant damage ($p < 0.05$), with a grade II response in four out of five tests and a grade III response in one test. At 10 µg/mL, Rituximab tested at 4 °C/0 h, 4 °C/48 h, and 37 °C/0 h showed a negative (grade I) response in all five tests. At 37 °C/48 h, a grade III response was observed in one test, and a grade II response was observed in one test of the five tests. At 40 °C/0 h, one test showed a grade III response, whereas the remaining four of the five tests showed a negative response. At 40 °C/48 h, a grade III response was observed in three tests, and a grade II response was observed in 1 out of five tests. Grade II positive responses were observed in response to OKT3.

The fact that Herceptin caused significant skin damage at both 1 and 10 µg/mL while Rituximab only caused significant damage at 10 µg/mL, suggested aggregated Herceptin and Rituximab have different immunotoxic potency profiles due to their aggregation status.

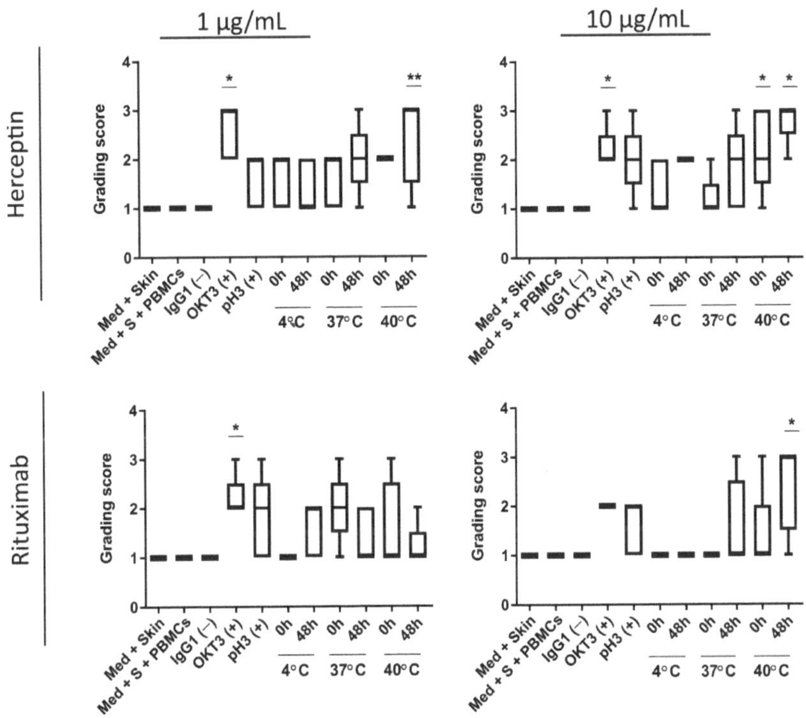

Figure 4. Skin explant assay results from exposure to heat-stressed mAb samples. Heat-stressed samples of Herceptin and Rituximab (1 and 10 μg/mL) were incubated with peripheral blood mononuclear cells (PBMCs) and autologous skin biopsies from healthy donors ($n = 5$) in the skin explant assay. The severity of the histopathological damage was graded using Lerner grades I–IV) Based on the damage observed, the boxes represent the median in the 25th and 75th percentiles with min and max whiskers. A significant increase (** $p < 0.005$ and * $p < 0.05$) in tissue damage response was observed at 1 μg/mL and 10 μg/mL Herceptin when compared to the negative control (skin in medium alone) at the 40 °C/48 h testing condition. Rituximab showed significant tissue damage at 40 °C/48 h when compared to the negative control (* $p < 0.05$). The data were assessed by one-way ANOVA and Bonferroni's post-hoc test.

Figure 5 shows representative H&E images of each of the treatment conditions for Rituximab and Herceptin. Black arrows indicate histological changes on the dermal/epidermal junction relevant to the grading process, namely epidermal disruption. T cell proliferation assays showed no significant increase in T cell proliferation upon exposure to aggregated Herceptin or Rituximab (Figure 6) in any of the testing conditions. The positive control, OKT3, elicited a T cell stimulation index above the log 2-fold threshold in both mAbs at 1 and 10 μg/mL.

Cytokine levels measured in cell culture supernatants from samples tested with 1 μg/mL (Figure 7) and 10 μg/mL (Figure 8) showed that IL-10, IFN-γ, and TNF-α levels were significantly elevated after exposure to OKT3 at both test concentrations ($p < 0.05$). Aggregated Rituximab and Herceptin tests, when compared to IgG1 controls, showed no significant increase in cytokine levels at both 1 and 10 μg/mL. However, increased yet non-significant IFN-γ levels were observed in response to 10 μg/mL in the Rituximab test when compared to the negative control. The positive control for aggregation (pH3) showed non significantly increased IL-10 levels when tested at 10 μg/mL (Figure 8) in response to Herceptin compared to a significant increased in response to Rituximab ($p < 0.05$).

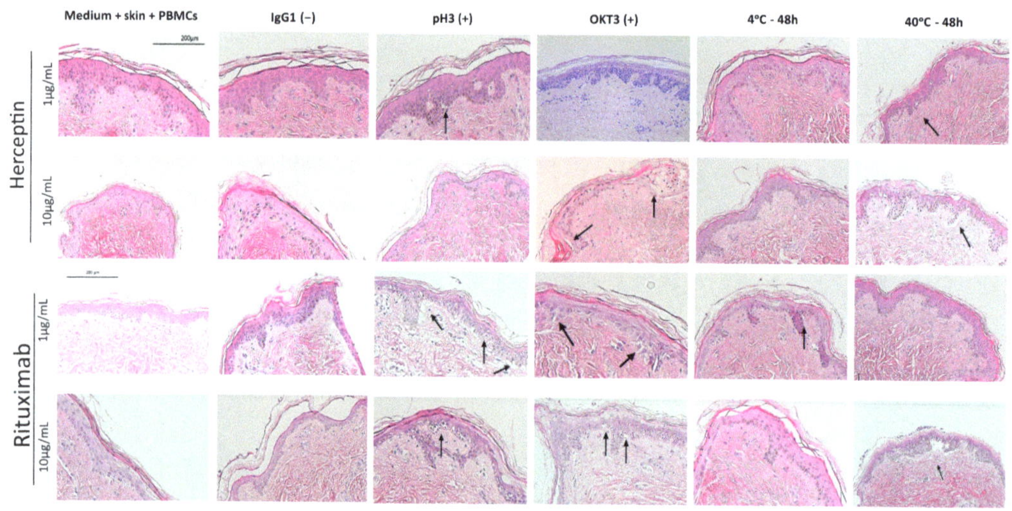

Figure 5. Histopathological damage from exposure to heat-stressed mAb samples. H&E staining of skin tissue after incubation with the heat-stressed samples (60 °C, pH3) of Herceptin and Rituximab at 1 and 10 µg/mL. Black arrows represent histopathological damage on the dermal/epidermal junction in the form of vacuolization of keratinocytes (Grade II positive reaction) and cleft formation leading to separation of the dermis and epidermis (Grade III positive reaction). Scale bars represent 200 µm.

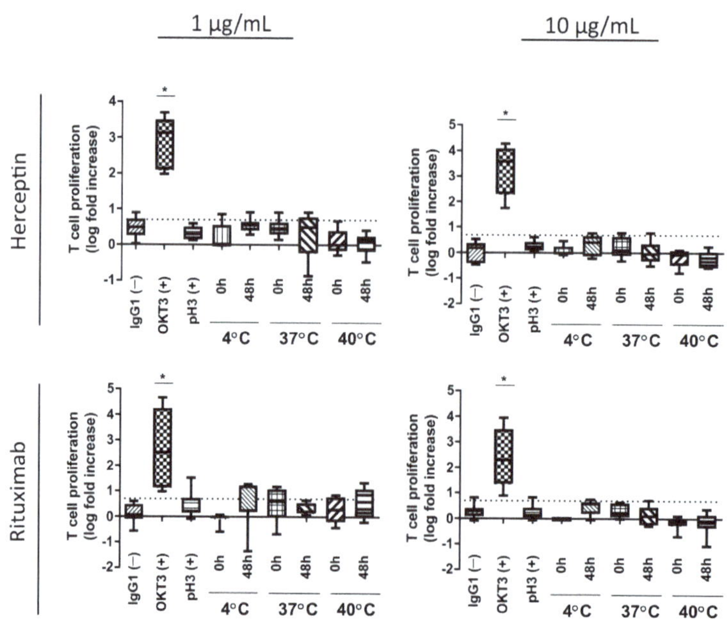

Figure 6. T-cell proliferation assessment resulting from exposure to heat-stressed mAb samples. Heat-stressed samples of Herceptin and Rituximab (1 and 10 µg/mL) were incubated with donor-derived PBMC (n = 5 for each condition). T-cell proliferation assessment alone was not capable of detecting an increased immunological profile induced through protein aggregation. The boxes represent the median with the 25th and 75th percentile with min and max whiskers. * $p < 0.05$ Bonferroni's post-hoc test after One-way ANOVA. The log 2-fold positive threshold is indicated by the dotted line.

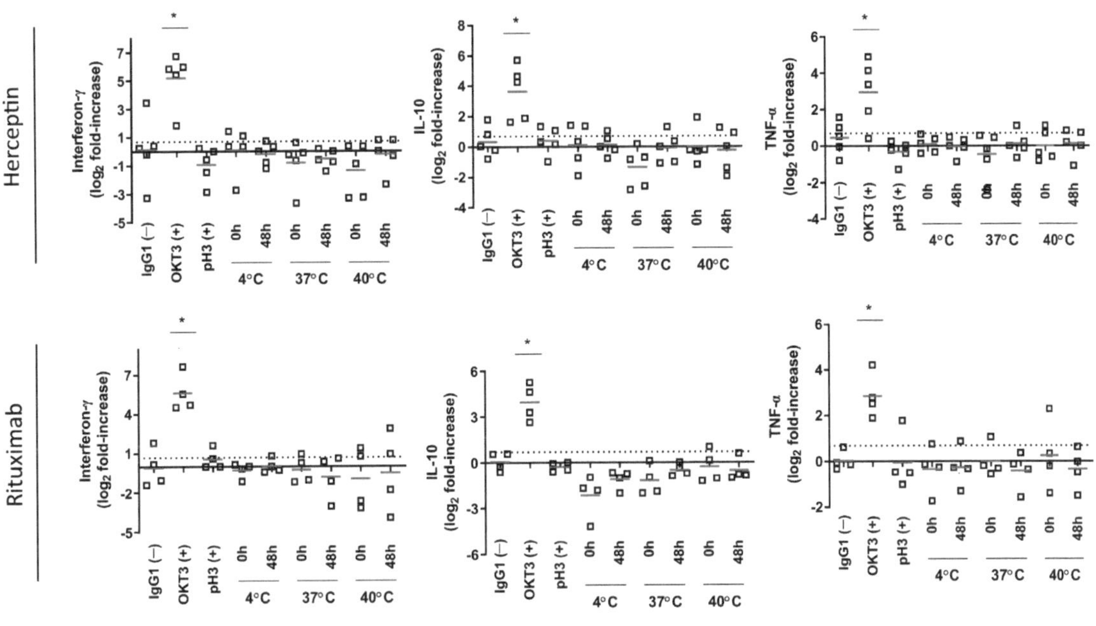

Figure 7. Multiplex cytokine panel analysis of experimental conditions tested at 1 μg/mL. For each condition ($n = 6$), the median is represented by a red line. The log 2-fold positive threshold is indicated by the dotted line. * $p < 0.05$ Bonferroni's post-hoc test after One-way ANOVA.

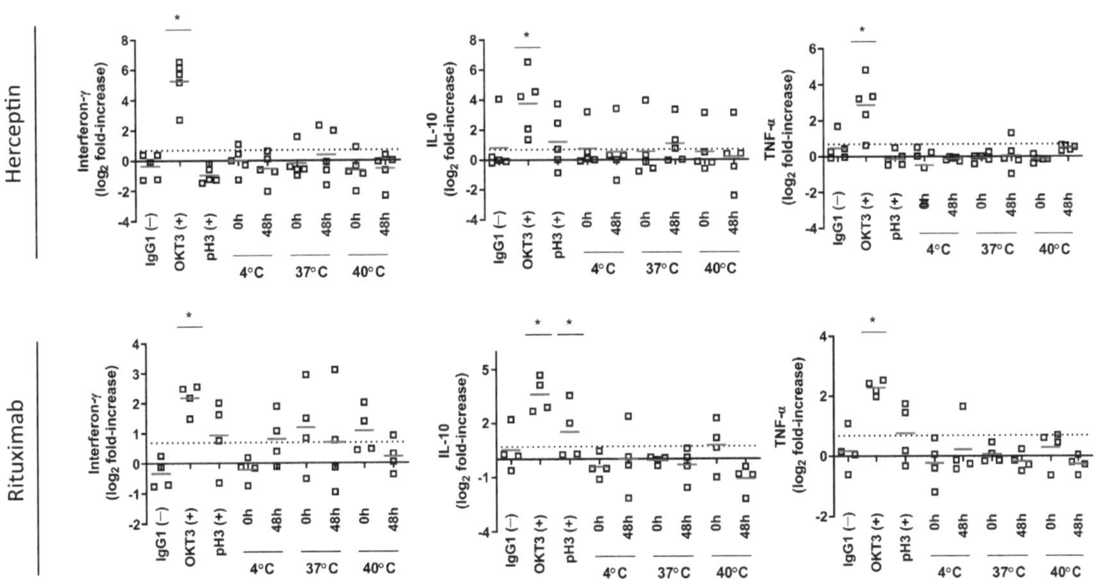

Figure 8. Multiplex cytokine panel analysis of experimental conditions tested at 10 μg/mL. For each condition ($n = 6$), the median is represented by a red line. The log 2-fold positive threshold is indicated by the dotted line. * $p < 0.05$ Bonferroni's post-hoc test after One-way ANOVA.

While no significant increases in T cell proliferation responses were observed upon exposure to heat-stressed mAbs, histopathological damage showing grade II and III positive responses were observed in the in vitro skin explant assay. To determine if apoptosis was occurring within the tissue sections, Heat Shock Protein (HSP) 70 staining was performed (Figure 9). Positive HSP 70 staining (bright green) was observed in the 37 °C and 40 °C conditions in both Herceptin and Rituximab at 10 µg/mL. Positive control for aggregation at 60 °C (pH3) also showed positive expression of HSP70.

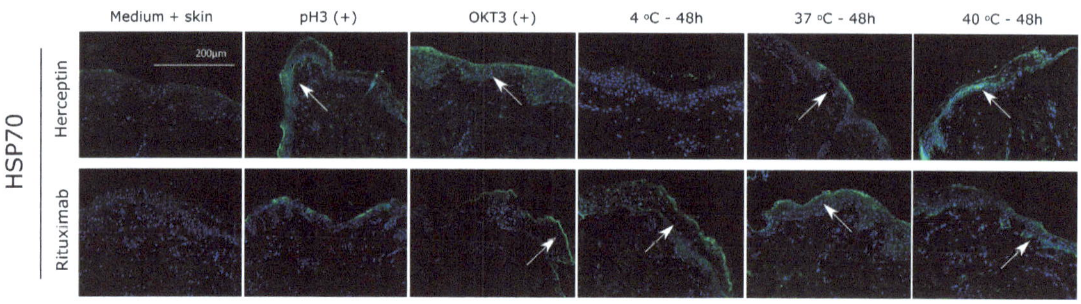

Figure 9. Apoptosis caused by mAb aggregation. Heat stressed aggregated samples of Herceptin and Rituximab were stained for heat shock protein 70 (HSP70) by immunofluorescence. DAPI was used as nuclear marker (blue). White arrows indicate areas of positive staining.

4. Discussion

Previous studies addressing the immunotoxic potential of mAb aggregation have used heat stress protocols with temperatures as high as 65 °C (25), 70 °C (26), or 80 °C [24,33]. In comparison, the maximum temperature condition set for this study was 40 °C. The temperature test points were chosen for their relevance to an in vivo scenario based on normal and fever-induced body temperatures. Our results demonstrated that temperature treatment of both Herceptin and Rituximab induced changes in the aggregative status of the tested samples. It could be suggested that a level of protein unfolding and consequent rearrangement of the mAb secondary structure was primarily responsible for the presence of larger molecules in the aggregated samples (as demonstrated by SV-AUC results). TEM analysis of aggregated mAbs demonstrated this effect visually, with clear changes in aggregation observed in heat-treated samples, especially at higher temperatures of 65 °C. These thermodynamic changes accounted for less than 4% aggregation of the total protein content, which is in accordance with previous reports of 3% aggregation [26] of the total protein content under heat stress conditions.

There is evidence that protein aggregation can occur at different stages of mAb production or during later storage, imposing a health risk to patients upon administration of therapeutic mAbs [38]. We have studied the potential immunotoxic effect of mAb aggregation by investigating histopathological assessment of cell damage in a novel in vitro skin explant assay, T cell proliferation, and cytokine release. Overall, our data shows heat stressed Herceptin and Rituximab samples did not enhance T cell proliferation rates. T cell proliferative induction is primarily stimulated through increased dendritic cell activity [28]. Our assay could be improved through the incorporation of dendritic cell activity alongside T cell proliferative responses. This would allow us to gain a more complete assessment of the ongoing processes within the assay and provide an additional data point to further improve the validity of our predictive outcome. Furthermore, the composition of PBMC populations can vary significantly across individuals. In this study, PBMC populations were not assessed and normalized prior to use in stimulatory experiments. Characterization of the PBMC population prior to use within the assay could allow for a level of data normalizing, which could further improve assay reliance and sensitivity.

A trend for elevated expression of IFN-γ was observed at 10 µg/mL in Rituximab test conditions. IL-10 is an immunosuppressive cytokine which inhibits expression of pro-inflammatory cytokines [38]. Elevated expression of IL-10 in the 10 µg/mL Rituximab positive control could suggest that cells are actively forming a protection mechanism from the damage caused by the aggregated mAbs, by dampening the immune response and sustaining normal tissue homeostasis.

Previously published data [31,32] based on the human tissue explant assay used in this study has resulted in convincing prediction of adverse immune events in response to chemical [31] and small molecule drug stimulation [32]. The histopathological damage assessment shown here suggests that heat-stressed Rituximab and Herceptin do cause some degree of immune activation. These results do not fully corroborate previously published studies, where it was shown that mAb aggregation by heat stress elicits in vitro innate and late-stage T cell responses [27], activates dendritic cells to stimulate T cells [24] and induces CD4+ T cell proliferation via pro-inflammatory cytokine release [25].

Heat shock protein 70 (HSP70) serves as a molecular chaperone in cases of cell stress (as heat shock or oxidative stress). HSP70 is crucial for cell survival after apoptotic stimuli [39,40], as shown by Mosser et al. [40] in which macrophages heat-stressed for up to two hours at 43 °C, showed protection of DNA fragmentation in the presence of HSP70 [39]. HSP70 was shown to be significantly upregulated in all treated samples and absent in negative controls. This suggests that some level of inflammatory stress was being imparted throughout the assay conditions. HSP70 is most likely involved in the protein refolding after exposure to the aggregated Herceptin and Rituximab, and its overexpression may rescue cells from apoptotic cell death. Skin explant assay samples were tested at 4 °C, 37 °C and 40 °C temperature conditions. Centrifugal analysis of aggregate formation at these temperatures only showed relatively moderate changes in aggregated antibody levels with the vast majority of mAbs still present in the monomer formation. This suggests that, at best, only a mild immune stimulatory effect would be observed and that this, alongside HSP70's anti-apoptotic effects, may partly explain why mild histological damage was observed in aggregated antibody-treated conditions.

Overall, only the highest testing temperature (40 °C) caused significant histopathological damage in comparison to the baseline condition whereby grade III skin damage was observed in most of the donors, while at 4 °C and 37 °C an overall grade II was observed. While the outcome is not a severe grade, these results still show positive skin damage caused by heat-stressed aggregated mAbs. The higher condition of heat stress (40 °C) did cause more severe histopathological damage than the positive control (pH3). We could hypothesize that highly aggregated mAbs (as in the case of the positive control for aggregation, pH = 3) can affect the normal binding of these antibodies to the antigen-presenting cells [24]. It has been shown that the structural mAb rearrangements that result from exposure to temperature shock can lead to neo-epitopes [41] (new epitopes formed during aggregation), glycosylation/pegylation changes [42] and other post-translational modifications that break tolerance towards these mAbs, thus causing immunotoxicity. Furthermore, it has been shown that only covalently modified aggregates can induce immunotoxicity [29]. However, for aggregated mAbs to cause an immunotoxic response, it is vital that conformation of the protein's active site remains preserved regardless of aggregation [43].

Thus, we can speculate that the low pH (pH3) internal positive control denatured the antibodies to such an extent that protein conformation was lost by denaturation during heat stress. The consequential aggregation occurring after denaturation impeded the preservation of the protein's active site. These results also highlight the fact that protein aggregation does not always directly imply immunotoxicity. While structural changes might occur due to aggregation, it does not mean that protein functionality is compromised. Previous studies showed that immunotoxicity is only achieved if the primary structure of the mAb protein is deeply modified after aggregation stress [23]. Therefore, it can be argued that the positive low pH control for aggregation did not cause a high level of immune activation/cell death due to the new structural rearrangement of the protein, hiding its active site.

Further work is required to better understand the potential use of the skin explant assay in the assessment of the immunotoxicity of aggregated antibodies. The data presented here represents a small study looking at relatively few aggregative conditions and only 2 mAbs. Nevertheless, it should still be emphasized that this novel assay was able to detect some immune activation through histological damage assessment directly brought about by immune cell activity within the assays. Further work is required to elucidate the mechanistic actions around aggregate formation and the effect of aggregate presence on immune cell activation. Future work involving this assay should focus on expanding the donor replicates within the explant assays and should include a greater range of temperatures to induce aggregate formation. Additionally, a larger sample size of mAbs based on different therapeutic areas and mechanistic action could elucidate additional areas of study interest.

5. Conclusions

This study was aimed at testing a novel in vitro test method for assessment of skin sensitization and adverse immune events relevant to human safety and has provided new insights for potentially reducing animal testing in the future. Our method was able to predict immune activation facilitated through temperature-mediated mAb aggregate formation in two antibody systems with a reasonable level of accuracy.

6. Patents

Alcyomics holds patents on the skin explant assay (trading as Skimune®) see table

Case Code	Applicant	Country	Status	Official Number	Title
P131663CH	Alcyomics Limited	Switzerland	Granted	'EP2524227	Skin Explant Assay
P131663DE	Alcyomics Limited	Germany	Granted	'602011004834	Skin Explant Assay
P131663DK	Alcyomics Limited	Denmark	Granted	'EP2524227	Skin Explant Assay
P131663ES	Alcyomics Limited	Spain	Granted	'EP2524227	Skin Explant Assay
P131663FR	Alcyomics Limited	France	Granted	'EP2524227	Skin Explant Assay
P131663GB1	Alcyomics Limited	United Kingdom	Granted	'EP2524227	Skin Explant Assay
P131663IT	Alcyomics Limited	Italy	Granted	'502014902254005	Skin Explant Assay
P131663US	Alcyomics Limited	USA	Granted	'9,651,544	Skin Explant Assay
P131663USD1	Alcyomics Limited	USA	Granted	'10,073,084	Skin Explant Assay

Author Contributions: A.M.-R. and A.K. performed the experiments, S.S.A. contributed to data analysis and experimental design, M.F. and A.I. contributed to reviewing the document, and J.G. and A.M.D. contributed to study design. All authors have read and agreed to the published version of the manuscript.

Funding: This work was funded by the European Union's Horizon 2020 research and innovation program under the Marie Skłodowska-Curie Actions grant agreement No. 643056 (BIORAPID). The authors gratefully acknowledge this financial support.

Institutional Review Board Statement: The study was conducted in accordance with the Declaration of Helsinki, and approved by the Ethics Committee of Newcastle and North Tyneside (UK) (reference no 10/H0906/58 and 2 October 2020) for studies involving humans.

Informed Consent Statement: Informed consent was obtained from all subjects involved in the study.

Data Availability Statement: The original data presented in the study are included in the article; further inquiries can be directed to the corresponding author.

Acknowledgments: The authors thank the NUPPA facility service from Newcastle University for the AUC-SV protein analysis.

Conflicts of Interest: Ana Martins-Ribeiro, Shaheda Sameena Ahmed, Abbas Ishaq, Matthew Freer, and Anne Mary Dickinson all were employed by Alcyomics Ltd. None of the authors has received or expects to receive research grants from funding agencies. None of the authors has received support from commercial sources of funding by companies that sell drugs, medical devices, or provide medical services. None of the authors has received honoraria for speaking at symposia associated with this work. None of the authors holds a position on advisory board associated with this work.

References

1. Vermeer, A.W.; Norde, W. The Thermal Stability of Immunoglobulin: Unfolding and Aggregation of a Multi-Domain Protein. *Biophys. J.* **2000**, *78*, 394–404. [CrossRef] [PubMed]
2. Joubert, M.K.; Luo, Q.; Nashed-Samuel, Y.; Wypych, J.; Narhi, L.O. Classification and Characterization of Therapeutic Antibody Aggregates. *J. Biol. Chem.* **2011**, *286*, 25118–25133. [CrossRef] [PubMed]
3. Menzen, T.; Friess, W. Temperature-Ramped Studies on the Aggregation, Unfolding, and Interaction of a Therapeutic Monoclonal Antibody. *J. Pharm. Sci.* **2014**, *103*, 445–455. [CrossRef] [PubMed]
4. Vermeer, A.W.; Bremer, M.G.; Norde, W. Structural changes of IgG induced by heat treatment and by adsorption onto a hydrophobic Teflon surface studied by circular dichroism spectroscopy. *Biochim. Biophys. Acta BBA Gen. Subj.* **1998**, *1425*, 1–12. [CrossRef] [PubMed]
5. Buchner, J.; Renner, M.; Lilie, H.; Hinz, H.J.; Jaenicke, R.; Kiefhaber, T.; Rudolph, R. Alternatively folded states of an immunoglobulin. *Biochemistry* **1991**, *30*, 6922–6929. [CrossRef]
6. Talley, K.; Alexov, E. On the pH-optimum of activity and stability of proteins. *Proteins* **2010**, *78*, 2699–2706. [CrossRef]
7. Kiese, S.; Papppenberger, A.; Friess, W.; Mahler, H.-C. Shaken, Not Stirred: Mechanical Stress Testing of an IgG1 Antibody. *J. Pharm. Sci.* **2008**, *97*, 4347–4366. [CrossRef]
8. Walchi, R.; Vermeire, P.-J.; Massant, J.; Arosi, P. Accelerated aggregation studies of Monoclonal Antibodies: Considerations for Storage Stability. *J. Pharm. Sci.* **2020**, *109*, 595–602. [CrossRef]
9. Peters, B.J.; Capelle, M.A.H.; Arvinte, T.; van de Garde, E.M.W. Validation of an automated method for compounding monoclonal antibody patient doses: Case studies of Avastin bevacizumab), Remicade (infliximab) and Herceptin (trastuzumab). *MAbs* **2013**, *5*, 162–170. [CrossRef]
10. Wafer, L.; Kloczewiak, M.; Luo, Y. Quantifying Trace Amounts of Aggregates in Biopharmaceuticals Using Analytical Ultracentrifugation Sedimentation Velocity: Bayesian Analyses and F Statistics. *AAPS J.* **2016**, *18*, 849–860. [CrossRef]
11. Ye, H. Simultaneous determination of protein aggregation, degradation, and absolute molecular weight by size exclusion chromatography-multiangle laser light scattering. *Anal. Biochem.* **2006**, *356*, 76–85. [CrossRef] [PubMed]
12. Fraunhofer, W.; Winter, G. The use of asymmetrical flow field-flow fractionation in pharmaceutics and biopharmaceutics. *Eur. J. Pharm. Biopharm.* **2004**, *58*, 369–383. [CrossRef] [PubMed]
13. Hemmig, R.; Rammage, P.; Johann, C. Asymmetric Flow Field-Flow Fractionation (AF4) with Multi-Angle Light Scattering (MALS) for High-Throughput Protein Refolding. *LCGC Eur.* **2005**, *18*, 532–538.
14. Berkowitz, S.A. Role of analytical ultracentrifugation in assessing the aggregation of protein biopharmaceuticals. *AAPS J.* **2006**, *8*, E590–E605. [CrossRef] [PubMed]
15. Cole, J.L.; Lary, J.W.; Moody, T.; Laue, T.M. Analytical ultracentrifugation: Sedimentation velocity and sedimentation equilibrium. *Methods Cell Biol.* **2008**, *84*, 143–179. [PubMed]
16. Secher, T.; Bodier-Montagutelli, E.; Parent, C.; Bouvart, L.; Cortes, M.; Ferreira, M.; MacLoughlin, R.; Ilango, G.; Schmid, O.; Respaud, R.; et al. Aggregates Associated with Instability of Antibodies during Aerosoliziation Induce Adverse Immunologival Effects. *Pharm. J.* **2022**, *14*, 671.
17. Ryan, M.; Lam, N.; Wright, K.; Siderov, J. Clinical Oncology Society of Australia Position Statement: 2022 update to the safe handling of monoclonal antibodies in healthcare settings. *Asia-Pac. J. Clin. Oncol.* **2023**, *19*, 723–730. [CrossRef] [PubMed]
18. Roberts, C.J. Therapeutic protein aggregation: Mechanisms, design, and control. *Trends Biotechnol.* **2014**, *32*, 372–380. [CrossRef] [PubMed]
19. Baert, F.; Noman, M.; Vermeire, S.; Van Assche, G.; D'Haens, G.; Carbonez, A.; Rutgeerts, P. Influence of immunogenicity on the long-term efficacy of infliximab in Crohn's disease. *N. Engl. J. Med.* **2003**, *348*, 601–608. [CrossRef]
20. West, R.L.; Zelinkova, Z.; Wolbink, G.J.; Kuipers, E.J.; Stokkers, P.C.F.; Van Der Woude, C.J. Immunogenicity negatively influences the outcome of adalimumab treatment in Crohn's disease. *Aliment. Pharmacol. Ther.* **2008**, *28*, 1122–1126. [CrossRef]
21. Swanson, M.D.; Rios, S.; Mittal, S.; Soder, G.; Jawa, V. Immunogenicity Risk Assessment of Spontaneously Occurring Therapeutic Monoclonal Antibody Aggregates. *Front. Immunol.* **2022**, *13*, 915412. [CrossRef] [PubMed]
22. Filipe, V.; Poole, R.; Kutscher, M.; Forier, K.; Braeckmans, K.; Jiskoot, W. Fluorescence Single Particle Tracking for the Characterization of Submicron Protein Aggregates in Biological Fluids and Complex Formulations. *Pharm. Res.* **2011**, *28*, 1112–1120. [CrossRef] [PubMed]
23. Filipe, V.; Poole, R.; Oladunjoye, O.; Braeckmans, K.; Jiskoot, W. Detection and Characterization of Subvisible Aggregates of Monoclonal IgG in Serum. *Pharm. Res.* **2012**, *29*, 2202–2212. [CrossRef] [PubMed]
24. Rombach-Riegraf, V.; Karle, A.C.; Wolf, B.; Sordé, L.; Koepke, S.; Gottlieb, S.; Krieg, J.; Djidja, M.-C.; Baban, A.; Spindeldreher, S.; et al. Aggregation of Human Recombinant Monoclonal Antibodies Influences the Capacity of Dendritic Cells to Stimulate Adaptive T-Cell Responses In Vitro. *PLoS ONE* **2014**, *9*, e86322. [CrossRef] [PubMed]
25. Ahmadi, M.; Bryson, C.J.; Cloake, E.A.; Welch, K.; Filipe, V.; Romeijn, S.; Hawe, A.; Jiskoot, W.; Baker, M.P.; Fogg, M.H. Small Amounts of Sub-Visible Aggregates Enhance the Immunogenic Potential of Monoclonal Antibody Therapeutics. *Pharm. Res.* **2014**, *32*, 1383–1394. [CrossRef] [PubMed]
26. Joubert, M.K.; Deshpande, M.; Yang, J.; Reynolds, H.; Bryson, C.; Fogg, M.; Baker, M.P.; Herskovitz, J.; Goletz, T.J.; Zhou, L.; et al. Use of In Vitro Assays to Assess Immunogenicity Risk of Antibody-Based Biotherapeutics. *PLoS ONE* **2016**, *11*, e0159328. [CrossRef] [PubMed]

27. Joubert, M.K.; Hokom, M.; Eakin, C.; Zhou, L.; Deshpande, M.; Baker, M.P.; Goletz, T.J.; Kerwin, B.A.; Chirmule, N.; Narhi, L.O.; et al. Highly Aggregated Antibody Therapeutics Can Enhance the in Vitro Innate and Late-stage T-cell Immune Responses. *J. Biol. Chem.* **2012**, *287*, 25266–25279. [CrossRef]
28. Chudnovskiy, A.; Pasqual, G.; Victora, G.D. Studying interactions between dendritic cells and T cells in vivo. *Curr. Opin. Immunol.* **2019**, *58*, 24–30. [CrossRef] [PubMed]
29. Bessa, J.; Boeckle, S.; Beck, H.; Buckel, T.; Schlicht, S.; Ebeling, M.; Kiialainen, A.; Koulov, A.; Boll, B.; Weiser, T.; et al. The Immunogenicity of Antibody Aggregates in a Novel Transgenic Mouse Model. *Pharm. Res.* **2015**, *32*, 2344–2359. [CrossRef]
30. Clair, J.B.S.; Detanico, T.; Aviszus, K.; Kirchenbaum, G.A.; Christie, M.; Carpenter, J.F.; Wysocki, L.J. Immunogenicity of Isogenic IgG in Aggregates and Immune Complexes. *PLoS ONE* **2017**, *12*, e0170556. [CrossRef]
31. Ahmed, S.S.; Wang, X.N.; Fielding, M.; Kerry, A.; Dickinson, I.; Munuswamy, R.; Kimber, I.; Dickinson, A.M. An in vitro human skin test for assessing sensitization potential. *J. Appl. Toxicol.* **2016**, *36*, 669–684. [CrossRef] [PubMed]
32. Ahmed, S.; Whritenour, J.; Ahmed, M.; Bibby, L.; Darby, L.; Wang, X.; Watson, J.; Dickinson, A. Evaluation of a human in vitro skin test for predicting drug hypersensitivity reactions. *Toxicol. Appl. Pharmacol.* **2019**, *369*, 39–48. [CrossRef] [PubMed]
33. Carpenter, P.A.; Pavlovic, S.; Tso, J.Y.; Press, O.W.; Gooley, T.; Yu, X.-Z.; Anasetti, C. Non-Fc Receptor-Binding Humanized Anti-CD3 Antibodies Induce Apoptosis of Activated Human T Cells. *J. Immunol.* **2000**, *165*, 6205–6213. [CrossRef] [PubMed]
34. Ahmed, S.S.; Ahmed, M.M.; Ashaq, A.; Freer, M.; Stebbings, R.; Dickinson, A.M. An In vitro human skin test for predicting skin sensitization and adverse immune reactions of biologics. *Toxics* **2024**, in press.
35. Lerner, K.G.; Kao, G.F.; Storb, R.; Buckner, C.D.; Clift, R.A.; Thomas, E.D. Histopathology of graft-vs.-host reaction (GvHR) in human recipients of marrow from HL-A-matched sibling donors. *Transplant. Proc.* **1974**, *6*, 367–371. [PubMed]
36. Ikwegbue, P.C.; Masamba, P.; Oyinloye, B.E.; Kappo, A.P. Roles of Heat Shock Proteins in Apoptosis, Oxidative Stress, Human Inflammatory Diseases, and Cancer. *Pharmaceuticals* **2018**, *11*, 2. [CrossRef] [PubMed]
37. Jarvis, M.; Marzolini, M.; Wang, X.N.; Jackson, G.; Sviland, L.; Dickinson, A.M. Heat shock protein 70: Correlation of expression with degree of graft-versus-host response and clinical graft-versus-host disease. *Transplantation* **2003**, *76*, 849–853. [CrossRef] [PubMed]
38. Obrezanova, O.; Arnell, A.; de la Cuesta, R.G.; Berthelot, M.E.; Gallagher, T.R.; Zurdo, J.; Stallwood, Y. Aggregation risk prediction for antibodies and its application to biotherapeutic development. *mAbs* **2015**, *7*, 352–363. [CrossRef]
39. Murray, P.J. The primary mechanism of the IL-10-regulated antiinflammatory response is to selectively inhibit transcription. *Proc. Natl. Acad. Sci. USA* **2005**, *102*, 8686–8691. [CrossRef]
40. Mosser, D.D.; Caron, A.W.; Bourget, L.; Denis-Larose, C.; Massie, B. Role of the Human Heat Shock Protein hsp70 in Protection against Stress-Induced Apoptosis. *Mol. Cell. Biol.* **1997**, *17*, 5317–5327. [CrossRef]
41. Klein, S.D.; Brune, B. Heat-shock protein 70 attenuates nitric oxide-induced apoptosis in RAW macrophages by preventing cytochrome c release. *Biochem. J.* **2002**, *362*, 635–641. [CrossRef] [PubMed]
42. Boll, B.; Bessa, J.; Folzer, E.; Quiroz, A.R.; Schmidt, R.; Bulau, P.; Finkler, C.; Mahler, H.-C.; Huwyler, J.; Iglesias, A.; et al. Extensive Chemical Modifications in the Primary Protein Structure of IgG1 Subvisible Particles Are Necessary for Breaking Immune Tolerance. *Mol. Pharm.* **2017**, *14*, 1292–1299. [CrossRef] [PubMed]
43. Hermeling, S.; Crommelin, D.J.A.; Schellekens, H.; Jiskoot, W. Structure-Immunogenicity Relationships of Therapeutic Proteins. *Pharm. Res.* **2004**, *21*, 897–903. [CrossRef] [PubMed]

toxics

MDPI

Review

Chemical-Specific T Cell Tests Aim to Bridge a Gap in Skin Sensitization Evaluation

Nele Fritsch [1,2,†], Marina Aparicio-Soto [1,*,†], Caterina Curato [1], Franziska Riedel [1], Hermann-Josef Thierse [1], Andreas Luch [1,3] and Katherina Siewert [1,*]

1 German Federal Institute for Risk Assessment (BfR), Department of Chemical and Product Safety, Dermatotoxicology Study Centre, 10589 Berlin, Germany; nele.fritsch@bfr.bund.de (N.F.); caterina.curato@bfr.bund.de (C.C.); franziska.riedel@bfr.bund.de (F.R.)
2 Institute of Biotechnology, Technical University of Berlin, 10115 Berlin, Germany
3 Institute of Pharmacy, Freie Universität Berlin, 14195 Berlin, Germany
* Correspondence: marina.aparicio-soto@bfr.bund.de (M.A.-S.); katherina.siewert@bfr.bund.de (K.S.); Tel.: +49-30-1841257003 (M.A.-S.); +49-30-1841257001 (K.S.)
† These authors contributed equally to this work.

Abstract: T cell activation is the final key event (KE4) in the adverse outcome pathway (AOP) of skin sensitization. However, validated new approach methodologies (NAMs) for evaluating this step are missing. Accordingly, chemicals that activate an unusually high frequency of T cells, as does the most prevalent metal allergen nickel, are not yet identified in a regulatory context. T cell reactivity to chemical sensitizers might be especially relevant in real-life scenarios, where skin injury, co-exposure to irritants in chemical mixtures, or infections may trigger the heterologous innate immune stimulation necessary to induce adaptive T cell responses. Additionally, cross-reactivity, which underlies cross-allergies, can only be assessed by T cell tests. To date, several experimental T cell tests are available that use primary naïve and memory CD4+ and CD8+ T cells from human blood. These include priming and lymphocyte proliferation tests and, most recently, activation-induced marker (AIM) assays. All approaches are challenged by chemical-mediated toxicity, inefficient or unknown generation of T cell epitopes, and a low throughput. Here, we summarize solutions and strategies to confirm in vitro T cell signals. Broader application and standardization are necessary to possibly define chemical applicability domains and to strengthen the role of T cell tests in regulatory risk assessment.

Keywords: adverse outcome pathway; allergic contact dermatitis; chemical-induced T cell epitopes; contact allergens; in vitro T cell test; key events; new approach methodologies; skin sensitization; T cell receptor

1. Introduction

In our modern industrialized society, humans are regularly exposed to numerous chemical compounds. Single or repeated skin contact with sensitizing chemicals may induce contact allergies in susceptible individuals. Upon re-exposure, an inflammatory skin condition known as allergic contact dermatitis (ACD) can develop [1–3]. Epidemic-like occurrence of contact allergies has been reported in recent decades [4,5]. Rough estimations claim that approximately 20% of the general population may be sensitized to at least one contact allergen [6]. The widespread occurrence of ACD implies significant economic and personal costs [7].

To prevent or reduce ACD, newly authorized chemicals must undergo skin sensitization testing to comply with national and international regulations on chemical safety. For instance, in the European Union (EU), for all chemicals manufactured or imported in quantities of ≥1 ton per year, information on skin sensitizing properties is required by the

REACH regulation (Registration, Evaluation, Authorization and Restriction of Chemicals; Regulation (EC) No. 1907/2006).

For decades, animal-based tests have been the gold standard to assess the skin sensitizing properties of chemicals (Figure 1). Guinea pig tests, which address both the sensitization and elicitation phase of ACD, have largely been replaced by the more animal-friendly murine local lymph node assay (LLNA) that only considers the sensitization phase (OECD TG 406; OECD 429) [8–10]. Early human patch tests designed to induce active sensitization have been banned under REACH [11,12]. However, human repeated insult patch tests (HRIPTs) are still used to confirm no-effect levels, mainly in the cosmetics, household products, and pharmaceutical sectors [13–15].

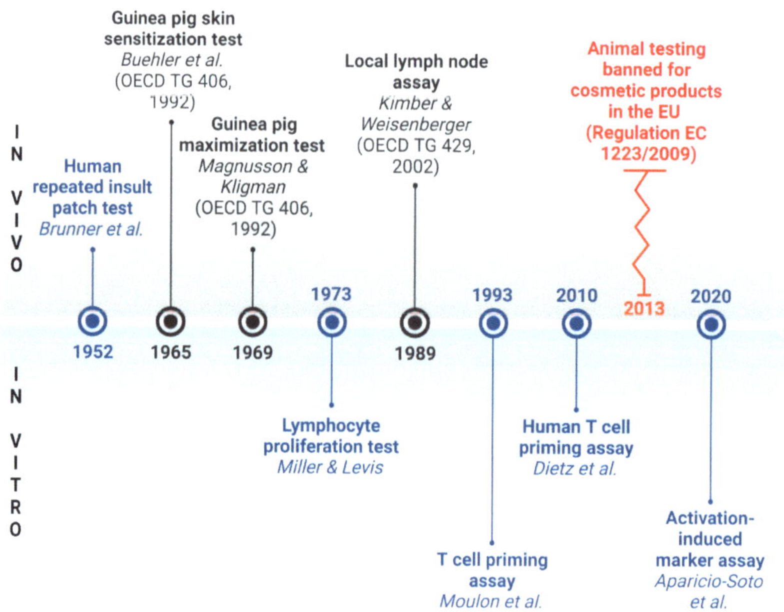

Figure 1. History of tests for assessing chemical-induced T cell responses. Traditional in vivo methods for assessing T cell responses in animals are represented in the OECD test guidelines (upper part) [8–10,12]. For a long time, these methods preceded the development of in vitro assays (lower part). Nevertheless, T cell assays have significantly advanced over time [16–19]. The necessity of alternative tests was demonstrated by the prohibition of animal testing for cosmetics in the EU in 2013 (red zigzag line). References were selected from among the earliest descriptions of these tests for studying contact allergens. The terms 'assay' and 'test' are used interchangeably. Blue text indicates human-based test systems, while dark gray denotes animal-based methods. EC—European Commission; EU—European Union; OECD—Organization for Economic Co-operation and Development; TG—test guideline. Created in BioRender.com. Siewert, K. (2024) https://BioRender.com/h57s252, accessed on 1 November 2024.

There is a growing need for alternative methods to replace or reduce in vivo testing, including skin sensitization assessment. In recent years, the concept of new approach methodologies (NAMs) has emerged to accommodate a higher throughput, reduce uncertainties resulting from interspecies differences, and provide insights into molecular mechanisms, while adhering to improved ethical standards [20–22]. Cutting-edge NAMs for skin sensitization testing have been incorporated into REACH and other chemical regulations, thereby allowing animal testing only when these alternative methods prove inconclusive [23,24]. The EU's implementation of a comprehensive ban on animal testing for cosmetics in 2013 underscores the ongoing evolution in the field of chemical regulation (Regulation (EC) No. 1223/2009).

NAMs for skin sensitization have been developed for a series of biological key events (KEs) considered essential to define the adverse outcome pathway (AOP) of skin sensitization [25] (Figure 2). The molecular initiating event (MIE) covers the binding of chemicals to endogenous skin proteins (KE1, haptenation) and can be addressed by in chemico peptide reactivity assays (OECD TG 442C). Subsequently, keratinocytes and dendritic cells (DCs) become activated (KE2 and 3, respectively), which can be analyzed by cellular in vitro tests (OECD TGs 442D and 442E). Activated DCs are then thought to migrate to the draining lymph nodes, where they present chemical-induced epitopes on their cell surface to prime naïve T cells (KE4). Subsequently, antigen-specific proliferation and differentiation into effector and memory T cells is induced, which terminates the sensitization phase [2,25,26]. For the final KE4, experimental assays are available, but have not yet been validated or accepted for regulatory use (Figures 1 and 2).

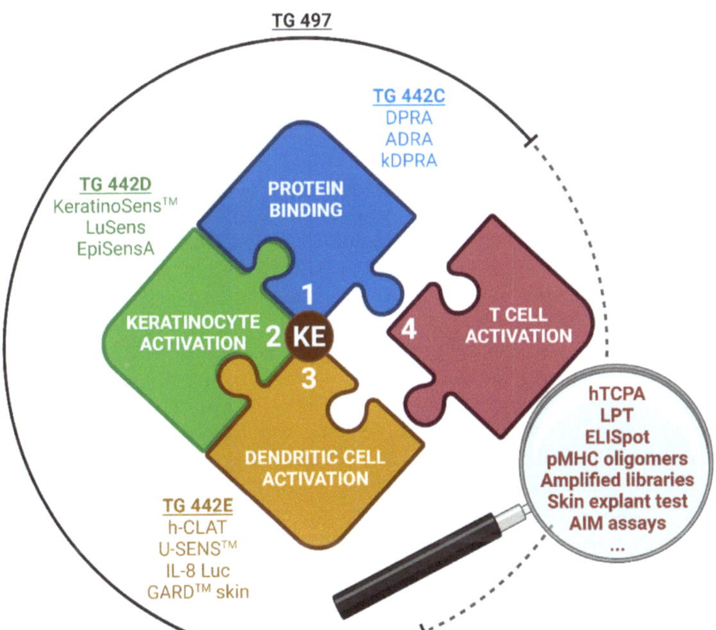

Figure 2. New approach methodologies (NAMs) address different key events (KEs) of the adverse outcome pathway (AOP) of human skin sensitization. Currently, NAMs are available for KE1 to 3, which are also covered by OECD guidelines. KE4 describes T cell activation as the final step of the AOP, but validated methods are still missing. This review describes approaches for measuring chemical-induced T cell responses, which may be further developed into a regulatory accepted predictive KE4 test method. ADRA—amino acid derivative reactivity assay; AIM—activation-induced marker; DPRA—direct peptide reactivity assay; ELISpot—enzyme-linked immunosorbent spot; EpiSensA—epidermal sensitization assay; GARD^{TM} skin—genomic allergen rapid detection for assessment of skin sensitizers; h-CLAT—human cell line activation test; hTCPA—human T cell priming assay; IL-8 Luc—interleukin 8 reporter gene assay; kDPRA—kinetic direct peptide reactivity assay; LPT—lymphocyte proliferation test; pMHC oligomers—peptide-major histocompatibility complex class I and II oligomers; U-SENS^{TM}—myeloid U937 skin sensitization test. Created in BioRender.com. Siewert, K. (2024). https://BioRender.com/a64x125, accessed on 1 November 2024.

So far, no individual NAM can replace in vivo testing for skin sensitization. Still, NAM combinations for KE1 to 3 that also include in silico tools have been implemented successfully as defined approaches (DAs) in OECD TG 497 and the related guidance document (GD 256). These DAs can replace animal testing and allow classification according to the globally harmonized system (GHS) of classification and labelling of chemicals within their applicability domains. Further research to improve the available methods is ongoing [27–29]. In this context, a validated T cell test is needed to fill a critical gap by resolving discrepancies among KE1 to KE3 NAMs, LLNA, human HRIPT, and clinical data, as well as to assess cross-reactivities. Still, T cell-mediated adaptive immunity exhibits unique characteristics including clonal and phenotypic diversity, as well as special antigen recognition mechanisms, as outlined in Section 2. These factors have influenced and delayed T cell test development.

In this review, we highlight key challenges due to the complexity of the underlying biology and provide potential solutions to the technical difficulties involved in assay development, expanding on previous work in this field [26,30–32].

2. Biological Basis of a Regulatory T Cell Test

A predictive T cell test for skin sensitizing properties is based on the assumption that chemical-specific T cells are also present in non-sensitized individuals, even without previous exposure. This phenomenon arises from the vast diversity of T cell receptors (TCRs) and their polyspecificity in humans. Each T cell expresses a unique TCR consisting of an α- and β-chain (or, more rarely, a γ- and δ-chain). The chains are generated in the thymus through recombination of V(D)J gene segments, and random insertions and deletions in the complementary determining region 3 (CDR3). These processes result in a vast theoretical variability of ~10^{15}–10^{61} possible TCRs [33,34]. From these theoretical numbers, a huge and mainly unique TCR repertoire is realized in each person, estimated to be about 100 million (10^8) TCRs [35].

TCR polyspecificity is also referred to as cross-reactivity. This property describes the ability of a single TCR to react to multiple epitopes [36–38]. Consequently, the human TCR repertoire covers virtually any epitope, but specific T cells for a certain allergen are only available at low levels in the naïve T cell pool due to the high diversity (~1 specific among $5 \cdot 10^4$ to 10^7 non-specific T cells) [39,40]. A naïve T cell with a given TCR (a clonotype) is represented by a low number of daughter cells (e.g., ~50 to 200 for murine CD8+ T cells) that are circulating between the blood and lymphatic system [41]. Therefore, cells from one naïve T cell clonotype are usually not detectable in several wells of an in vitro culture, a phenomenon known as a sampling effect. In contrast, the effector and memory T cell pools contain a broad range of individual clonotype frequencies, including some that are highly expanded [42,43].

Epitopes recognized by TCRs consist of a peptide presented by a protein of the highly polymorphic major histocompatibility complex (MHC), also known as human leukocyte antigen (HLA) [44]. For most chemical-induced epitopes, the identity of the HLA allele, the presented self-peptide, and the exact haptenation site and mechanism remain unknown [45–48]. Covalent chemical binding via electrophile/nucleophile or radical mechanisms (haptenation) as well as complex formation (i.e., chelation, also termed pharmacological interaction, or p-i concept) are possible (see also Section 4.1) [3,49,50].

So far, reliable in vitro T cell signals have been observed for only a limited number of contact allergens and drugs [45,47]. A prominent example is nickel (Ni^{2+} ions), which astoundingly activates approximately ~1 in 1000 CD4+ T cells (at ~200 μM), due to the interaction with certain common TCR antigen binding sites [19]. Similar observations were published for cobalt (Co^{2+}) and palladium (Pd^{2+}) [51]. In an analogous manner, the experimental contact allergen 2,4,6-trinitrobenzene-1-sulfonic acid (TNBS) has been shown to activate many CD4+ and CD8+ T cells [52,53]. Here, the chemical moiety (trinitrophenyl group) dominates the TCR interactions at specific positions, while the sequence of the underlying MHC-presented peptides can vary. This mechanism allows the involvement of many TCRs, possibly with distinct CDR3 motifs [52,54]. While TCR repertoires are largely unique among individuals, bias not only in thymic selection but also among reactive TCRs exists. This encompasses the selection of common gene segments, amino acid motifs in the CDR3, or a combination thereof [34,51,55–58].

A T cell test is indicated to determine the extent to which a chemical sensitizer activates a broader range of T cells, as well as to distinguish signals from heterologous immune stimulation. Sufficient stimulation of the innate immune system, required for an adaptive T cell response, is thought to be mediated by the reactivity of the chemical and addressed by KE1 to 3 methods [59]. In real-life scenarios, heterologous stimulation from skin injury, barrier defects, infections, or co-exposure to other irritants in mixtures may contribute to innate immune activation [60–62]. A prominent example is nickel, predicted to exhibit a weak to moderate potency in available tests and displaying also some irritant characteristics in vitro [63–65]. However, nickel allergy is strongly associated with skin injury which is exceptionally immunogenic and could also provide the required heterologous stimulation. Thus, broad T cell reactivity, combined with skin injury, may contribute to the high prevalence of nickel allergy [66,67]. Heterologous immune stimulation is currently not considered in any predictive testing. All reference data on skin sensitizing properties are derived from human or animal models with an intact skin barrier, without the presence of additional chemicals (mixtures) or pathogens, with the exception of sodium dodecyl sulfate (SDS), which was used sometimes in the past. Taken together, the contribution of heterologous immune stimulation to skin sensitization under real-life scenarios remains unclear. However, T cell tests may help to dissect the contributions of innate and adaptive immune activation to skin sensitization for a given sensitizing chemical.

3. Available T Cell-Based Assays

Human T cell tests use blood as a readily accessible source of primary T cells and APCs [45]. All assays rely on autologous systems, where T cells and APCs derive from the same donor to prevent reactions against polymorphic HLA proteins (mixed lymphocyte reactions). This setting ensures that the detection of chemical-specific responses will not be hindered by allogeneic stimulation. After antigen exposure, different readouts can be used and the results are compared to cells treated with an unrelated chemical (specificity control) and/or untreated cells (background control). The available tests can be categorized based on their capacity to interrogate naïve, memory, or both T cell populations (Table 1).

Table 1. In vitro assays used for the identification of human chemical-specific T cells. The remark column mentions general features and example references. In addition, variations, e.g., additional readouts or key experimental details are listed from individual exemplary publications.

Assay	Main Read-Out	Remarks and Example References	Limitations	Example Chemicals
Naïve T cells				
Priming assays • hTCPA	proliferation	phenotyping is possible (CD4/CD8) APCs: cultured LCs [17] APCs: MoDCs [68] depletion of CD25+ regulatory T cells [69] cytokine detection (IFN-γ, TNF-α) [18,70] clone characterization [71]	requires professional APCs to overcome activation threshold of naïve T cells, time-intensive	BB, DNCB, DNBS, Eugenol, FITC, HCA, Isoeugenol, PPD, TNBS
Memory T cells				
LPT	proliferation	cytokines (inverse relation of IL-5 and IL-8) [72] phenotyping is possible (CD4/CD8) [75]	radioactivity is the most sensitive read-out but prevents down-stream analysis [73,74]	Mercury, Nickel

Table 1. *Cont.*

Assay	Main Read-Out	Remarks and Example References	Limitations	Example Chemicals
Cytokine-based assays • ELISA • Bead assays	production of cytokines in supernatant	frequently analyzed cytokines: IFN-γ, IL-10 single cytokines (ELISA) or multiple soluble analytes (multiplex-bead assays) [76,77]	not all specific T cells produce a given cytokine no resolution at single-cell level	BB, Nickel, PPD
• ELISpot		cytokine detection on single-cell level [78], quantitative	prevents down-stream analysis, no phenotyping possible	
• ICS	cell-based	Parallel analysis of multiple cytokines and phenotyping of responsive cells on single cell level, quantitative	fixation and permeabilization required, thus preventing down-stream analysis of living cells [79]	
• Capture assays		viable single cell analysis enables down-stream analysis [80]	limited cytokine analysis [81]	
Naïve and memory T cells				
pMHC oligomers		phenotyping and downstream analysis of viable, antigen-specific cells [82]	epitope knowledge required for (chemically modified) peptide-MHC, oligomer production	Nickel (mimotope)
Amplified T cells libraries	proliferation	identification of low frequency clones, HTS possible [83]	work- and time intensive, amplification of single clones may deviate from real-life conditions	
Autologous skin explant test	histo-pathological analysis of human skin	histopathological score for damage induced in skin explants added to chemical-treated MoDCs—T cell co-cultures [84,85]	time intensive, limited availability of skin samples	Cinnamic alcohol, oxazolone
AIM assay	activation-induced surface protein expression	fast, sensitive, quantitative, downstream analysis of life specific T cells (further phenotyping, TCR repertoire analysis) [19,51,52]	less established for antigen-specific CD8 T cells, fluorescence interferences may limit analysis of high chemical concentrations	TNBS, Nickel

AIM—activation-induced marker; APC—antigen presenting cell; BB—Brandowsky's base; CD—cluster of differentiation; DNBS—2,4-dinitrobenzenesulfonic acid; DNCB—2,4-dinitrochlorobenzene; ELISA—enzyme-linked immunosorbent assay; ELISpot—enzyme-linked immunoSpot assay; FITC—fluorescein isothiocyanate; HCA—α-hexylcinnamaldehyde; hTCPA—human T cell priming assay; HTS—high-throughput sequencing; ICS—intracellular cytokine staining; IFN—interferon; IL—interleukin; LC—Langerhans cells; LPT—lymphocyte proliferation test; MoDCs—monocyte-derived dendritic cells; pMHC—peptide major histocompatibility complex; PPD—p-phenylenediamine; TCR—T cell receptor; TNBS—2,4,6-trinitrobenzenesulfonic acid; TNF—tumor necrosis factor.

3.1. Detection of Naïve T Cells

Unlike memory T cells, naïve T cells require further signals beyond TCR engagement, including co-stimulation via CD28 and cytokines like interleukin (IL)-2, for full activation and proliferation [86–88]. Professional APCs like DCs supply essential co-stimulatory molecules like CD86 or cytokines, which also drive the shift of T cells towards effector and memory phenotypes including Th1, Th2, or Th17 subpopulations [89].

Priming assays, among the first tests developed for skin sensitizers, are currently the only ones that mimic the sensitization phase in vitro. These assays involve the co-culture of naïve human T cells with APCs in the presence of a chemical, integrating the necessary co-stimulatory signals from the innate immune system [31,90]. This approach not only activates naïve T cells but also potentially allows for the assessment of chemical-induced shifts in immune responses [17]. After the initial priming, cells are restimulated at one or several time points.

T cell responses are primarily evaluated by cell proliferation, traditionally using [^{3}H]-thymidine incorporation into DNA [17,18,70]. Results are expressed as counts per minute (cpm) and typically converted into a stimulation index. The stimulation index quantifies cell proliferation by comparing chemical-treated cells to unstimulated control cells under identical conditions, thus accounting for biological variations [69,70]. Alternative techniques have been established to replace radioactive thymidine due to safety concerns and limitations in single-cell resolution. The use of fluorescent dyes (e.g., carboxyfluorescein succinimidyl ester (CFSE) or Oregon Green) as well as proliferation markers like 5-bromo-2′-deoxyuridine (BrDU) or Ki67, along with the detection of cytokines provides better specificity for cell identification and allows for detailed phenotyping [18,30,32]. However, these alternatives offer a lower sensitivity and make them less suitable for the identification of small populations of proliferating cells.

Various adjustments have been implemented to increase assay sensitivity, including modification of T cell and APC populations. Blood monocyte-derived DCs (abbreviated MDDCs or MoDCs) or monocyte-derived Langerhans cells (MoLCs) are commonly used APCs [17,18,91,92]. In the human T cell priming assay (hTCPA), additional stimulation to support effective T cell priming has been used, e.g., the addition of CD28 antibodies for T cell co-stimulation or maturation of immature MoDCs with stimulators like lipopolysaccharide (LPS) [18,70]. However, it is important to note that once MoDCs mature, they exhibit a significant reduced phagocytic capacity, which may limit their ability to effectively take up and process haptenated proteins [92]. Other modifications in experimental protocols, such as the addition of cytokines for improved T cell survival, the depletion of several immune-regulatory cells (e.g., CD56+ and CD25+ cells), and/or blocking of inhibitory checkpoint receptors (e.g., programmed cell death ligand-1 (PD-L1) or cytotoxic T-lymphocyte-associated protein 4 (CTLA4)) may contribute to an improved assay sensitivity [32,70,91].

Despite advancements, human priming assays still face difficulties in the reliable identification of sensitizing chemicals. Multiple restimulations may increase background levels due to uneven proliferation of individual T cell clones in samples without antigen stimulation. Optimization for individual parameters within the complex protocol remains labor-intensive and time consuming with prolonged cultivation periods, thus challenging the high-throughput implementation of priming assays [32].

3.2. Detection of Memory T Cells

Assays that evaluate the secondary immune response mediated by expanded effector or memory T cells usually aim to detect chemical-specific T cells in previously sensitized individuals. Naïve T cells do not typically proliferate in PBMC cultures unless they receive both antigen-specific TCR engagement and co-stimulatory signals along with cytokine support due to insufficient priming stimulation [86–88,92]. On the other hand, effector or memory T cells can be efficiently detected. However, in a regulatory context it is crucial to reach the detection limit in non-allergic donors. For some chemicals, like metals, specific T cells can easily be identified in vitro. In contrast, other chemicals are more challenging, e.g., due to low frequencies of specific T cells and/or difficulties in epitope generation (see Sections 4.1 and 4.4).

3.2.1. Lymphocyte Proliferation Test (LPT)

The lymphocyte proliferation test (LPT), also termed lymphocyte transformation test (LTT) or lymphocyte stimulation test (LST), measures the antigen-specific response of effector or memory T cell populations based on their in vitro proliferation [16,74,93–95]. It is also referred to as a lymphocyte activation test (LAT) in some literature [96,97]. PBMCs, including both T cells and mainly monocytes and B cells as APCs, are cultured with the chemical of interest (or otherwise chemical-induced epitopes, see Section 4.1), usually using replicates. Cell proliferation is measured after several days, typically by [^{3}H]-thymidine incorporation and subsequent determination of the stimulation index (see Section 3.1).

Alternative readouts, like cytokine production or the expression of several activation markers, can also be combined with the proliferation protocol [98,99]. The addition of several cytokines in cell culture medium has been proposed to enhance assay sensitivity, but their use remains controversial as it may skew overall T cell responses by selectively activating or suppressing specific T cell subsets [99,100].

One main challenge of the LPT is that it does not provide information on original specific cell frequencies unless performed in the setting of limiting dilution cultures. In addition, the results of the LPT may reflect altered frequencies of T cell subpopulations due to varying division speeds [101].

3.2.2. Cytokine-Based Tests

Cytokine-based assays aim to quantify cytokine production by TCR-activated effector or memory T cells and are commonly used as an additional readout of LPT or priming assays [18,77,78].

Among these methods, the enzyme-linked immunosorbent assay (ELISA) or bead-based immunoassays are widely employed to measure cytokine concentrations in cell supernatants. Depending on the applied method, a single cytokine (ELISA), or a variety of soluble analytes (e.g., cytometric bead array (CBA)) can be detected at a time [72]. However, supernatant analysis does not identify cytokine production at single-cell level. In contrast, ELISpot assays visualize cytokine-releasing cells as colored spots, enabling the quantification of responsive individual T cells, though without preserving cell viability [78,102]. In most studies, IFN-γ and IL-10 are the main redout, although ELISpot enables the detection of a variety of cytokines in parallel, e.g., IL-2, IL-4, IL-5 or IL-13, to improve qualification of the induced response [103]. Capture assays, such as the cytokine secretion assay (CSA), also termed cytokine catch assay, detect cytokines at individual cell level while maintaining cell viability [80]. Another approach is intracellular cytokine staining (ICS) using multiparameter flow cytometry. This method employs protein transport inhibitors to prevent extracellular cytokine migration, followed by cell fixation and permeabilization to allow the penetration of fluorochrome-conjugated antibodies. While ICS enables analysis of multiple cytokines and the phenotyping of responsive cells, it does not preserve cell viability [79,104]

The isolated quantification of cytokines as a single readout is associated with certain limitations. The complex nature of cytokine production makes it difficult to efficiently capture antigen-specific responses at any given time point, as not all chemical-specific T cells produce all cytokines [105]. Additionally, elevated or altered baseline cytokine levels, especially during active inflammatory processes, can affect assay interpretation. Furthermore, the detection of T cell subsets with low cytokine profiles can be challenging, leading to incomplete characterization of certain T cell responses [106].

3.3. Detection of Both Naïve and Memory T Cells

3.3.1. Peptide-Major Histocompatibility Complex Class I and II Oligomers

The peptide-MHC oligomer technique analyzes the binding of T cells to specific antigens via the interaction between TCRs and recombinantly generated soluble peptide-MHC complexes. Fluorescent tetramers, as well as dimers, pentamers, and higher-order oligomers of MHC I and II molecules loaded with specific peptides, bind to CD8+ or CD4+ T cells, respectively. Antigen-specific T cells are then identified using multiparameter flow cytometry, which also allows for T cell phenotyping and isolation of specific viable T cells [82,107].

The multimer approach is necessary to overcome the low avidity of TCRs, particularly for CD4+ T cells, and stabilize their interaction with MHC [108–110]. However, it remains unclear whether MHC oligomers can efficiently capture all relevant T cells, regardless of their activation state or epitope avidity [111,112]. Nevertheless, the adaptability of the assay for detecting chemical-specific T cells has been demonstrated by the development of a nickel mimotope tetramer [107].

Optimized protocols with magnetic-bead enrichment improved assay detection capabilities, allowing for the analysis of rare antigen-specific T cells even in unexposed individuals [110,112]. Besides, a combination with mass cytometry is possible. In this approach, antibodies labeled with isotopically purified metal conjugates are measured by a time-of-flight mass spectrometer. This offers a wider range of parameter detection than flow cytometry and fewer variations in signal intensity between parameters [113,114].

The main bottleneck of the peptide-MHC oligomer technique is that all relevant contributors need to be known, including the presented peptide, the presenting MHC protein and, in the case of chemical sensitizers, the haptenated entity. Moreover, the complex production of peptide-MHC tetramer complexes, especially for MHC II, restricts its widespread use [113,115,116].

3.3.2. Amplified T Cell Libraries

The in vitro generation of amplified T cell libraries enables the quantitative and functional analysis of naïve and memory antigen-specific CD4+ and CD8+ T cells [83]. Initially, T cells are isolated based on their naïve or memory phenotype, followed by polyclonal expansion and subsequent stimulation of fractions of the expanded pools with autologous monocytes, in the presence of an antigen. T cell proliferation of expanded antigen-specific T cell clones is assessed via [^{3}H]-thymidine incorporation or flow cytometry and can be combined with further TCR repertoire analysis and cytokine detection assays [83,117,118].

This method offers high sensitivity in the detection of rare antigen-specific T cells. However, it involves a labor-intensive process with extended culture times that requires advanced expertise. Additionally, the in vitro expansion of T cells may not accurately reflect in vivo scenarios due to possible outgrowth or loss of certain specific clones. Moreover, potential phenotypic or functional alterations in T cells induced by in vitro manipulation may also limit reproducibility of real-life settings [40,117]. To date, this method has not been applied to chemical allergens.

3.3.3. Autologous Skin Explant Test

The human in vitro skin explant assay presents an alternative approach to assess the skin sensitization potential of chemicals through a co-culture of MoDCs and T cells from healthy individuals combined with a skin explant from the same donor [84,85,119]. Not unlike priming assay, fast MoDCs are generated (day 1), matured with maturation stimuli (day 2), modified with chemical allergens (day 3), and then co-cultured with T cells (days 4–7). Here, a comprehensive T cell pool is used (CD14- PBMC fraction), which include all T cell subpopulations. Thus, the T cell responses observed in the assay may consist of either primed naïve T cells or (cross-reactive) memory responses and includes CD4+ as well as CD8+ T cells. Then, skin biopsies are co-cultured for another 3 days (days 8–10).

By including a skin explant besides dendritic cell models and T cells, this assay mimics the human in vivo relevant cutaneous immune response more closely compared to other tests. Histopathological damage is the major readout and evaluated according to the Lerner criteria, using hematoxylin and eosin staining [120]. The visible damage in histological sections mainly reflects the response of keratinocytes to the inflammatory environment generated by the in vitro culture, likely dominated by the proliferated, activated chemical-specific T cells. Additional readouts from parallel MoDC-T cell co-cultures include cytokine (IFN-γ) assessment in the cell culture supernatants and T cell proliferation via [^{3}H]-thymidine incorporation. The assay has also been applied to biologicals and small-molecule drugs, besides a comparatively large pool of sensitizing chemicals [85,119]. However, the complexity of the in vitro system may also pose some challenges, such as the necessary logistics to obtain autologous skin samples.

3.3.4. Activation-Induced Marker (AIM) Assays

AIM assays offer a short-term functional approach for the detection, isolation, and potential expansion of antigen-specific T cells, even from low original precursor frequencies. These assays are based on the upregulation of surface proteins (activation markers) upon TCR stimulation [40,121,122].

The background expression of the activation markers can complicate the accurate distinction of low-frequency T cell subsets, which compromises assay sensitivity [123,124]. Ideally, these markers should be expressed only after TCR interaction, independently of cytokine secretion or cell differentiation stages, with fast kinetics sustained for a reliable period [40,125]. Some early activation markers like CD69 are intrinsically expressed on several T cell subsets or can be induced by TCR-independent signals [124,126]. Late activation markers, including HLA-DR (HLA-DR) and CD38, show high expression variability and are also prone to TCR-independent bystander activation [127–129]. In contrast, CD154 (CD40L) is transiently expressed after 5–7 h on activated CD4+, and to a lesser extent among CD8+ T cells, upon TCR engagement, with minimal bystander activation within this time window [130]. CD154 is rapidly internalized after interacting with CD40, so its detection is only possible intracellularly or with the concomitant addition of a CD40-blocking antibody [106,123,131,132]. Another well-known marker is CD137 (4–1BB), which is upregulated primarily in CD8+ but also CD4+ T cells. Its expression generally peaks around 24 h, with regulatory CD4+ T cells showing peak upregulation at approximately 5 h, and it remains detectable for up to 5 days [125,133–135]. Other surface proteins described in the literature include CD134 (OX40) and programmed death-ligand 1 (PD-L1, CD274), which are typically upregulated after 18–24 h [105,133]. It often remains unclear whether the expression of these activation markers varies between cell types or in response to different antigens. The lack of comparative data hinders a comprehensive understanding and the accurate characterization of immune responses, particularly concerning low-frequency T cell subsets [122].

The co-staining of activation markers, e.g., CD154 or CD137 with CD69 and Nur77, an intracellular marker for TCR-mediated activation, can further confirm T cell specificity [40,126,136]. Other markers allow for simultaneous analyses, such as cell proliferation, single-cell cytokine detection, or T cell phenotyping, which also informs on in vivo relevance [19,40,51,52,105]. For instance, HLA-DR, a late activation marker, and Ki-67, a nuclear protein only expressed during active cell division, are often detected among antigen-specific T cells [19,137]. Increased co-expression of cutaneous leukocyte-associated antigen (CLA) on activated cells indicates skin-homing responses [138,139]. However, if the frequency of antigen-specific T cells is low, the analysis of further markers may not be feasible.

Using flow cytometry, only a limited number of cells can be acquired and computationally processed within a reasonable timeframe, which contrasts with the need for a sufficient number of target events for statistically robust analysis, particularly for certain subpopulations [140]. To address these limitations, the antigen-reactive T cell enrichment (ARTE) protocol has been developed as a modification of the AIM assay. This protocol involves magnetic pre-enrichment of activation marker-positive cells, allowing for the acquisition and analysis of rare antigen-specific viable cells from large blood samples [121,140]. Currently, ARTE has successfully been established for CD154 and CD137 [124,141]. Another approach for flow cytometry is to acquire events specifically from target T cell subpopulations to reduce data volume [19,51]. In both options, a small fraction of cells is analyzed in detail to capture original frequencies within parent populations, providing information about cell phenotype and functionality.

Further optimization of AIM assays, e.g., in analogy to priming assays, is possible and it has to be tested whether this leads to improved chemical-specific T cell detection, e.g., using co-stimulation with CD28 or depletion of regulatory T cells [142].

Overall, AIM assays offer promising advantages for both laboratory and clinical applications. Their shorter experimental time leads to fast results and facilitates their large-scale application. Because T cells do not divide during short-term stimulation, T cell frequencies, phenotypes, and TCR profiles obtained remain unbiased, accurately reflecting the in vivo repertoire and frequencies [143]. Specifically, the CD154-based CD4+ AIM assay has proven to be highly effective in the identification of chemical-specific T cells, underscoring its value for monitoring chemical-related immune responses [19,36,52].

4. Critical Steps for T Cell Assay Development

Several crucial steps are involved in the development of T cell tests (Figure 3), which are discussed in the following sections.

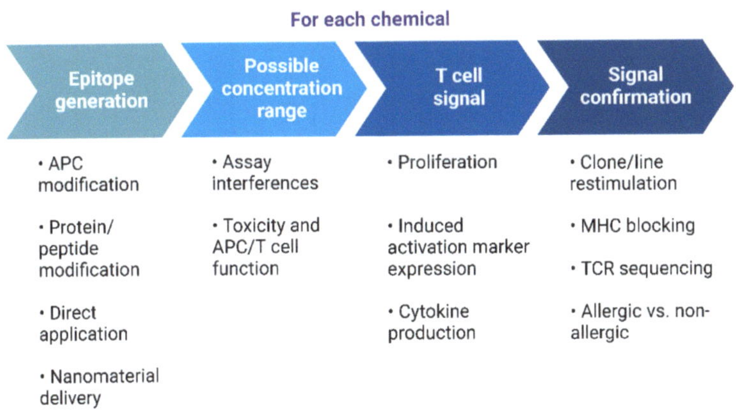

Figure 3. Key steps in the development of in vitro T cell assays. For each key step (large blue arrows), several experimental options can support the analysis of a new chemical (see also Sections 4.1–4.4 below). APC—antigen presenting cell; MHC—major histocompatibility complex; TCR—T cell receptor. Created in BioRender.com. Siewert, K. (2024) https://BioRender.com/j18a303, accessed on 1 November 2024.

4.1. Generation of T Cell Epitopes

For most sensitizers, the identity of the induced T cell epitope and the conditions required for efficient in vitro generation are not well understood. This renders the development of T cell tests for chemical sensitizers more challenging compared to those for protein antigens. The physicochemical properties of the sensitizer drive the molecular mechanisms of its interaction with cellular components. Different aspects need to be considered, including protein binding affinity at different sites, chemical reactivity (including pre- or pro-haptens), or solubility [45,144]. Covalent protein binding (haptenation) can occur on extracellular proteins, which are subsequently processed by APCs, resulting in hapten-derived peptides being presented on MHC II molecules to CD4+ T cells [145–147]. Cross-presentation enables the activation of CD8+ T cells due to presentation of haptenized peptides on MHC I molecules. Some haptens can penetrate into the cell and modify intracellular proteins to serve as an epitope source, which is subsequently shuffled into the MHC I presentation pathway [146,147]. Moreover, some chemicals have been shown to covalently modify peptides bound to MHC molecules, leading to the formation of TCR epitopes in a processing-independent way [145,148]. Others may alter intracellular processes to produce neo-epitopes without chemical contact of the TCR [36,149,150]. As a second main mechanism for epitope formation, some chemical-derived epitopes are formed in a processing-independent way [151]. This has been described for metal allergens where epitopes are formed via complex formation (coordination bonds), also termed pharmacological interaction (or p-i) in the context of drug hypersensitivity [36,49,152].

In real life, proteins in the skin are typically modified. Such skin proteins are rarely available in in vitro assays, but this may not pose a problem. According to the concept of T cell polyspecificity, sufficiently similar modified proteins or peptides can be processed and presented by the APCs in the assay, thus engaging a representative fraction of the relevant chemical-specific TCR repertoire, similar to the in vivo skin responses [95,153,154].

Considering the possible molecular mechanism for epitope formation in vivo, the following four major in vitro approaches can be distinguished in the literature: separate modification of the APCs, modification of model proteins, which are provided to the APCs, direct addition of a chemical to the culture medium, and chemical delivery by nanomaterials (Figure 4).

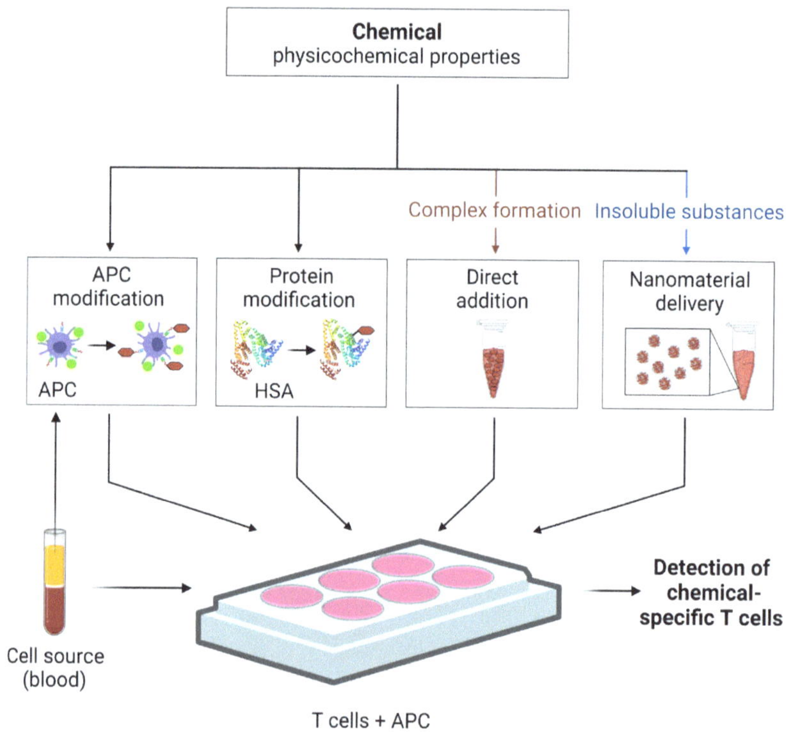

Figure 4. Approaches for the in vitro generation of chemical-induced T cell epitopes. The physicochemical properties of a substance influence which epitope formation approach may be the most suitable. For instance, certain features, like binding via complex formation (chelation) or insolubility, may require direct addition into the culture medium or benefit from nanomaterial-mediated delivery, respectively. Human blood samples serve as the cell source (e.g., for PBMCs, indicated here as white cellular layer after blood density gradient centrifugation), which includes fresh blood donations, cord blood, buffy coats, and leukocyte depletion filters. APC—antigen presenting cells; HSA—human serum albumin. Created in BioRender.com. Siewert, K. (2024). https://BioRender.com/b98e522, accessed on 1 November 2024.

For sensitizers that interact via covalent protein binding (haptenation), APC modification or the addition of modified proteins (or peptides) to the APCs have been described. APC modification, also termed pulsing, facilitates efficient antigen processing under culture conditions. Additionally, pre-haptens, which require prior oxidation, or pro-haptens, which require metabolic activation, can convert into their immunogenic reactive forms. In contrast, a second way comprises a short-term incubation with a chemical for ~10 min in PBS which may limit toxic effects due to the shorter exposure time. This latter approach allows for a targeted modification of pre-existing surface peptide-MHC complexes, due to the lack of reactive proteins in the cell culture medium [52,70]. To determine if a chemical-induced epitope requires antigen processing, APCs can be fixed with glutaraldehyde, which leads to an immediate stop of antigen uptake and processing [18,70].

The use of haptenated model proteins provides an excess source of extracellular protein for APC processing. Human serum albumin (HSA) is often used as a model protein due to its several lysine and cysteine residues, which readily react with many sensitizers, including nickel and 2,4-dinitrobenzenesulfonic acid [70,155]. Moreover, HSA is the most abundant plasma protein and a common carrier protein in the skin [156,157]. The modification of a single protein provides a limited number of functional groups and consequently, a restricted range of potential T cell epitopes [153,158]. To better mimic the extensive protein diversity encountered during skin sensitization, the modification of a well-defined mix of model proteins (e.g., skin) may improve in vitro epitope formation [159]. Each haptenated model protein or peptide should also be tested without chemical modification to determine background T cell reactivity. Successful protein modifications can be determined analytically, e.g., by mass spectrometry.

The direct addition of the chemicals to the cell culture medium is a straightforward approach to induce T cell responses but requires careful controls for toxicity and cell functions (see Section 4.2). For chemicals that form T cell epitopes via complex formation, such a direct application is essential to maintain the reaction equilibrium.

As an additional approach, the encapsulation within nanomaterials potentially offers an improved epitope formation for insoluble chemicals by maintaining availability, dispersibility and stability in aqueous solutions. Nanomaterials may be taken up by the APC and deliver chemicals for intracellular protein modification. This approach potentially boosts T cell activation, further preventing chemical aggregation and protecting against metabolic or oxidative degradation. It also enables the controlled release of haptens over time, thus mimicking in vivo conditions [160–162]. As the use of nanomaterials is already a dynamic field in drug delivery research, an adaptation to the assessment of chemical sensitizers in T cell tests seems promising.

4.2. Determination of Suitable In Vitro Assay Conditions

For each chemical and epitope generation approach, suitable chemical concentration ranges should be determined by performing a concentration-dependent stimulation to assess assay interferences and potential toxic effects.

Autofluorescence, which is an inherent property of several chemicals, can complicate the detection of fluorescence-based parameters. To address this issue, it is possible to either select autofluorescence-free flow cytometry channels or to employ alternative detection methods, such as cytometry by time-of-flight (CyTOF) or through [^{3}H]-thymidine incorporation. Moreover, chemical binding to the cell membrane can interfere with the binding of viability stains or antibodies, thereby affecting flow cytometry results (unpublished own observation) [163,164]. However, assay interferences due to reduced antibody binding are unlikely to occur when modified APCs or proteins are used as the bases of epitope formation.

Monitoring cell viability and functionality is essential for the accuracy and reliability of T cell assays. Multiparameter flow cytometry is widely used for this purpose, as it allows for the discrimination between viable, apoptotic, and necrotic cells across different cell types and subpopulations [165]. Depending on the kind of epitope formation approach, monitoring the functionality of the T cells and/or the APCs, often a more sensitive parameter than cell viability, is necessary. Proper cell function can be assessed by concurrent incubation of a positive control with the chemical concentrations of interest. The positive control should trigger widespread activation in a processing-independent manner to determine if T cell function is affected. A decrease in control-mediated T cell activation in response to increasing chemical concentrations will indicate functional defects. For this purpose, polyclonal stimulation controls, such as the superantigen staphylococcal enterotoxin B (SEB), have been employed recently in AIM assays [51,52]. It is also imperative to monitor the viability and function of APCs, particularly with regard to the processing-dependent formation of epitopes. Co-exposure to a universal, processing-dependent antigen may help to evaluate APCs' function (Seifert et al., in preparation). Of note, a preserved APC function during the initial hours of antigen incubation can be sufficiently effective to generate chemical-induced epitopes and specific T cell activation. In contrast, chemicals that act as pre- or pro-haptens may require functional APCs for extended periods.

If solvents like dimethyl sulfoxide (DMSO) are used, their effects on cell viability and function should be carefully monitored. To exclude solvent-related effects, a consistent concentration of the solvent should be maintained throughout the experimental protocol [166,167].

The tolerable concentration of a chemical in the assay can differ depending on the incubation time and the type of cells used. For example, PBMCs in AIM assays may tolerate higher chemical concentrations compared to PBMCs in proliferation assays, due to the shorter experimental time. Careful adjustment of all the described parameters is essential to ensure that assay conditions accurately reflect biological responses. Accordingly, the concentration selected for the T cell assay should be the highest non-toxic level, for example preserving 75–80% of the signal detected in the assay readout. Subsequently, lower concentrations are expected to have no toxic effects. These adjustments will guarantee the reliability and reproducibility of assays while minimizing potential confounding effects on T cell function and assay outcomes.

4.3. Confirmation of Chemical-Specific T Cell Responses

Once a signal is obtained from a T cell test, it should be verified to distinguish between true TCR-mediated responses and bystander activation. For instance, higher nickel concentrations were previously interpreted as nonspecific mitogenic effects [168,169]. However, AIM assay data confirmed that nickel-specific T cell responses are induced via the interaction of Ni^{2+} ions with certain features of the TCR repertoire [19,51].

In most assays, the confirmation of T cell signals has traditionally relied on the generation of clones or lines from activated cells, followed by restimulation with the original antigen. Since all T cell clones originate from a single T cell, they share the same TCR sequence and antigen specificity. If these clones react positively upon re-exposure to the original antigen, it confirms specific TCR involvement. These clones can be further characterized for research purposes through cytokine and chemokine analysis or HLA typing. In addition, this method allows for the assessment of potential cross-reactivities by demonstrating positive responses to non-original but related antigens [40,51]. However, the experimental protocol for generating clones is labor-intensive and time-consuming, involving several steps such as culture with feeder cells, expansion, and splitting if necessary. Furthermore, clone restimulation requires autologous APCs, like CD3-depleted PBMCs or Epstein–Barr virus (EBV)-transformed B cells. This can be challenging due to the limited availability of blood samples or the biosafety level 2 requirements, respectively [40,121,139].

To verify MHC restrictions, MHC-blocking antibodies or isotype controls can be employed. These antibodies prevent T cell activation by targeting specific MHC molecules, whether class II (HLA-DR, -DQ or DP) or class I (HLA-A, -B and -C) [40,51,52,170]. This approach is most effective with T cell clones or lines, because blocking antibodies may not be completely effective, particularly due to differences in HLA alleles. As a result, small signals, such as those frequently obtained in assays using bulk PBMCs, may be less reliably controlled by MHC blocking [40,51,52].

Distinctive TCR repertoire features or CDR3 motifs can also serve as indicators of chemical specificity in activated T cells. For example, in nickel-specific T cells, an overrepresentation of the TRAV9-2 gene segments (TAV9-2) and histidine in the CDR3 region indicates TCR-mediated activation [19]. Similar results have been observed for cobalt and palladium and the experimental allergen TNBS, but these features remain to be determined for other chemicals [51,52].

Lastly, differences in cytokine production, phenotypes, proliferation, or frequencies of specific T cells between allergic and non-allergic individuals can indicate specific T cell activation. Increased signals in allergic individuals also demonstrate that the in vitro test accurately reflects in vivo immune responses, as has been shown for nickel or p-phenylenediamine (PPD) [19,171].

4.4. T Cell Test as Diagnostic Tool

The gold standard for diagnosing ACD in clinical practice is the patch test which, despite its widespread use, has several limitations. The patch test is an epicutaneous provocation test in which the allergic skin responses of a patient to small amounts of potential allergens on the back are evaluated at repeated time points. One difficulty is that patients with active skin conditions may be unable to undergo testing. In addition, there is a lack of test preparations for less common allergens, which are not included in the predefined patch test panel [172,173]. Furthermore, some allergens may not penetrate the skin efficiently. This phenomenon has been observed in studies examining the reactivity of palladium, a metal allergen, and in investigations of tattoo ink reactions [174,175].

Blood-based in vitro T cell tests, particularly LPT and AIM assays, provide several advantages over the current diagnostic standard [99,176]. They eliminate the discomfort associated with patch testing and reduce the need for multiple visits in the clinic. Particularly, AIM assays offer short processing times, which enables the evaluation of the allergic state within ~48 h after blood sampling. Additionally, in vitro readouts provide a less subjective result compared to patch tests readings and a wider range of potential allergens beyond those applicable in patch test preparations can be tested. Since in vitro tests do not rely on skin penetration, systemic reactions, like adverse drug reactions or implant failure due to allergic reactions, may be more readily detectable with a blood-based assay. Documented half-lives for antigen-specific memory T cells, as seen with vaccinations or infections, and general clonal persistence in longitudinal studies indicate that increased T cell frequencies can be detected over time [42,177–179]. In addition, acute eczema and the related ongoing T cell responses may be more clearly identified and linked to the causative allergen. For example, patients who recently underwent patch tests exhibited a marked increase in nickel-specific CD154+CD4+ memory T cells that were also frequently CLA-positive, likely associated with active skin eczema, compared to patients tested in the distant past [19,180]. Thus, in vitro T cell tests would offer an accurate, comprehensive, and more patient-friendly diagnosis of contact allergies. In addition, improved allergy diagnosis can also provide data to further improve regulatory risk assessment.

Still, current challenges limit the replacement of patch tests by in vitro tests, which are not yet accepted as diagnostic tools [181]. One key issue is how well blood reflects the immune status of the skin, given that some T cells circulate between the two compartments while others remain resident [43,182]. Other limitations include the need to better understand and optimize the in vitro epitope generation and the selection of appropriate concentrations for each chemical (see Sections 4.1 and 4.2). Particularly, the CD154- and CD137-based AIM assays for antigen-specific CD4+ and CD8+ T cell detection have potential for further development as a diagnostic tool [19,125,140]. However, studies comparing current clinical diagnostic methods, including patch testing, medical history, and relevance assessment, with T cell test results are needed.

In general, positive results from diagnostic tests indicate sensitization, whereas the actual manifestation of allergic symptoms, referred to as in vivo relevance, may vary. Consequently, a positive patch test or induced T cell activation may not always accurately reflect in vivo conditions. In vitro, excessively high chemical concentrations may lead to extended T cell activation that may not reflect real-life scenarios, as described for palladium [51,174]. Therefore, determining the range of local in vivo concentrations of chemicals in the skin and/or draining lymph nodes, known as exposure assessment in risk assessment, would be crucial to interpret T cell test results. This topic, however, lies beyond the scope of this review.

5. Remaining Challenges and Future Directions

The regulatory implementation of predictive T cell tests for sensitizing chemicals requires a strategic approach that overcomes several challenges. Ideally, chemical-specific T cells will be detected using PBMCs from blood donations and thus likely non-allergic individuals, which requires clarification of ethical issues. The number of required blood samples that is needed to cover both intra- and inter-donor variations, arising from, e.g., sampling effects, differences in TCR repertoires, HLA haplotypes, or immune status and age, still needs to be determined. If rarer alleles are involved, a test with a limited number of donors will not reliably detect T cell responses associated with specific HLA haplotypes. While no HLA associations have been identified for contact allergies, certain drug hypersensitivities are linked to specific HLA haplotypes [47].

Another relevant question is whether to interrogate the naïve or memory T cell pool. The naïve T cell pool is more diverse, but the memory pool may still contain a sufficiently extensive TCR repertoire to detect chemical-specific T cells [19,26,45]. Memory T cells are easier to handle, as they do not require priming. They have already been preselected for antigen reactivity, which means they potentially comprise a higher proportion of responsive TCRs. Pre-existing memory T cells may even be recruited during chemical sensitization in vivo, as shown analogously for memory B cells [183]. However, it is difficult to trace their specific involvement because TCRs do not undergo somatic hypermutation like B cells. Moreover, memory T cells have already undergone differentiation and may not fully translate innate signals from chemical-treated APCs into phenotypic changes as effectively as naïve T cells when used in priming assays.

The assessment of T cell cross-reactivity to predict potential cross-allergies is a major unmet regulatory need. Currently, the most commonly used method involves the generation, expansion, and restimulation of chemical-specific T cell clones or lines, which is highly labor-intensive. In silico methods are not yet capable of predicting TCR reactivity [184]. Recently, bulk high-throughput TCR sequencing has emerged as a promising solution to assess cross-reactivity among activated expanded memory T cell clonotypes identified by the AIM assay. Cross-reactivity must be analyzed within the T cell pool from the same individual, as TCR repertoires are unique to each person. The assessment of the β-chain is sufficient to identify cross-reactive clonotypes, but additional α-chain analysis may reveal unique binding sites [19,51].

A T cell assay could offer valuable insights in a tiered approach for hazard identification, particularly in the assessment of chemicals that show discrepancies or unclear results in current KE1-3 NAM, LLNA, and human data. However, a negative T cell test result remains inconclusive until more experience is gained with known skin sensitizing chemicals. In this context, in vitro T cell assays may not always provide sufficient predictions of human clinical outcomes but rather complement existing approaches. T cell tests alone may underestimate severe adverse reactions, if they do not effectively mimic in vivo conditions for antigen presentation. This emphasizes the need for more comprehensive and integrated approaches for safety assessment beyond in vitro T cells tests alone. As a final objective, the practical application of in vitro T cell assays within regulatory frameworks would be highly beneficial. Currently, the state of T cell tests for regulatory purposes could be assigned a regulatory readiness level (RLL) of 2 (mid research stage), as reviewed and outlined by Haase et al. [185] The scientific relevance should be further illustrated by case studies. Standardization and validation of these assays must follow a structured pathway to ensure their acceptance and implementation. First, internal validation must be achieved with a clear protocol (SOP; standard operating procedure) with a well-defined endpoint and applicability domain [32,185]. This preliminary process also includes the validation of the accuracy and reproducibility of the assay. Next, large-scale inter-laboratory validation studies should be conducted to further confirm the transferability, reproducibility, and reliability of the assays. In these trials, participating laboratories will adhere to the established protocol while conducting independent assays with their own setups, to assess the effectiveness of the assays across diverse environments. To the best of our knowledge, there is currently no established standardization or inter-laboratory collaboration specific to a regulatory T cell test. However, upon successful completion and data analysis to determine accuracy, predictive capacity, applicability domain, and performance standards of an assay under these settings, a final report can be submitted to a recognized validation organism, such as the European Union Reference Laboratory for Alternatives to Animal Testing (EURL ECVAM), which will evaluate whether the method meets the necessary criteria for formal validation. After successful validation and potential approval, the assays can be proposed for inclusion in the OECD test guidelines, facilitating their acceptance in regulatory frameworks throughout the EU and other OECD member countries [186,187]. Nevertheless, since not all NAMs can be prioritized for inclusion in the OECD test guidelines, a potential alternative which is currently in evaluation is the establishment of a qualification system. This system is assessed by experts to ensure that NAMs are used effectively in suitable contexts and to help optimize them for regulatory applications [185]. Besides, continuous monitoring of the assay protocol is essential to maintain ongoing effectiveness and relevance. Looking ahead, it will be important also to assess respiratory sensitizers and drugs, including biologicals [85,119]. Further, NAMs should aim to replace or reduce the use of animal-derived products, such as bovine serum or animal-derived antibodies, with several options already available [188,189].

An important emerging field comprises efforts towards a precise assessment of the potency of skin sensitizers, i.e., the dose per cm^2 of skin associated with a certain degree of sensitization within a fraction of individuals, optimally measured on a continuous scale [190–192]. This information is relevant as it should help to define safe levels and enable risk assessment beyond hazard identification. Potency data are derived from dose-response curves (hazard characterization). A combination of methods from the different KEs in the skin sensitization AOP is likely needed to obtain reliable values. Integrating findings from various testing methods, such as T cell assays, could lead to better-defined applicability domains for chemicals, ultimately enhancing regulatory risk assessment. It is still unknown to which extent immunological responses, which may include tolerance and threshold mechanisms, follow conventional dose-response curves [193,194]. Concentration-dependent T cell activation has been shown for metal allergens binding via complex formation [19,51]. However, how and whether frequencies of in vitro reactive T cells can be translated into potency data is still uncertain and further experience with additional chemicals is needed [26,31,32].

6. Conclusions

Overall, T cell assays remain demanding experimental methods due to the intricacies of chemical-induced epitope generation, combined with the need to identify specific T cells among numerous irrelevant bystanders. Recently, AIM assays have been adapted to investigate skin sensitizers [19,51,52]. Particularly, CD154-based AIM assays are fast, highly sensitive, comprehensive, and directly quantitative methods for antigen-specific CD4+ T cell responses, supporting an efficient optimization of experimental conditions without prior knowledge of epitopes or cytokine secretion patterns. When combined with bulk TCR high-throughput sequencing, AIM assays can assess T cell cross-reactivity more efficiently than restimulation of specific T cell clones or lines [51].

The in vitro T cell tests presented here highlight the advantages and potential of NAMs by being directly relevant to human conditions, offering ethical benefits, and providing new insights not available from animal models. However, more resources are needed to test more chemicals and facilitate regulatory implementation. While not all chemicals or substances may require T cell testing, bridging the T cell-related gap in NAM-based sensitizing chemical hazard identification could significantly complement and enhance regulatory frameworks.

Author Contributions: Conceptualization and original draft preparation: K.S., M.A.-S. and N.F.; review and editing, K.S., M.A.-S., N.F., C.C., F.R., H.-J.T. and A.L.; supervision and project administration, K.S.; funding acquisition, K.S. and M.A.-S. All authors have read and agreed to the published version of the manuscript.

Funding: This work was supported by BfR internal grant BfR-CPS-23-1322-848 (to M. Aparicio Soto, C. Curato, F. Riedel, and K. Siewert) and by DFG grant 500312706 (to K. Siewert).

Institutional Review Board Statement: Not applicable.

Informed Consent Statement: Not applicable.

Data Availability Statement: No new data were created or analyzed in this study. Data sharing is not applicable to this article.

Acknowledgments: We would like to thank our BfR colleagues Anna Sonnenburg, Matthias Herzler, Michael Oelgeschläger, Tanja Burgdorf, and Ulrike Bernauer for their thoughtful comments and discussion on this review.

Conflicts of Interest: The authors declare no conflicts of interest.

Abbreviations

ACD—allergic contact dermatitis; AIM—activation-induced marker; AOP—adverse outcome pathway; APC—antigen presenting cell; CD—cluster of differentiation; DC—dendritic cell; KE—key event; LPT—lymphocyte proliferation test; MHC—major histocompatibility complex; NAM—new approach methodology; OECD—Organization for Economic Co-operation and Development; TCR—T cell receptor; TG—test guideline; TNBS—2,4,6-trinitrobenzenesulfonic acid.

References

1. Kimber, I.; Basketter, D.A.; Gerberick, G.F.; Dearman, R.J. Allergic contact dermatitis. *Int. Immunopharmacol.* **2002**, *2*, 201–211. [CrossRef] [PubMed]
2. Martin, S.F. Contact dermatitis: From pathomechanisms to immunotoxicology. *Exp. Dermatol.* **2012**, *21*, 382–389. [CrossRef] [PubMed]
3. Scheinman, P.L.; Vocanson, M.; Thyssen, J.P.; Johansen, J.D.; Nixon, R.L.; Dear, K.; Botto, N.C.; Morot, J.; Goldminz, A.M. Contact dermatitis. *Nat. Rev. Dis. Primers* **2021**, *7*, 38. [CrossRef] [PubMed]
4. Thyssen, J.P.; Johansen, J.D.; Menne, T. Contact allergy epidemics and their controls. *Contact Dermat.* **2007**, *56*, 185–195. [CrossRef]
5. Uter, W.; Aalto-Korte, K.; Agner, T.; Andersen, K.E.; Bircher, A.J.; Brans, R.; Bruze, M.; Diepgen, T.L.; Foti, C.; Gimenez Arnau, A.; et al. The epidemic of methylisothiazolinone contact allergy in Europe: Follow-up on changing exposures. *J. Eur. Acad. Dermatol. Venereol.* **2020**, *34*, 333–339. [CrossRef]
6. Alinaghi, F.; Bennike, N.H.; Egeberg, A.; Thyssen, J.P.; Johansen, J.D. Prevalence of contact allergy in the general population: A systematic review and meta-analysis. *Contact Dermat.* **2019**, *80*, 77–85. [CrossRef]
7. Uter, W.; Werfel, T.; Lepoittevin, J.P.; White, I.R. Contact Allergy-Emerging Allergens and Public Health Impact. *Int. J. Environ. Res. Public. Health* **2020**, *17*, 2404. [CrossRef]
8. Buehler, E.V. Delayed Contact Hypersensitivity in the Guinea Pig. *Arch. Dermatol.* **1965**, *91*, 171–177. [CrossRef]
9. Magnusson, B.; Kligman, A.M. The identification of contact allergens by animal assay. The guinea pig maximization test. *J. Investig. Dermatol.* **1969**, *52*, 268–276. [CrossRef]
10. Kimber, I.; Weisenberger, C. A murine local lymph node assay for the identification of contact allergens. Assay development and results of an initial validation study. *Arch. Toxicol.* **1989**, *63*, 274–282. [CrossRef]
11. Strickland, J.; Haugabrooks, E.; Allen, D.G.; Balottin, L.B.; Hirabayashi, Y.; Kleinstreuer, N.C.; Kojima, H.; Nishizawa, C.; Prieto, P.; Ratzlaff, D.E.; et al. International regulatory uses of acute systemic toxicity data and integration of new approach methodologies. *Crit. Rev. Toxicol.* **2023**, *53*, 385–411. [CrossRef] [PubMed]
12. Brunner, M.J.; Smiljanic, A. Procedure for evaluating skin-sensitizing power of new materials. *AMA Arch. Derm. Syphilol.* **1952**, *66*, 703–705. [CrossRef] [PubMed]
13. Bialas, I.; Zelent-Kraciuk, S.; Jurowski, K. The Skin Sensitisation of Cosmetic Ingredients: Review of Actual Regulatory Status. *Toxics* **2023**, *11*, 392. [CrossRef] [PubMed]
14. Basketter, D.A. The human repeated insult patch test in the 21st century: A commentary. *Cutan. Ocul. Toxicol.* **2009**, *28*, 49–53. [CrossRef] [PubMed]

15. Politano, V.T.; Api, A.M. The Research Institute for Fragrance Materials' human repeated insult patch test protocol. *Regul. Toxicol. Pharmacol.* **2008**, *52*, 35–38. [CrossRef]
16. Miller, A.E.; Levis, W.R. Lymphocyte transformation during dinitrochlorobenzene contact sensitization. An in vitro and in vivo evaluation of the primary immune response in man. *J. Clin. Investig.* **1973**, *52*, 1925–1930. [CrossRef]
17. Moulon, C.; Peguet-Navarro, J.; Courtellemont, P.; Redziniak, G.; Schmitt, D. Human in vitro T cell sensitization using hapten-modified epidermal Langerhans cells. *Adv. Exp. Med. Biol.* **1993**, *329*, 209–212. [CrossRef]
18. Richter, A.; Schmucker, S.S.; Esser, P.R.; Traska, V.; Weber, V.; Dietz, L.; Thierse, H.J.; Pennino, D.; Cavani, A.; Martin, S.F. Human T cell priming assay (hTCPA) for the identification of contact allergens based on naive T cells and DC--IFN-gamma and TNF-alpha readout. *Toxicol. Vitr.* **2013**, *27*, 1180–1185. [CrossRef]
19. Aparicio-Soto, M.; Riedel, F.; Leddermann, M.; Bacher, P.; Scheffold, A.; Kuhl, H.; Timmermann, B.; Chudakov, D.M.; Molin, S.; Worm, M.; et al. TCRs with segment TRAV9-2 or a CDR3 histidine are overrepresented among nickel-specific CD^{4+} T cells. *Allergy* **2020**, *75*, 2574–2586. [CrossRef]
20. Zaunbrecher, V.; Beryt, E.; Parodi, D.; Telesca, D.; Doherty, J.; Malloy, T.; Allard, P. Has Toxicity Testing Moved into the 21st Century? A Survey and Analysis of Perceptions in the Field of Toxicology. *Environ. Health Perspect.* **2017**, *125*, 087024. [CrossRef]
21. Carnesecchi, E.; Langezaal, I.; Browne, P.; Batista-Leite, S.; Campia, I.; Coecke, S.; Dagallier, B.; Deceuninck, P.; Dorne, J.L.C.; Tarazona, J.V.; et al. OECD harmonised template 201: Structuring and reporting mechanistic information to foster the integration of new approach methodologies for hazard and risk assessment of chemicals. *Regul. Toxicol. Pharmacol.* **2023**, *142*, 105426. [CrossRef] [PubMed]
22. Allen, T.E.; Goodman, J.M.; Gutsell, S.; Russell, P.J. Defining molecular initiating events in the adverse outcome pathway framework for risk assessment. *Chem. Res. Toxicol.* **2014**, *27*, 2100–2112. [CrossRef] [PubMed]
23. Stucki, A.O.; Barton-Maclaren, T.S.; Bhuller, Y.; Henriquez, J.E.; Henry, T.R.; Hirn, C.; Miller-Holt, J.; Nagy, E.G.; Perron, M.M.; Ratzlaff, D.E.; et al. Use of new approach methodologies (NAMs) to meet regulatory requirements for the assessment of industrial chemicals and pesticides for effects on human health. *Front. Toxicol.* **2022**, *4*, 964553. [CrossRef] [PubMed]
24. Gadarowska, D.; Kalka, J.; Daniel-Wojcik, A.; Mrzyk, I. Alternative Methods for Skin-Sensitization Assessment. *Toxics* **2022**, *10*, 740. [CrossRef] [PubMed]
25. OECD. *The Adverse Outcome Pathway for Skin Sensitisation Initiated by Covalent Binding to Proteins*; Part 1: Scientific Evidence. ENV/JM/MONO(2012)10/PART1; OECD: Paris, France, 2012.
26. Esser, P.R.; Kimber, I.; Martin, S.F. Correlation of contact sensitizer potency with T cell frequency and TCR repertoire diversity. *Exp. Suppl.* **2014**, *104*, 101–114. [CrossRef]
27. Strickland, J.; Zang, Q.; Kleinstreuer, N.; Paris, M.; Lehmann, D.M.; Choksi, N.; Matheson, J.; Jacobs, A.; Lowit, A.; Allen, D.; et al. Integrated decision strategies for skin sensitization hazard. *J. Appl. Toxicol.* **2016**, *38*, 432. [CrossRef]
28. MacKay, C.; Davies, M.; Summerfield, V.; Maxwell, G. From pathways to people: Applying the adverse outcome pathway (AOP) for skin sensitization to risk assessment. *ALTEX* **2013**, *30*, 473–486. [CrossRef]
29. Gilmour, N.; Kimber, I.; Williams, J.; Maxwell, G. Skin sensitization: Uncertainties, challenges, and opportunities for improved risk assessment. *Contact Dermat.* **2019**, *80*, 195–200. [CrossRef]
30. Martin, S.F.; Esser, P.R.; Schmucker, S.; Dietz, L.; Naisbitt, D.J.; Park, B.K.; Vocanson, M.; Nicolas, J.F.; Keller, M.; Pichler, W.J.; et al. T-cell recognition of chemicals, protein allergens and drugs: Towards the development of in vitro assays. *Cell Mol. Life. Sci.* **2010**, *67*, 4171–4184. [CrossRef]
31. Kimber, I.; Maxwell, G.; Gilmour, N.; Dearman, R.J.; Friedmann, P.S.; Martin, S.F. Allergic contact dermatitis: A commentary on the relationship between T lymphocytes and skin sensitising potency. *Toxicology* **2012**, *291*, 18–24. [CrossRef]
32. van Vliet, E.; Kuhnl, J.; Goebel, C.; Martinozzi-Teissier, S.; Alepee, N.; Ashikaga, T.; Blomeke, B.; Del Bufalo, A.; Cluzel, M.; Corsini, E.; et al. State-of-the-art and new options to assess T cell activation by skin sensitizers: Cosmetics Europe Workshop. *Altern. Anim. Exp.* **2018**, *35*, 179–192. [CrossRef] [PubMed]
33. Davis, M.M.; Bjorkman, P.J. T-cell antigen receptor genes and T-cell recognition. *Nature* **1988**, *334*, 395–402. [CrossRef] [PubMed]
34. Dash, P.; Fiore-Gartland, A.J.; Hertz, T.; Wang, G.C.; Sharma, S.; Souquette, A.; Crawford, J.C.; Clemens, E.B.; Nguyen, T.H.O.; Kedzierska, K.; et al. Quantifiable predictive features define epitope-specific T cell receptor repertoires. *Nature* **2017**, *547*, 89–93. [CrossRef] [PubMed]
35. Robins, H.S.; Campregher, P.V.; Srivastava, S.K.; Wacher, A.; Turtle, C.J.; Kahsai, O.; Riddell, S.R.; Warren, E.H.; Carlson, C.S. Comprehensive assessment of T-cell receptor beta-chain diversity in alphabeta T cells. *Blood* **2009**, *114*, 4099–4107. [CrossRef] [PubMed]
36. Riedel, F.; Aparicio-Soto, M.; Curato, C.; Thierse, H.J.; Siewert, K.; Luch, A. Immunological Mechanisms of Metal Allergies and the Nickel-Specific TCR-pMHC Interface. *Int. J. Environ. Res. Public. Health* **2021**, *18*, 10867. [CrossRef]
37. Sewell, A.K. Why must T cells be cross-reactive? *Nat. Rev. Immunol.* **2012**, *12*, 669–677. [CrossRef]
38. Baker, B.M.; Scott, D.R.; Blevins, S.J.; Hawse, W.F. Structural and dynamic control of T-cell receptor specificity, cross-reactivity, and binding mechanism. *Immunol. Rev.* **2012**, *250*, 10–31. [CrossRef]
39. Su, L.F.; Davis, M.M. Antiviral memory phenotype T cells in unexposed adults. *Immunol. Rev.* **2013**, *255*, 95–109. [CrossRef]
40. Bacher, P.; Scheffold, A. Flow-cytometric analysis of rare antigen-specific T cells. *Cytom. A* **2013**, *83*, 692–701. [CrossRef]
41. Williams, M.A.; Bevan, M.J. Effector and memory CTL differentiation. *Annu. Rev. Immunol.* **2007**, *25*, 171–192. [CrossRef]

42. Britanova, O.V.; Shugay, M.; Merzlyak, E.M.; Staroverov, D.B.; Putintseva, E.V.; Turchaninova, M.A.; Mamedov, I.Z.; Pogorelyy, M.V.; Bolotin, D.A.; Izraelson, M.; et al. Dynamics of Individual T Cell Repertoires: From Cord Blood to Centenarians. *J. Immunol.* **2016**, *196*, 5005–5013. [CrossRef] [PubMed]
43. Poon, M.M.L.; Caron, D.P.; Wang, Z.; Wells, S.B.; Chen, D.; Meng, W.; Szabo, P.A.; Lam, N.; Kubota, M.; Matsumoto, R.; et al. Tissue adaptation and clonal segregation of human memory T cells in barrier sites. *Nat. Immunol.* **2023**, *24*, 309–319. [CrossRef] [PubMed]
44. Rudolph, M.G.; Stanfield, R.L.; Wilson, I.A. How TCRs bind MHCs, peptides, and coreceptors. *Annu. Rev. Immunol.* **2006**, *24*, 419–466. [CrossRef] [PubMed]
45. Aparicio-Soto, M.; Curato, C.; Riedel, F.; Thierse, H.J.; Luch, A.; Siewert, K. In Vitro Monitoring of Human T Cell Responses to Skin Sensitizing Chemicals-A Systematic Review. *Cells* **2021**, *11*, 83. [CrossRef] [PubMed]
46. Weltzien, H.U.; Moulon, C.; Martin, S.; Padovan, E.; Hartmann, U.; Kohler, J. T cell immune responses to haptens. Structural models for allergic and autoimmune reactions. *Toxicology* **1996**, *107*, 141–151. [CrossRef]
47. Sullivan, A.; Watkinson, J.; Waddington, J.; Park, B.K.; Naisbitt, D.J. Implications of HLA-allele associations for the study of type IV drug hypersensitivity reactions. *Expert. Opin. Drug. Metab. Toxicol.* **2018**, *14*, 261–274. [CrossRef]
48. Ndreu, L.; Erber, L.N.; Tornqvist, M.; Tretyakova, N.Y.; Karlsson, I. Characterizing Adduct Formation of Electrophilic Skin Allergens with Human Serum Albumin and Hemoglobin. *Chem. Res. Toxicol.* **2020**, *33*, 2623–2636. [CrossRef]
49. Pichler, W.J. Immune pathomechanism and classification of drug hypersensitivity. *Allergy* **2019**, *74*, 1457–1471. [CrossRef]
50. Vileno, B.; Port-Lougarre, Y.; Gimenez-Arnau, E. Electron paramagnetic resonance and spin trapping to detect free radicals from allergenic hydroperoxides in contact with the skin: From the molecule to the tissue. *Contact Dermat.* **2022**, *86*, 241–253. [CrossRef]
51. Riedel, F.; Aparicio-Soto, M.; Curato, C.; Munch, L.; Abbas, A.; Thierse, H.J.; Peitsch, W.K.; Luch, A.; Siewert, K. Unique and common TCR repertoire features of Ni^{2+}-, Co^{2+}-, and Pd^{2+}-specific human CD154 + CD4+ T cells. *Allergy* **2023**, *78*, 270–282. [CrossRef]
52. Curato, C.; Aparicio-Soto, M.; Riedel, F.; Wehl, I.; Basaran, A.; Abbas, A.; Thierse, H.J.; Luch, A.; Siewert, K. Frequencies and TCR Repertoires of Human 2,4,6-Trinitrobenzenesulfonic Acid-specific T Cells. *Front. Toxicol.* **2022**, *4*, 827109. [CrossRef] [PubMed]
53. Martin, S.; Delattre, V.; Leicht, C.; Weltzien, H.U.; Simon, J.C. A high frequency of allergen-specific $CD8^+$ Tc1 cells is associated with the murine immune response to the contact sensitizer trinitrophenyl. *Exp. Dermatol.* **2003**, *12*, 78–85. [CrossRef] [PubMed]
54. Martin, S.; Ruh, H.; Hebbelmann, S.; Pflugfelder, U.; Rude, B.; Weltzien, H.U. Carrier-reactive hapten-specific cytotoxic T lymphocyte clones originate from a highly preselected T cell repertoire: Implications for chemical-induced self-reactivity. *Eur. J. Immunol.* **1995**, *25*, 2788–2796. [CrossRef] [PubMed]
55. Miles, J.J.; Douek, D.C.; Price, D.A. Bias in the alphabeta T-cell repertoire: Implications for disease pathogenesis and vaccination. *Immunol. Cell Biol.* **2011**, *89*, 375–387. [CrossRef]
56. Ronel, T.; Harries, M.; Wicks, K.; Oakes, T.; Singleton, H.; Dearman, R.; Maxwell, G.; Chain, B. The clonal structure and dynamics of the human T cell response to an organic chemical hapten. *Elife* **2021**, *10*, e54747. [CrossRef]
57. Li, H.; Ye, C.; Ji, G.; Han, J. Determinants of public T cell responses. *Cell Res.* **2012**, *22*, 33–42. [CrossRef]
58. Ruiz Ortega, M.; Spisak, N.; Mora, T.; Walczak, A.M. Modeling and predicting the overlap of B- and T-cell receptor repertoires in healthy and SARS-CoV-2 infected individuals. *PLoS Genet.* **2023**, *19*, e1010652. [CrossRef]
59. Esser, P.; Sautier, L.; Sarkar, S.; Schilling, G.; Bokemeyer, C.; Koch, U.; Rose, M.; Friedrich, M.; Nolte, S.; Walter, O.; et al. Correction to: Evaluation of an electronic psycho-oncological adaptive screening program (EPAS) with immediate patient feedback: Findings from a German cluster intervention study. *J. Cancer Surviv.* **2023**, *17*, 859. [CrossRef]
60. Bonefeld, C.M.; Nielsen, M.M.; Rubin, I.M.; Vennegaard, M.T.; Dabelsteen, S.; Gimenez-Arnau, E.; Lepoittevin, J.P.; Geisler, C.; Johansen, J.D. Enhanced sensitization and elicitation responses caused by mixtures of common fragrance allergens. *Contact Dermat.* **2011**, *65*, 336–342. [CrossRef]
61. Bonefeld, C.M.; Nielsen, M.M.; Vennegaard, M.T.; Johansen, J.D.; Geisler, C.; Thyssen, J.P. Nickel acts as an adjuvant during cobalt sensitization. *Exp. Dermatol.* **2015**, *24*, 229–231. [CrossRef]
62. Cumberbatch, M.; Scott, R.C.; Basketter, D.A.; Scholes, E.W.; Hilton, J.; Dearman, R.J.; Kimber, I. Influence of sodium lauryl sulphate on 2,4-dinitrochlorobenzene-induced lymph node activation. *Toxicology* **1993**, *77*, 181–191. [CrossRef] [PubMed]
63. Basketter, D. Nickel: Intrinsic Skin Sensitization Potency and Relation to Prevalence of Contact Allergy. *Dermatitis* **2021**, *32*, 71–77. [CrossRef] [PubMed]
64. Höper, T.; Siewert, K.; Dumit, V.I.; von Bergen, M.; Schubert, K.; Haase, A. The Contact Allergen NiSO(4) Triggers a Distinct Molecular Response in Primary Human Dendritic Cells Compared to Bacterial LPS. *Front. Immunol.* **2021**, *12*, 644700. [CrossRef] [PubMed]
65. Schmidt, M.; Raghavan, B.; Muller, V.; Vogl, T.; Fejer, G.; Tchaptchet, S.; Keck, S.; Kalis, C.; Nielsen, P.J.; Galanos, C.; et al. Crucial role for human Toll-like receptor 4 in the development of contact allergy to nickel. *Nat. Immunol.* **2010**, *11*, 814–819. [CrossRef] [PubMed]
66. Liu, L.; Zhong, Q.; Tian, T.; Dubin, K.; Athale, S.K.; Kupper, T.S. Epidermal injury and infection during poxvirus immunization is crucial for the generation of highly protective T cell-mediated immunity. *Nat. Med.* **2010**, *16*, 224–227. [CrossRef] [PubMed]
67. Schuttelaar, M.L.A.; Ofenloch, R.F.; Bruze, M.; Cazzaniga, S.; Elsner, P.; Goncalo, M.; Naldi, L.; Svensson, A.; Diepgen, T.L. Prevalence of contact allergy to metals in the European general population with a focus on nickel and piercings: The EDEN Fragrance Study. *Contact Dermat.* **2018**, *79*, 1–9. [CrossRef]

68. Guironnet, G.; Dalbiez-Gauthier, C.; Rousset, F.; Schmitt, D.; Peguet-Navarro, J. In vitro human T cell sensitization to haptens by monocyte-derived dendritic cells. *Toxicol. Vitr.* **2000**, *14*, 517–522. [CrossRef]
69. Vocanson, M.; Cluzel-Tailhardat, M.; Poyet, G.; Valeyrie, M.; Chavagnac, C.; Levarlet, B.; Courtellemont, P.; Rozieres, A.; Hennino, A.; Nicolas, J.F. Depletion of human peripheral blood lymphocytes in $CD25^+$ cells allows for the sensitive in vitro screening of contact allergens. *J. Investig. Dermatol.* **2008**, *128*, 2119–2122. [CrossRef]
70. Dietz, L.; Esser, P.R.; Schmucker, S.S.; Goette, I.; Richter, A.; Schnolzer, M.; Martin, S.F.; Thierse, H.J. Tracking human contact allergens: From mass spectrometric identification of peptide-bound reactive small chemicals to chemical-specific naive human T-cell priming. *Toxicol. Sci.* **2010**, *117*, 336–347. [CrossRef]
71. Gibson, A.; Kim, S.H.; Faulkner, L.; Evely, J.; Pirmohamed, M.; Park, K.B.; Naisbitt, D.J. In Vitro Priming of Naïve T-cells with p-Phenylenediamine and Bandrowski's Base. *Chem. Res. Toxicol.* **2015**, *28*, 2069–2077. [CrossRef]
72. Summer, B.; Stander, S.; Thomas, P. Cytokine patterns in vitro, in particular IL-5/IL-8 ratio, to detect patients with nickel contact allergy. *J. Eur. Acad. Dermatol. Venereol.* **2018**, *32*, 1542–1548. [CrossRef] [PubMed]
73. Schopf, E.; Schulz, K.H.; Isensee, I. Lymphocyte transformation in mercury hypersensitivity. Non-specific transformation due to mercury compounds. *Arch. Klin. Exp. Dermatol.* **1969**, *234*, 420–433. [PubMed]
74. Pichler, W.J.; Tilch, J. The lymphocyte transformation test in the diagnosis of drug hypersensitivity. *Allergy* **2004**, *59*, 809–820. [CrossRef] [PubMed]
75. Koponen, M.; Pichler, W.J.; de Weck, A.L. T cell reactivity to penicillin: Phenotypic analysis of in vitro activated cell subsets. *J. Allergy Clin. Immunol.* **1986**, *78*, 645–652. [CrossRef] [PubMed]
76. Coulter, E.M.; Jenkinson, C.; Wu, Y.; Farrell, J.; Foster, B.; Smith, A.; McGuire, C.; Pease, C.; Basketter, D.; King, C.; et al. Activation of T-cells from allergic patients and volunteers by p-phenylenediamine and Bandrowski's base. *J. Investig. Dermatol.* **2008**, *128*, 897–905. [CrossRef] [PubMed]
77. Coulter, E.M.; Jenkinson, C.; Farrell, J.; Lavergne, S.N.; Pease, C.; White, A.; Aleksic, M.; Basketter, D.; Williams, D.P.; King, C.; et al. Measurement of $CD4^+$ and $CD8^+$ T-lymphocyte cytokine secretion and gene expression changes in p-phenylenediamine allergic patients and tolerant individuals. *J. Investig. Dermatol.* **2010**, *130*, 161–174. [CrossRef]
78. Bordignon, V.; Palamara, F.; Altomonte, G.; Sperduti, I.; Pietravalle, M.; Cavallotti, C.; Cordiali-Fei, P.; Fuggetta, M.P.; Cristaudo, A.; Ensoli, F. A laboratory test based on determination of cytokine profiles: A promising assay to identify exposition to contact allergens and predict the clinical outcome in occupational allergic contact dermatitis. *BMC Immunol.* **2015**, *16*, 4. [CrossRef]
79. Jung, T.; Schauer, U.; Heusser, C.; Neumann, C.; Rieger, C. Detection of intracellular cytokines by flow cytometry. *J. Immunol. Methods* **1993**, *159*, 197–207. [CrossRef]
80. Saxena, A.; Dagur, P.K.; Desai, A.; McCoy, J.P., Jr. Ultrasensitive Quantification of Cytokine Proteins in Single Lymphocytes From Human Blood Following ex-vivo Stimulation. *Front. Immunol.* **2018**, *9*, 2462. [CrossRef]
81. Campbell, J.D. Detection and enrichment of antigen-specific $CD4^+$ and $CD8^+$ T cells based on cytokine secretion. *Methods* **2003**, *31*, 150–159. [CrossRef]
82. Altman, J.D.; Moss, P.A.; Goulder, P.J.; Barouch, D.H.; McHeyzer-Williams, M.G.; Bell, J.I.; McMichael, A.J.; Davis, M.M. Phenotypic analysis of antigen-specific T lymphocytes. *Science* **1996**, *274*, 94–96. [CrossRef] [PubMed]
83. Geiger, R.; Duhen, T.; Lanzavecchia, A.; Sallusto, F. Human naive and memory $CD4^+$ T cell repertoires specific for naturally processed antigens analyzed using libraries of amplified T cells. *J. Exp. Med.* **2009**, *206*, 1525–1534. [CrossRef] [PubMed]
84. Ahmed, S.S.; Wang, X.N.; Fielding, M.; Kerry, A.; Dickinson, I.; Munuswamy, R.; Kimber, I.; Dickinson, A.M. An in vitro human skin test for assessing sensitization potential. *J. Appl. Toxicol.* **2016**, *36*, 669–684. [CrossRef] [PubMed]
85. Ahmed, S.S.; Whritenour, J.; Ahmed, M.M.; Bibby, L.; Darby, L.; Wang, X.N.; Watson, J.; Dickinson, A.M. Evaluation of a human in vitro skin test for predicting drug hypersensitivity reactions. *Toxicol. Appl. Pharmacol.* **2019**, *369*, 39–48. [CrossRef] [PubMed]
86. Croft, M.; Dubey, C. Accessory Molecule and Costimulation Requirements for CD4 T Cell Response. *Crit. Rev. Immunol.* **2017**, *37*, 261–290. [CrossRef]
87. Dubey, C.; Croft, M.; Swain, S.L. Naive and effector CD4 T cells differ in their requirements for T cell receptor versus costimulatory signals. *J. Immunol.* **1996**, *157*, 3280–3289. [CrossRef]
88. Bour-Jordan, H.; Esensten, J.H.; Martinez-Llordella, M.; Penaranda, C.; Stumpf, M.; Bluestone, J.A. Intrinsic and extrinsic control of peripheral T-cell tolerance by costimulatory molecules of the CD28/ B7 family. *Immunol. Rev.* **2011**, *241*, 180–205. [CrossRef]
89. Kunzli, M.; Masopust, D. $CD4^+$ T cell memory. *Nat. Immunol.* **2023**, *24*, 903–914. [CrossRef]
90. Martin-Fontecha, A.; Sebastiani, S.; Hopken, U.E.; Uguccioni, M.; Lipp, M.; Lanzavecchia, A.; Sallusto, F. Regulation of dendritic cell migration to the draining lymph node: Impact on T lymphocyte traffic and priming. *J. Exp. Med.* **2003**, *198*, 615–621. [CrossRef]
91. Vocanson, M.; Achachi, A.; Mutez, V.; Cluzel-Tailhardat, M.; Varlet, B.L.; Rozieres, A.; Fournier, P.; Nicolas, J.F. Human T cell priming assay: Depletion of peripheral blood lymphocytes in $CD25^+$ cells improves the in vitro detection of weak allergen-specific T cells. *Exp. Suppl.* **2014**, *104*, 89–100. [CrossRef]
92. Sallusto, F.; Lanzavecchia, A. Efficient presentation of soluble antigen by cultured human dendritic cells is maintained by granulocyte/macrophage colony-stimulating factor plus interleukin 4 and downregulated by tumor necrosis factor alpha. *J. Exp. Med.* **1994**, *179*, 1109–1118. [CrossRef] [PubMed]
93. Sarkany, I. Lymphocyte transformation in drug hypersensitivity. *Lancet* **1967**, *1*, 743–745. [CrossRef] [PubMed]

94. Ganesan, N.; Ronsmans, S.; Hoet, P. Methods to Assess Proliferation of Stimulated Human Lymphocytes In Vitro: A Narrative Review. *Cells* **2023**, *12*, 386. [CrossRef] [PubMed]
95. Friedmann, P.S.; Haddadeen, C.; Lai, C.; Healy, E. In vitro human T cell responses to diphencyprone. *Contact Dermat.* **2017**, *76*, 251–253. [CrossRef]
96. Kneilling, M.; Caroli, U.; Grimmel, C.; Fischer, J.; Eichner, M.; Wieder, T.; Maier, F.C.; Rocken, M.; Biedermann, T. Para-phenylenediamine-specific lymphocyte activation test: A sensitive in vitro assay to detect para-phenylenediamine sensitization in patients with severe allergic reactions. *Exp. Dermatol.* **2010**, *19*, 435–441. [CrossRef]
97. Martin, S.; Barbaud, A.; Schmutz, J.L.; Gobert, B.; Faure, G.; Bene, M.C. Polyclonal T lymphocyte proliferation in drug lymphocyte activation test. *Ann. Dermatol. Venereol.* **2000**, *127*, 268–272.
98. Sachs, B.; Fatangare, A.; Sickmann, A.; Glassner, A. Lymphocyte transformation test: History and current approaches. *J. Immunol. Methods* **2021**, *493*, 113036. [CrossRef]
99. Popple, A.; Williams, J.; Maxwell, G.; Gellatly, N.; Dearman, R.J.; Kimber, I. The lymphocyte transformation test in allergic contact dermatitis: New opportunities. *J. Immunotoxicol.* **2016**, *13*, 84–91. [CrossRef]
100. Rustemeyer, T.; von Blomberg, B.M.; van Hoogstraten, I.M.; Bruynzeel, D.P.; Scheper, R.J. Analysis of effector and regulatory immune reactivity to nickel. *Clin. Exp. Allergy* **2004**, *34*, 1458–1466. [CrossRef]
101. Kretschmer, L.; Flossdorf, M.; Mir, J.; Cho, Y.L.; Plambeck, M.; Treise, I.; Toska, A.; Heinzel, S.; Schiemann, M.; Busch, D.H.; et al. Differential expansion of T central memory precursor and effector subsets is regulated by division speed. *Nat. Commun.* **2020**, *11*, 113. [CrossRef]
102. Tanguay, S.; Killion, J.J. Direct comparison of ELISPOT and ELISA-based assays for detection of individual cytokine-secreting cells. *Lymphokine Cytokine Res.* **1994**, *13*, 259–263. [PubMed]
103. Masjedi, K.; Ahlborg, N.; Gruvberger, B.; Bruze, M.; Karlberg, A.T. Methylisothiazolinones elicit increased production of both T helper (Th)1- and Th2-like cytokines by peripheral blood mononuclear cells from contact allergic individuals. *Br. J. Dermatol.* **2003**, *149*, 1172–1182. [CrossRef] [PubMed]
104. Lovelace, P.; Maecker, H.T. Multiparameter Intracellular Cytokine Staining. *Methods Mol. Biol.* **2018**, *1678*, 151–166. [CrossRef] [PubMed]
105. Poloni, C.; Schonhofer, C.; Ivison, S.; Levings, M.K.; Steiner, T.S.; Cook, L. T-cell activation-induced marker assays in health and disease. *Immunol. Cell Biol.* **2023**, *101*, 491–503. [CrossRef] [PubMed]
106. Chattopadhyay, P.K.; Yu, J.; Roederer, M. A live-cell assay to detect antigen-specific $CD4^+$ T cells with diverse cytokine profiles. *Nat. Med.* **2005**, *11*, 1113–1117. [CrossRef] [PubMed]
107. Zhang, Y.; Wang, Y.; Anderson, K.; Novikov, A.; Liu, Z.; Pacheco, K.; Dai, S. Using DR52c/Ni^{2+} mimotope tetramers to detect Ni^{2+} reactive $CD4^+$ T cells in patients with joint replacement failure. *Toxicol. Appl. Pharmacol.* **2017**, *331*, 69–75. [CrossRef]
108. Adair, K.; Meng, X.; Naisbitt, D.J. Drug hapten-specific T-cell activation: Current status and unanswered questions. *Proteomics* **2021**, *21*, e2000267. [CrossRef]
109. Pastore, G.; Carraro, M.; Pettini, E.; Nolfi, E.; Medaglini, D.; Ciabattini, A. Optimized Protocol for the Detection of Multifunctional Epitope-Specific $CD4^+$ T Cells Combining MHC-II Tetramer and Intracellular Cytokine Staining Technologies. *Front. Immunol.* **2019**, *10*, 2304. [CrossRef]
110. Vollers, S.S.; Stern, L.J. Class II major histocompatibility complex tetramer staining: Progress, problems, and prospects. *Immunology* **2008**, *123*, 305–313. [CrossRef]
111. Dileepan, T.; Malhotra, D.; Kotov, D.I.; Kolawole, E.M.; Krueger, P.D.; Evavold, B.D.; Jenkins, M.K. MHC class II tetramers engineered for enhanced binding to CD4 improve detection of antigen-specific T cells. *Nat. Biotechnol.* **2021**, *39*, 943–948. [CrossRef]
112. Pan, Y.G.; Su, L.F. Identification of Human Antigen-Specific T Cells Using Class II MHC Tetramer Staining and Enrichment. *Methods Mol. Biol.* **2022**, *2574*, 31–40. [CrossRef] [PubMed]
113. Newell, E.W. Higher throughput methods of identifying T cell epitopes for studying outcomes of altered antigen processing and presentation. *Front. Immunol.* **2013**, *4*, 430. [CrossRef] [PubMed]
114. Christophersen, A. Peptide-MHC class I and class II tetramers: From flow to mass cytometry. *HLA* **2020**, *95*, 169–178. [CrossRef] [PubMed]
115. Kwok, W.W.; Tan, V.; Gillette, L.; Littell, C.T.; Soltis, M.A.; LaFond, R.B.; Yang, J.; James, E.A.; DeLong, J.H. Frequency of epitope-specific naive $CD4^+$ T cells correlates with immunodominance in the human memory repertoire. *J. Immunol.* **2012**, *188*, 2537–2544. [CrossRef] [PubMed]
116. Nepom, G.T. MHC class II tetramers. *J. Immunol.* **2012**, *188*, 2477–2482. [CrossRef]
117. Theaker, S.M.; Rius, C.; Greenshields-Watson, A.; Lloyd, A.; Trimby, A.; Fuller, A.; Miles, J.J.; Cole, D.K.; Peakman, M.; Sewell, A.K.; et al. T-cell libraries allow simple parallel generation of multiple peptide-specific human T-cell clones. *J. Immunol. Methods* **2016**, *430*, 43–50. [CrossRef]
118. Ogura, H.; Preston-Hurlburt, P.; Perdigoto, A.L.; Amodio, M.; Krishnaswamy, S.; Clark, P.; Yu, H.; Egli, D.; Fouts, A.; Steck, A.K.; et al. Identification and Analysis of Islet Antigen-Specific $CD8^+$ T Cells with T Cell Libraries. *J. Immunol.* **2018**, *201*, 1662–1670. [CrossRef]
119. Ahmed, S.S.; Ahmed, M.M.; Ishaq, A.; Freer, M.; Stebbings, R.; Dickinson, A.M. An In Vitro Human Skin Test for Predicting Skin Sensitization and Adverse Immune Reactions to Biologics. *Toxics* **2024**, *12*, 401. [CrossRef]

120. Lerner, K.G.; Kao, G.F.; Storb, R.; Buckner, C.D.; Clift, R.A.; Thomas, E.D. Histopathology of graft-vs.-host reaction (GvHR) in human recipients of marrow from HL-A-matched sibling donors. *Transpl. Proc.* **1974**, *6*, 367–371.
121. Bacher, P.; Schink, C.; Teutschbein, J.; Kniemeyer, O.; Assenmacher, M.; Brakhage, A.A.; Scheffold, A. Antigen-reactive T cell enrichment for direct, high-resolution analysis of the human naive and memory Th cell repertoire. *J. Immunol.* **2013**, *190*, 3967–3976. [CrossRef]
122. Lemieux, A.; Sannier, G.; Nicolas, A.; Nayrac, M.; Delgado, G.G.; Cloutier, R.; Brassard, N.; Laporte, M.; Duchesne, M.; Sreng Flores, A.M.; et al. Enhanced detection of antigen-specific T cells by a multiplexed AIM assay. *Cell Rep. Methods* **2024**, *4*, 100690. [CrossRef] [PubMed]
123. Reiss, S.; Baxter, A.E.; Cirelli, K.M.; Dan, J.M.; Morou, A.; Daigneault, A.; Brassard, N.; Silvestri, G.; Routy, J.P.; Havenar-Daughton, C.; et al. Comparative analysis of activation induced marker (AIM) assays for sensitive identification of antigen-specific CD4 T cells. *PLoS ONE* **2017**, *12*, e0186998. [CrossRef] [PubMed]
124. Elias, G.; Ogunjimi, B.; Van Tendeloo, V. Activation-induced surface proteins in the identification of antigen-responsive CD4 T cells. *Immunol. Lett.* **2020**, *219*, 1–7. [CrossRef] [PubMed]
125. Wolfl, M.; Kuball, J.; Ho, W.Y.; Nguyen, H.; Manley, T.J.; Bleakley, M.; Greenberg, P.D. Activation-induced expression of CD137 permits detection, isolation, and expansion of the full repertoire of $CD8^+$ T cells responding to antigen without requiring knowledge of epitope specificities. *Blood* **2007**, *110*, 201–210. [CrossRef]
126. Bremser, A.; Brack, M.; Izcue, A. Higher Sensitivity of $Foxp3^+$ Treg Compared to Foxp3- Conventional T Cells to TCR-Independent Signals for CD69 Induction. *PLoS ONE* **2015**, *10*, e0137393. [CrossRef]
127. Evans, R.L.; Faldetta, T.J.; Humphreys, R.E.; Pratt, D.M.; Yunis, E.J.; Schlossman, S.F. Peripheral human T cells sensitized in mixed leukocyte culture synthesize and express Ia-like antigens. *J. Exp. Med.* **1978**, *148*, 1440–1445. [CrossRef]
128. Munoz, P.; Mittelbrunn, M.; de la Fuente, H.; Perez-Martinez, M.; Garcia-Perez, A.; Ariza-Veguillas, A.; Malavasi, F.; Zubiaur, M.; Sanchez-Madrid, F.; Sancho, J. Antigen-induced clustering of surface CD38 and recruitment of intracellular CD38 to the immunologic synapse. *Blood* **2008**, *111*, 3653–3664. [CrossRef]
129. Ghosh, A.; Khanam, A.; Ray, K.; Mathur, P.; Subramanian, A.; Poonia, B.; Kottilil, S. CD38: An ecto-enzyme with functional diversity in T cells. *Front. Immunol.* **2023**, *14*, 1146791. [CrossRef]
130. Frentsch, M.; Stark, R.; Matzmohr, N.; Meier, S.; Durlanik, S.; Schulz, A.R.; Stervbo, U.; Jurchott, K.; Gebhardt, F.; Heine, G.; et al. CD40L expression permits $CD8^+$ T cells to execute immunologic helper functions. *Blood* **2013**, *122*, 405–412. [CrossRef]
131. Skov, S.; Bonyhadi, M.; Odum, N.; Ledbetter, J.A. IL-2 and IL-15 regulate CD154 expression on activated CD4 T cells. *J. Immunol.* **2000**, *164*, 3500–3505. [CrossRef]
132. Frentsch, M.; Arbach, O.; Kirchhoff, D.; Moewes, B.; Worm, M.; Rothe, M.; Scheffold, A.; Thiel, A. Direct access to $CD4^+$ T cells specific for defined antigens according to CD154 expression. *Nat. Med.* **2005**, *11*, 1118–1124. [CrossRef] [PubMed]
133. Watts, T.H. TNF/TNFR family members in costimulation of T cell responses. *Annu. Rev. Immunol.* **2005**, *23*, 23–68. [CrossRef] [PubMed]
134. Cannons, J.L.; Lau, P.; Ghumman, B.; DeBenedette, M.A.; Yagita, H.; Okumura, K.; Watts, T.H. 4-1BB ligand induces cell division, sustains survival, and enhances effector function of CD4 and CD8 T cells with similar efficacy. *J. Immunol.* **2001**, *167*, 1313–1324. [CrossRef] [PubMed]
135. Schoenbrunn, A.; Frentsch, M.; Kohler, S.; Keye, J.; Dooms, H.; Moewes, B.; Dong, J.; Loddenkemper, C.; Sieper, J.; Wu, P.; et al. A converse 4-1BB and CD40 ligand expression pattern delineates activated regulatory T cells (Treg) and conventional T cells enabling direct isolation of alloantigen-reactive natural $Foxp3^+$ Treg. *J. Immunol.* **2012**, *189*, 5985–5994. [CrossRef]
136. Cohen, G.B.; Kaur, A.; Johnson, R.P. Isolation of viable antigen-specific CD4 T cells by CD40L surface trapping. *J. Immunol. Methods* **2005**, *302*, 103–115. [CrossRef]
137. Gerdes, J.; Lemke, H.; Baisch, H.; Wacker, H.H.; Schwab, U.; Stein, H. Cell cycle analysis of a cell proliferation-associated human nuclear antigen defined by the monoclonal antibody Ki-67. *J. Immunol.* **1984**, *133*, 1710–1715. [CrossRef]
138. Santamaria Babi, L.F.; Picker, L.J.; Perez Soler, M.T.; Drzimalla, K.; Flohr, P.; Blaser, K.; Hauser, C. Circulating allergen-reactive T cells from patients with atopic dermatitis and allergic contact dermatitis express the skin-selective homing receptor, the cutaneous lymphocyte-associated antigen. *J. Exp. Med.* **1995**, *181*, 1935–1940. [CrossRef]
139. Cavani, A.; Mei, D.; Guerra, E.; Corinti, S.; Giani, M.; Pirrotta, L.; Puddu, P.; Girolomoni, G. Patients with allergic contact dermatitis to nickel and nonallergic individuals display different nickel-specific T cell responses. Evidence for the presence of effector $CD8^+$ and regulatory $CD4^+$ T cells. *J. Investig. Dermatol.* **1998**, *111*, 621–628. [CrossRef]
140. Bacher, P.; Scheffold, A. New technologies for monitoring human antigen-specific T cells and regulatory T cells by flow-cytometry. *Curr. Opin. Pharmacol.* **2015**, *23*, 17–24. [CrossRef]
141. Saggau, C.; Scheffold, A.; Bacher, P. Flow Cytometric Characterization of Human Antigen-Reactive T-Helper Cells. *Methods Mol. Biol.* **2021**, *2285*, 141–152. [CrossRef]
142. Cavani, A.; Nasorri, F.; Ottaviani, C.; Sebastiani, S.; De Pita, O.; Girolomoni, G. Human $CD25^+$ regulatory T cells maintain immune tolerance to nickel in healthy, nonallergic individuals. *J. Immunol.* **2003**, *171*, 5760–5768. [CrossRef] [PubMed]
143. Cook, L.; Munier, C.M.L.; Seddiki, N.; Hardy, M.Y.; Anderson, R.P.; Zaunders, J.; Tye-Din, J.A.; Kelleher, A.D.; van Bockel, D. Circulating gluten-specific, but not CMV-specific, $CD39^+$ regulatory T cells have an oligoclonal TCR repertoire. *Clin. Transl. Immunol.* **2020**, *9*, e1096. [CrossRef] [PubMed]
144. Toholka, R.; Nixon, R. Allergic contact dermatitis to chlorhexidine. *Australas J. Dermatol.* **2013**, *54*, 303–306. [CrossRef] [PubMed]

145. Weltzien, H.U.; Hebbelmann, S.; Pflugfelder, U.; Ruh, H.; Ortmann, B.; Martin, S.; Iglesias, A. Antigen contact sites in class I major histocompatibility complex-restricted, trinitrophenyl-specific T cell receptors. *Eur. J. Immunol.* **1992**, *22*, 863–866. [CrossRef] [PubMed]
146. Rock, K.L.; Reits, E.; Neefjes, J. Present Yourself! By MHC Class I and MHC Class II Molecules. *Trends. Immunol.* **2016**, *37*, 724–737. [CrossRef]
147. Pishesha, N.; Harmand, T.J.; Ploegh, H.L. A guide to antigen processing and presentation. *Nat. Rev. Immunol.* **2022**, *22*, 751–764. [CrossRef]
148. Ortmann, B.; Martin, S.; von Bonin, A.; Schiltz, E.; Hoschutzky, H.; Weltzien, H.U. Synthetic peptides anchor T cell-specific TNP epitopes to MHC antigens. *J. Immunol.* **1992**, *148*, 1445–1450. [CrossRef]
149. Fontenot, A.P.; Falta, M.T.; Kappler, J.W.; Dai, S.; McKee, A.S. Beryllium-Induced Hypersensitivity: Genetic Susceptibility and Neoantigen Generation. *J. Immunol.* **2016**, *196*, 22–27. [CrossRef]
150. Griem, P.; Panthel, K.; Kalbacher, H.; Gleichmann, E. Alteration of a model antigen by Au(III) leads to T cell sensitization to cryptic peptides. *Eur. J. Immunol.* **1996**, *26*, 279–287. [CrossRef]
151. Thierse, H.J.; Moulon, C.; Allespach, Y.; Zimmermann, B.; Doetze, A.; Kuppig, S.; Wild, D.; Herberg, F.; Weltzien, H.U. Metal-protein complex-mediated transport and delivery of $Ni2^{+}$ to TCR/MHC contact sites in nickel-specific human T cell activation. *J. Immunol.* **2004**, *172*, 1926–1934. [CrossRef]
152. Budinger, L.; Hertl, M. Immunologic mechanisms in hypersensitivity reactions to metal ions: An overview. *Allergy* **2000**, *55*, 108–115. [CrossRef] [PubMed]
153. Jenkinson, C.; Jenkins, R.E.; Aleksic, M.; Pirmohamed, M.; Naisbitt, D.J.; Park, B.K. Characterization of p-phenylenediamine-albumin binding sites and T-cell responses to hapten-modified protein. *J. Investig. Dermatol.* **2010**, *130*, 732–742. [CrossRef] [PubMed]
154. Pickard, C.; Smith, A.M.; Cooper, H.; Strickland, I.; Jackson, J.; Healy, E.; Friedmann, P.S. Investigation of mechanisms underlying the T-cell response to the hapten 2,4-dinitrochlorobenzene. *J. Investig. Dermatol.* **2007**, *127*, 630–637. [CrossRef] [PubMed]
155. Thierse, H.J.; Gamerdinger, K.; Junkes, C.; Guerreiro, N.; Weltzien, H.U. T cell receptor (TCR) interaction with haptens: Metal ions as non-classical haptens. *Toxicology* **2005**, *209*, 101–107. [CrossRef]
156. Oakes, T.; Popple, A.L.; Williams, J.; Best, K.; Heather, J.M.; Ismail, M.; Maxwell, G.; Gellatly, N.; Dearman, R.J.; Kimber, I.; et al. The T Cell Response to the Contact Sensitizer Paraphenylenediamine Is Characterized by a Polyclonal Diverse Repertoire of Antigen-Specific Receptors. *Front. Immunol.* **2017**, *8*, 162. [CrossRef]
157. Wicks, K.; Stretton, C.; Popple, A.; Beresford, L.; Williams, J.; Maxwell, G.; Gosling, J.P.; Kimber, I.; Dearman, R.J. T lymphocyte phenotype of contact-allergic patients: Experience with nickel and p-phenylenediamine. *Contact Dermat.* **2019**, *81*, 43–53. [CrossRef]
158. Eilstein, J.; Gimenez-Arnau, E.; Duche, D.; Rousset, F.; Lepoittevin, J.P. Mechanistic studies on the lysine-induced N-formylation of 2,5-dimethyl-p-benzoquinonediimine. *Chem. Res. Toxicol.* **2007**, *20*, 1155–1161. [CrossRef]
159. Parkinson, E.; Aleksic, M.; Cubberley, R.; Kaur-Atwal, G.; Vissers, J.P.C.; Skipp, P. Determination of Protein Haptenation by Chemical Sensitizers Within the Complexity of the Human Skin Proteome. *Toxicol. Sci.* **2018**, *162*, 429–438. [CrossRef]
160. Cortial, A.; Nosbaum, A.; Rozieres, A.; Baeck, M.; de Montjoye, L.; Grande, S.; Briancon, S.; Nicolas, J.F.; Vocanson, M. Encapsulation of hydrophobic allergens into nanoparticles improves the in vitro immunological diagnosis of allergic contact dermatitis. *Nanomedicine* **2015**, *11*, 1029–1033. [CrossRef]
161. Di Felice, G.; Colombo, P. Nanoparticle-allergen complexes for allergen immunotherapy. *Int. J. Nanomed.* **2017**, *12*, 4493–4504. [CrossRef]
162. Pateiro, M.; Gomez, B.; Munekata, P.E.S.; Barba, F.J.; Putnik, P.; Kovacevic, D.B.; Lorenzo, J.M. Nanoencapsulation of Promising Bioactive Compounds to Improve Their Absorption, Stability, Functionality and the Appearance of the Final Food Products. *Molecules* **2021**, *26*, 1547. [CrossRef] [PubMed]
163. Hulspas, R.; O'Gorman, M.R.; Wood, B.L.; Gratama, J.W.; Sutherland, D.R. Considerations for the control of background fluorescence in clinical flow cytometry. *Cytom. B Clin. Cytom.* **2009**, *76*, 355–364. [CrossRef] [PubMed]
164. Sober, S.A.; Darmani, H.; Alhattab, D.; Awidi, A. Flow cytometric characterization of cell surface markers to differentiate between fibroblasts and mesenchymal stem cells of different origin. *Arch. Med. Sci.* **2023**, *19*, 1487–1496. [CrossRef] [PubMed]
165. Robinson, J.P.; Ostafe, R.; Iyengar, S.N.; Rajwa, B.; Fischer, R. Flow Cytometry: The Next Revolution. *Cells* **2023**, *12*, 1875. [CrossRef] [PubMed]
166. Timm, M.; Saaby, L.; Moesby, L.; Hansen, E.W. Considerations regarding use of solvents in in vitro cell based assays. *Cytotechnology* **2013**, *65*, 887–894. [CrossRef]
167. Holthaus, L.; Lamp, D.; Gavrisan, A.; Sharma, V.; Ziegler, A.G.; Jastroch, M.; Bonifacio, E. $CD4^{+}$ T cell activation, function, and metabolism are inhibited by low concentrations of DMSO. *J. Immunol. Methods* **2018**, *463*, 54–60. [CrossRef]
168. Cederbrant, K.; Anderson, C.; Andersson, T.; Marcusson-Stahl, M.; Hultman, P. Cytokine production, lymphocyte proliferation and T-cell receptor Vbeta expression in primary peripheral blood mononuclear cell cultures from nickel-allergic individuals. *Int. Arch. Allergy Immunol.* **2003**, *132*, 373–379. [CrossRef]
169. Al-Tawil, N.G.; Marcusson, J.A.; Moller, E. Lymphocyte transformation test in patients with nickel sensitivity: An aid to diagnosis. *Acta. Derm. Venereol.* **1981**, *61*, 511–515. [CrossRef]

170. Bacher, P.; Heinrich, F.; Stervbo, U.; Nienen, M.; Vahldieck, M.; Iwert, C.; Vogt, K.; Kollet, J.; Babel, N.; Sawitzki, B.; et al. Regulatory T Cell Specificity Directs Tolerance versus Allergy against Aeroantigens in Humans. *Cell* **2016**, *167*, 1067–1078.e16. [CrossRef]
171. Sieben, S.; Kawakubo, Y.; Al Masaoudi, T.; Merk, H.F.; Blomeke, B. Delayed-type hypersensitivity reaction to paraphenylene-diamine is mediated by 2 different pathways of antigen recognition by specific alphabeta human T-cell clones. *J. Allergy Clin. Immunol.* **2002**, *109*, 1005–1011. [CrossRef]
172. Uter, W.; Wilkinson, S.M.; Aerts, O.; Bauer, A.; Borrego, L.; Brans, R.; Buhl, T.; Dickel, H.; Dugonik, A.; Filon, F.L.; et al. Patch test results with the European baseline series, 2019/20-Joint European results of the ESSCA and the EBS working groups of the ESCD, and the GEIDAC. *Contact Dermat.* **2022**, *87*, 343–355. [CrossRef] [PubMed]
173. Brans, R.; Skudlik, C. Patch testing in occupational dermatology: Practical aspects in relation to the conditions in Germany. *Allergol. Select.* **2024**, *8*, 82–89. [CrossRef] [PubMed]
174. Simon, K.; Reichardt, P.; Luch, A.; Roloff, A.; Siewert, K.; Riedel, F. Less efficient skin penetration of the metal allergen $Pd2^{+}$ compared to $Ni2^{+}$ and $Co2^{+}$ from patch test preparations. *Contact Dermat.* **2024**, *91*, 11–21. [CrossRef] [PubMed]
175. Serup, J.; Hutton Carlsen, K. Patch test study of 90 patients with tattoo reactions: Negative outcome of allergy patch test to baseline batteries and culprit inks suggests allergen(s) are generated in the skin through haptenization. *Contact Dermat.* **2014**, *71*, 255–263. [CrossRef]
176. Stander, S.; Oppel, E.; Thomas, P.; Summer, B. Evaluation of lymphocyte transformation tests as compared with patch tests in nickel allergy diagnosis. *Contact Dermat.* **2017**, *76*, 228–234. [CrossRef]
177. Walker, J.M.; Slifka, M.K. Longevity of T-cell memory following acute viral infection. *Adv. Exp. Med. Biol.* **2010**, *684*, 96–107. [CrossRef]
178. Hammarlund, E.; Lewis, M.W.; Hansen, S.G.; Strelow, L.I.; Nelson, J.A.; Sexton, G.J.; Hanifin, J.M.; Slifka, M.K. Duration of antiviral immunity after smallpox vaccination. *Nat. Med.* **2003**, *9*, 1131–1137. [CrossRef]
179. Minervina, A.A.; Komech, E.A.; Titov, A.; Bensouda Koraichi, M.; Rosati, E.; Mamedov, I.Z.; Franke, A.; Efimov, G.A.; Chudakov, D.M.; Mora, T.; et al. Longitudinal high-throughput TCR repertoire profiling reveals the dynamics of T-cell memory formation after mild COVID-19 infection. *eLife* **2021**, *10*, e63502. [CrossRef]
180. Clark, R.A.; Chong, B.; Mirchandani, N.; Brinster, N.K.; Yamanaka, K.; Dowgiert, R.K.; Kupper, T.S. The vast majority of CLA^{+} T cells are resident in normal skin. *J. Immunol.* **2006**, *176*, 4431–4439. [CrossRef]
181. Renz, H.; Biedermann, T.; Bufe, A.; Eberlein, B.; Jappe, U.; Ollert, M.; Petersen, A.; Kleine-Tebbe, J.; Raulf-Heimsoth, M.; Saloga, J.; et al. In-vitro-Allergiediagnostik—Leitlinie der Deutschen Gesellschaft für Allergologie und klinische Immunologie (DGAKI) unter Beteiligung des Ärzteverbandes Deutscher Allergologen (ÄDA), der Gesellschaft für Pädiatrische Allergologie und Umweltmedizin (GPA) und der Deutschen Dermatologische Gesellschaft (DDG). *Allergo J. Int.* **2010**, *19*, 110–128.
182. Watanabe, R.; Gehad, A.; Yang, C.; Scott, L.L.; Teague, J.E.; Schlapbach, C.; Elco, C.P.; Huang, V.; Matos, T.R.; Kupper, T.S.; et al. Human skin is protected by four functionally and phenotypically discrete populations of resident and recirculating memory T cells. *Sci. Transl. Med.* **2015**, *7*, 279ra39. [CrossRef] [PubMed]
183. Giesecke, C.; Meyer, T.; Durek, P.; Maul, J.; Preiss, J.; Jacobs, J.F.M.; Thiel, A.; Radbruch, A.; Ullrich, R.; Dorner, T. Simultaneous Presence of Non- and Highly Mutated Keyhole Limpet Hemocyanin (KLH)-Specific Plasmablasts Early after Primary KLH Immunization Suggests Cross-Reactive Memory B Cell Activation. *J. Immunol.* **2018**, *200*, 3981–3992. [CrossRef] [PubMed]
184. Hudson, D.; Fernandes, R.A.; Basham, M.; Ogg, G.; Koohy, H. Can we predict T cell specificity with digital biology and machine learning? *Nat. Rev. Immunol.* **2023**, *23*, 511–521. [CrossRef] [PubMed]
185. Haase, A.; Barroso, J.; Bogni, A.; Bremer-Hoffmann, S.; Fessard, V.; Gutleb, A.C.; Mast, J.; McVey, E.; Mertens, B.; Oomen, A.G.; et al. Proposal for a qualification system for New Approach Methodologies (NAMs) in the food and feed sector: Example of implementation for nanomaterial risk assessment. *EFSA Support. Publ.* **2024**, *21*, 9008E. [CrossRef]
186. Bas, A.; Burns, N.; Gulotta, A.; Junker, J.; Drasler, B.; Lehner, R.; Aicher, L.; Constant, S.; Petri-Fink, A.; Rothen-Rutishauser, B. Understanding the Development, Standardization, and Validation Process of Alternative In Vitro Test Methods for Regulatory Approval from a Researcher Perspective. *Small* **2021**, *17*, e2006027. [CrossRef]
187. Hartung, T.; Bremer, S.; Casati, S.; Coecke, S.; Corvi, R.; Fortaner, S.; Gribaldo, L.; Halder, M.; Hoffmann, S.; Roi, A.J.; et al. A modular approach to the ECVAM principles on test validity. *Altern. Lab. Anim.* **2004**, *32*, 467–472. [CrossRef]
188. van der Valk, J.; Brunner, D.; De Smet, K.; Fex Svenningsen, A.; Honegger, P.; Knudsen, L.E.; Lindl, T.; Noraberg, J.; Price, A.; Scarino, M.L.; et al. Optimization of chemically defined cell culture media--replacing fetal bovine serum in mammalian in vitro methods. *Toxicol. Vitr.* **2010**, *24*, 1053–1063. [CrossRef]
189. Duarte, A.C.; Costa, E.C.; Filipe, H.A.L.; Saraiva, S.M.; Jacinto, T.; Miguel, S.P.; Ribeiro, M.P.; Coutinho, P. Animal-derived products in science and current alternatives. *Biomater. Adv.* **2023**, *151*, 213428. [CrossRef]
190. Herzler, M.; Abedini, J.; Allen, D.G.; Germolec, D.; Gordon, J.; Ko, H.S.; Matheson, J.; Reinke, E.; Strickland, J.; Thierse, H.J.; et al. Use of human predictive patch test (HPPT) data for the classification of skin sensitization hazard and potency. *Arch. Toxicol.* **2024**, *98*, 1253–1269. [CrossRef]
191. Irizar, A.; Bender, H.; Griem, P.; Natsch, A.; Vey, M.; Kimber, I. Reference Chemical Potency List (RCPL): A new tool for evaluating the accuracy of skin sensitisation potency measurements by New Approach Methodologies (NAMs). *Regul. Toxicol. Pharmacol.* **2022**, *134*, 105244. [CrossRef]

192. Donthamsetty, S.; Forreryd, A.; Sterchele, P.; Huang, X.; Gradin, R.; Johansson, H.; Mattson, U.; Lee, I.; Api, A.M.; Ladics, G. GARDskin dose-response assay and its application in conducting Quantitative Risk Assessment (QRA) for fragrance materials using a Next Generation Risk Assessment (NGRA) framework. *Regul. Toxicol. Pharmacol.* **2024**, *149*, 105597. [CrossRef] [PubMed]
193. Artik, S.; Haarhuis, K.; Wu, X.; Begerow, J.; Gleichmann, E. Tolerance to nickel: Oral nickel administration induces a high frequency of anergic T cells with persistent suppressor activity. *J. Immunol.* **2001**, *167*, 6794–6803. [CrossRef] [PubMed]
194. Cavani, A. Breaking tolerance to nickel. *Toxicology* **2005**, *209*, 119–121. [CrossRef] [PubMed]

MDPI AG
Grosspeteranlage 5
4052 Basel
Switzerland
Tel.: +41 61 683 77 34

Toxics Editorial Office
E-mail: toxics@mdpi.com
www.mdpi.com/journal/toxics

www.ingramcontent.com/pod-product-compliance
Lightning Source LLC
Chambersburg PA
CBHW041355210926
51850CB00013B/3092
* 9 7 8 3 7 2 5 8 4 3 8 3 1 *